D0712518

Introduction to OCEAN ENGINEERING

Introduction to OCEAN ENGINEERING

Edited by

Hilbert Schenck, Jr.

Professor of Ocean and
Mechanical Engineering
University of Rhode Island

McGRAW-HILL BOOK COMPANY

New York St. Louis San Francisco Auckland Düsseldorf
Johannesburg Kuala Lumpur London Mexico Montreal New Delhi
Panama Paris São Paulo Singapore Sydney Tokyo Toronto

This book was set in Press Roman by Scripta Graphica.
The editors were B. J. Clark and J. W. Maisel;
the cover was designed by J. E. O'Connor;
the production supervisor was Leroy A. Young.
The drawings were done by Vantage Art, Inc.
Kingsport Press, Inc., was printer and binder.

INTRODUCTION TO OCEAN ENGINEERING

1 2 3 4 5 6 7 8 9 0 KPKP 7 9 8 7 6 5

Library of Congress Cataloging in Publication Data

Schenck, Hilbert van Nydeck, date
Introduction to ocean engineering.

1. Ocean engineering. I. Title.
TC1645.S3 620'.416'2 74-14577
ISBN 0-07-055240-1

Contents

Preface

The expansion of ocean engineering curricula in the United States has followed the increasing interest and legislation in areas of environmental quality, oil recovery, and waterborne recreation. In the United Kingdom, the North Sea oil developments have been an important catalyst, but all these and other concerns have stirred modest academic activity in other areas of the world. In general, ocean engineering programs have been at graduate level and have accepted students from all the engineering disciplines and often from physics, geology, and oceanography as well. A primary difficulty with such curricula, especially in the typical 30-credit-hour master's degree program, is their rather elementary character. Since most degree candidates come to the areas of underwater soil mechanics, ocean wave theory, hyperbaric physiology, underwater sound, and corrosion theory with little or no background, they are constrained to take five or more courses, involving half or more of their master's level load, at a beginning-course level. This makes any significant specialization difficult and tends to produce master's candidates who are little advanced in engineering sophistication beyond typical bachelor of science candidates in the traditional engineering disciplines.

One solution to this present difficulty is to provide course materials that will enable an undergraduate engineer (or a beginning graduate student) to cover rapidly the fundamentals of the several important ocean engineering topics in, say, a six- or even three-credit-hour course. *Introduction to Ocean Engineering* is intended to provide these materials.

Ideally, the first 10 chapters can be covered in a two-semester three-credit-hour course involving undergraduate seniors in engineering. Chapter 11 is intended for schools where some associated lab work may be possible. Because of the "do-it-yourself" character of Chap. 11, it should adapt quite well to senior "project" labs now so popular in the United States and elsewhere. Where only a single semester is available, some selection and compression will be necessary. We suggest for such a course that Chaps. 2, 3, 4, 5, and 7 be attempted, since these areas are the ones least likely to be covered in traditional undergraduate engineering curricula. Alternatively, a lab-oriented course where project lab time is available might consider a schedule of Chaps. 2, 3, 5, 10, and 11, since this group involves the student most heavily in the making and use of measurements in water.

The scheduling problem at the graduate level is more complex, for it is desirable for the student to have completed the elementary material in a chapter here before starting on a full, specialized course. This could be arranged where part-time students are the primary enrollees, but would be difficult to arrange in a one-year 30-credit-hour master's program.

A third and, we believe, important application of this book lies in summer and extension courses for graduate engineers whose firms or schools are moving into ocean-related engineering.

Although various books and course materials exist in ocean engineering, none of these are really suitable for a formal, first course in an academic setting. These materials often lack numerical examples, end-chapter problems, and an organized relation between the disparate topics of ocean technology. For ocean engineering to become a significant elective or specialization in undergraduate engineering, it will be necessary for professors in traditional engineering to teach in the many different areas noted here, albeit at an elementary level. In effect, they will have to learn this new field with their students. Although such activity is never painless, *Introduction to Ocean Engineering* is intended to minimize such transitions and guide the newcomer whether he be student or professor.

This book is entirely an effort of members of the College of Engineering at Rhode Island. All the authors either are professors in this college or, in two cases, were graduate students here when the work was done. Course materials are generally what these teachers would wish to see in their students' backgrounds in advanced, specialized courses or what they themselves teach in elementary courses. The entire book is used in a two-semester senior course in mechanical engineering, a core course required of students obtaining a joint BS degree in mechanical and ocean engineering.

As to whether engineering education needs yet another new area of this sort, the authors will make no large arguments. Claims of "relevance" and "survival" have fueled every program from moon journeys to mind alteration by psychosurgery. We say only that students can fulfill themselves in ocean technology, as engineers and as human beings.

Hilbert Schenck, Jr.

Introduction to OCEAN ENGINEERING

1

The Scope of Ocean Engineering

Hilbert Schenck, Jr.

1-1 BACKGROUND

During most of the nineteenth century, the major engineering disciplines in both the United States and Europe were civil or mechanical in character. Railroads, industrial plants, bridges, steel-framed buildings, and machine tools occupied the majority of engineering graduates from those few colleges specializing in technical education.

By 1880, a new and exciting specialty was beginning to take form, electric power generation. Now one of the more tradition-bound subjects, electric power in the eighties caught the imagination of young technical graduates in the same way that nuclear engineering became the "glamour subject" in the 1950s. By the end of the century all four major engineering branches—civil, mechanical, electrical, and chemical—were well established, chemical engineering having boomed after the discovery of oil in Pennsylvania and the rapid development of the petrochemical industry.

These four major specialties are mainly organized around clusters of basic scientific principles—mechanical engineering drawing from hydraulics and structural statics, and so on. Overlapping areas are frequent in most curricula and all four disciplines have common interests in applied mathematics.

During the 1920s the first engineering departments organized around a specific application, air transport, were formed. Aeronautical engineering, although heavily biased toward fluid mechanics, also involved the student in structural (airframe) analysis, the design and analysis of engines, instrumentation development, and related areas. Aeronautical engineering was often a graduate curriculum. In effect, this was an early attempt at systems engineering, the idea being to synthesize material and ideas from the four basic engineering areas into a complete, mission-oriented package.

After the Second World War and the Hiroshima bomb, nuclear engineering was projected as the technology of the future. Like the four traditional disciplines, nuclear technology was based on a scientific area (nuclear physics), and like the older areas, it appeared to have important roles in a variety of applications: stationary power generation, excavation and landscape alteration, industrial gamma-ray uses, tracer chemistry, various propulsion systems, and a host of unguessed new fields. The projected growth of nuclear engineering in the early 1950s did not come true. The health hazards of fallout and radioactive contamination were more dangerous and intractable and their prevention more expensive than early, optimistic spokesmen realized. Strict government regulation, the growing concerns of ecology, and the continual, close tie-in of all nuclear engineering with weapons of drastic power have served to limit the field of application of this modern discipline.

If nuclear engineering was the style of the 1950s, so aerospace engineering was surely the big new area of the 1960s. Often derived from aeronautical departments and heavily tied to NASA largesse, aerospace departments had systems-oriented curricula and drew heavily from fields both inside and outside the traditional engineering areas; biology, materials science, combustion, astronomy, and orbital mechanics were some of the diverse disciplines considered by the astronautic system designer.

The nuclear engineering boom floundered on the problems of health and safety, and aerospace engineering appears to be withering because of problems of relevance. For whereas the generation of safe and nonpolluting power remains an urgent and central problem, the flying of the very rich to Paris in 3 h or the placing of a few men for a few hours on the moon seem increasingly meaningless, at least to the taxpayers who must buy such toys and sports. In spite of vast efforts, NASA and the aerospace industry have been unable to identify critical and urgent technological needs for which their sophisticated and elegant systems have answers.

Ocean engineering and aerospace engineering have one common aspect: they are both organized around a particular environmental condition rather than around a series of scientific principles. But whereas space is mostly empty and its exploitation for human purposes severely limited, the oceans cover much of the earth, regulate its climate, produce food, and hide large amounts of important raw materials. We can thus hope that the variety of possible ocean missions and the human scale of its operations will give ocean engineering a stable and fruitful future, less critically dependent on a single principle or a single task than nuclear and aerospace technologies are.

Ocean engineering will not be the only new technology for the 1970s. It is easy to foresee departments of arctic engineering, urban engineering, and health engineering springing up as money and need appear. By 1980, the great engineering efforts of the 1960s may seem rather perverse and aberrant. Sampling water in a murky estuary is less exciting than digging rocks from the moon's crust; history will decide which is the more essential task.

1-2 TOPICS IN THE OCEAN

In some schools, ocean engineering is an expansion of that old and honorable discipline, marine architecture. Things that float and move across the water surface are used by engineers but, in addition, other devices must sink stably, or move as submarines, or rise from the bottom in controlled ways. All these modes of operation involve us in propulsion and stability, engineering concepts as ancient as human recorded history.

Fixed structures along the shore or those tied to the bottom by piers or piles involve the engineer in "wet civil engineering." The dry civil engineer is only mildly interested in wind loading, but the wet civil engineer is centrally concerned with waves and their gigantic forces and, thus, with aspects of physical oceanography.

Engineering constructions that sit on, or sink into, the bottom involve us in soils and sediments underwater, again a "wet analog" of dry-land soils engineering. Cables, pipes, moorings, and footings will sink into the bottom and the rate must be predicted, as must be the retrieval forces in many cases. Drilling and mining on the bottom require sediment and geological studies.

Buoys, surface and underwater vehicles, moorings, cables, pipes, breakwaters, piles, and every other wet device must be built of materials of appropriate strength and appropriate cost. They must also resist corrosion and fouling, or else be made large enough to accept these ubiquitous environmental attacks over a projected lifetime.

Pipes are laid in the water to carry or discharge material. Large underwater constructions in estuaries are often there because seawater is an excellent coolant. The driving of piles to form a marina implies a future population of people in the area living on the water. Oil, sewerage, waste, and power pollute the estuary. Only recently have we cared very much about it. Estuaries are multiple-use systems. Swimming, shellfishing, boating, game fishing, and sailing must coexist with the industrial and municipal burdens. Clearly they have at least equal rights with the large corporate users. The engineer wishing to avoid controversy and debate should stay away from the ocean.

Because the ocean is murky, cold, and turbulent and the pressure in the deep regions hard to deal with, remote sensing in the ocean is an important art. Underwater sound is the best method yet discovered for probing the bottom sediments for oil structures or for communicating with a deep-diving research sub. The generation, direction, and transmission of sound in water is basic to ocean engineering and is a tool used by all its practitioners.

Since human manipulation is indispensable in many ocean operations, diving and life support is an important part of our subject. It is also a particularly unique and interesting area, for it draws heavily from biomedical topics, especially respiration mechanics and body heat flow. An understanding of what the diver can do is as important to the ocean engineer as an understanding of human motion and reactions is to the production engineer.

The putting together of these special arts and technologies to perform an ocean task is the end product of an ocean engineering education. And what makes the field intriguing is that even simple and basic ocean systems may require a knowledge of every topic in this book. Consider a simple bottom-anchored buoy implanted to measure the vertical temperature profile in an estuary. To anchor it, we must know about the bottom sediment and the forces on the buoy and line. But forces can be computed only from a knowledge of local currents and wave characteristics and their effect on the floating buoy. Does the buoy have more than one stable position? What kind of material should we use to connect the anchor and float? Will it corrode, and what is its life? What about retrieval forces, the sinking of the mooring, and the reduced strength due to corrosion of the tether? How will we read the data, by sound telemetry? How will we deploy the system? Must we use a diver to inspect the mooring to ensure that it has taken the correct attitude, check the instruments, or work on the float in the water? And what do the daily readings mean in regard to the state of our water column? Are we thermally polluting in this area, and what will be the effect on marine life of any temperature changes? Should we monitor the temperatures at some particular time of day (peak load at the nearby nuclear steam station, perhaps)?

What we are suggesting here is that ocean engineering has two components: a "primary" part encompassing basic engineering physical and oceanographic knowledge, and a "secondary" or "systems" part whereby these separate bits and pieces are combined into an effective and harmonious whole.

1-3 A NOTE ABOUT UNIT SYSTEMS

In common with most technologists, the ocean engineer faces a variety of unit and dimensional systems in his daily work. Added to the usual English and metric systems are the purely marine units such as the fathom and the nautical mile. Areas closely allied to oceanography or physics (such as water-wave theory, sound propagation, submarine geology) tend to use mainly the meter-kilogram-second system. The more engineering-based subjects (such as ocean structures, marine propulsion, or power-plant siting) tend to stay with the English system of feet-pounds-British thermal units-seconds. Medically trained diving physiologists speak in terms of liters, calories, and centimeters. Diving engineers run calculations in cubic feet, foot-pounds, Btu, and inches. The knot, or nautical mile per hour, is basic in marine architecture, but oceanographers measure wind speed in kilometers per hour and the U.S. Weather Bureau records miles per hour.

These nuisances are little more than that, and the working engineer soon learns to think in several unit systems. The concept that the metric system is a "universal" or "the best" system has been thoroughly sold by world scientists and efforts are afoot to replace the English (and other) systems altogether. This will not be a complete blessing. For example, for measuring inshore ocean depth, the foot is a dimension of handy size. Meters are really too large or coarse, centimeters absurdly small.

Student engineers resent mixed-unit problems, and it would be irresponsible to write a handbook with indiscriminate combinations of unit systems. A textbook, on the other hand, is intended to teach practitioners in as realistic and general a way as possible. When the job, for example, is to place a buoy in a seaway, certain information must be gathered from various sources. Tidal currents, found from charts, will usually have velocities given in knots. Wavelengths and heights, say for a given sea state, may be found in meters, but the depth of water under the buoy will be charted in feet. The tensile strength of the wire-rope mooring is given in pounds per square inch (psi or lb/in^2), but the bearing strength of the soil on which the mooring rests may be computed in kilograms per square centimeter (kg/cm^2). The divers' tanks contain 70 ft^3 of standard air, but their demand at depth on this job is tabulated in liters per minute. And so on.

Thus the several authors of this material have tried to use *realistic* (rather than consistent) units, the same units that they would use in professional design work in their specialty and the units a reader would find if he were to use the specialized literature in the given area. We wish this were all a little less chaotic, but every interdisciplinary area suffers from the same problems.

1-4 CAREERS AND RELEVANCE

As with most other human endeavors, greed and ethical honesty are in continual struggle within the discipline of ocean engineering. The fires and oil spills due to negligence and corner cutting by the offshore oil producers, themselves a major employer of ocean engineers, are one side of the coin. Selfless efforts by fish-gear and systems engineers in improving the national fisheries of small countries is the other. The young practitioner may have basic moral decisions to make, whatever his special field in the ocean.

Compared to many giant manufacturing firms, ocean engineering companies are mostly small and often broke. In the later 1960s federal funds of one sort or another were the main source of income. Expensive tools, such as the oceanographic vessels and midget research subs, operated when federal largesse permitted. The question was, and is, does ocean engineering have sufficient diversity and sufficient basic importance to take its place as a major force in the technology of the world? We believe the answer can only be yes. The free use of estuaries, lakes, and seas by companies and cities cannot continue. Sewerage and spoil dumping, thermal cooling, and chemical disposal are actually services rendered by all the citizens abutting the estuary and these services must be paid for. The payment will be in the form of control, processing, and monitoring of the dumped material or operation, and it is in these new areas that many ocean engineers will be employed in future years. Offshore oil in deeper and deeper water will surely continue—with far more effective and careful engineering management, it is hoped, than has been noted. Control of the movement of tankers, and detection and prevention of spills and tank pump-outs, the development of systems to mop up after an accident are all large areas of activity. In a rational society they should be a basic part of the expense of doing business in the ocean.

We see recreation as a central job source for many engineers, involving new products, safety, control of traffic in estuaries, sport fishing improvement, and a host of other tasks. Mining in the deep and shallow ocean is only now under study, and it is hard to say how it may develop because we do not know exactly what lies in the crushing depths of the sea. The design of vessels of all sorts will surely continue as will the building of harbors and shore protection structures. Tsunamis and currents, sea ice, hurricanes, internal waves, and the intricate biological machines of the ocean will continue to attract scientific study and many of the tools and methods will be created by young ocean engineers.

We thus foresee plenty to do, much of it challenging at both a human and a technical level, much of it conducted from the rolling decks of small ships far from the great football-field-sized "bull pens" of the automotive or aerospace giants.

REFERENCES

American Institute of Mining, Metallurgical, and Petroleum Engineers (1969, 1970, 1971, 1972): "Preprints of the Annual Offshore Technology Conference," Dallas and Houston, Texas.

Brantz, J. F. (ed.) (1968): "Ocean Engineering," Wiley, New York.

Craven, J. P. (1971): "Ocean Engineering Systems," M.I.T. Sea Grant Program, Project GH-88, Cambridge, Mass.

Marine Technology Society (1965 through 1972): "Conference Reprints, Annual Meeting," Washington.

Myers, J. (ed.) (1969): "Handbook of Ocean and Underwater Engineering," McGraw-Hill, New York.

Paddleford, N. J. (1968): "Public Policy and the Use of the Seas," M.I.T. Sea Grant Program, Project GH-1, Cambridge, Mass.

——, and J. E. Cook (1971): New Dimensions of U.S. Marine Policy," M.I.T. Sea Grant Program, Project GH-88, Cambridge, Mass.

Sheets, H. E., and V. T. Boatwright (eds.) (1970): "Hydronautics," Academic, New York and London.

Society for Underwater Technology (1972): "Oceanology International '72 Conference Papers," 1 Birdcage Walk, London, S.W. 1.

2
Hydrodynamics of Coastal Areas

F. Mangrem White

2-1 INTRODUCTION: NATURAL, NONDETERMINISTIC HYDRODYNAMICS

As far as we know the motions of coastal waters—waves, currents, tides, surges, and seiches—always satisfy the basic laws of conservation of mass, momentum, energy, and species that are developed in engineering textbooks on fluid mechanics. Thus the computation of a coastal motion is, in principle, a straightforward task. However, ocean flows have three special characteristics which few other than civil engineers have had occasion to associate with flow problems. They are (1) "natural" as opposed to "forced"; (2) "nondeterministic" as opposed to "deterministic"; and (3) "hydrodynamic" as opposed to "aerodynamic."

Most engineers are familiar with forced motions, which are under the direct control of the designer, through pumps, spillways, thrusting engines, and other prime movers. Ocean flows are "natural," being caused by giant forces—the moon, the sun, the earth's rotation, the wind, the seasons, earthquakes—far outside the engineer's control. We cannot make a tide or a storm surge; we must accept their existence and seek to modify or otherwise take account of their effects. For example, the theory of waves (Sec. 2-3) does not specify their cause but simply computes their propagation characteristics if and when they exist.

We are also used to deterministic motion: a given pressure drop applied to a given pipe causes a given flow rate, day in and day out. In the ocean, only the tides lead to such predictable effects. Typically, sea motions are chaotic, disorderly, random–sometime things which encompass wide frequency and magnitude spectra and obey statistical rather than deterministic laws. Lord Rayleigh once remarked that "the only law of the seaway is the apparent lack of any law." Therefore, in many cases, we must be content with statistical information about currents and waves and forces on ocean structures.

Finally, ocean flows are hydrodynamic: they are liquid, not gas flows. There is always a "free surface," and compressibility effects, so important in acoustic propagation problems (Chap. 7), are entirely negligible in coastal flows. The single most important effect we consider here is the interaction between the atmosphere (pressure and wind) and the free surface of the ocean. This is discussed in some detail in Sec. 2-4.

2-2 FUNDAMENTAL EQUATIONS OF OCEAN HYDRODYNAMICS

The equations we discuss here are nothing more than the four laws of conservation which any pure substance (not undergoing chemical or nuclear reactions) must satisfy:

a. Conservation of mass: $m = \text{constant}$

b. Conservation of momentum: $\mathbf{F} = m\mathbf{a}$ (2-1)

c. Conservation of energy: $\delta E = \delta Q + \delta W$

d. Conservation of species: $S = \text{constant}$

It is merely necessary to relate these laws directly to the physical properties of the ocean. Also, since ocean properties vary from point to point, it is useless, for example, to take m as the mass of the ocean. Rather, m will be the mass of a single elemental particle of seawater, and it follows that the "fundamental" equations will in fact be partial differential equations.

2-2.1 Physical properties of the ocean

The properties of a fluid can be divided into three classes for a hydrodynamic study: (1) thermodynamic properties: pressure, temperature, density, and salinity; (2) the kinematic properties: wave height, velocity, and acceleration; and (3) the transport properties: viscosity, conductivity, and diffusion coefficient. Each has a particular "dimension" which can be written in terms of no more than four "primary" dimensions, which we take here to be mass M, length L, time T, and temperature Θ. Let us list these properties, their symbols, dimensions, and typical average values for seawater:

Property	*Symbol*	*Dimension*	*Typical average value*
1. Thermodynamic:			
Pressure	p	$ML^{-1}T^{-2}$	1–1,000 atm
Temperature	T	Θ	20°C
Density	ρ	ML^{-3}	1.025 g/cm³
Salinity	S	none	35 ppt
2. Kinematic:			
Wave height	η	L	0–2,000 cm
Velocity	$\mathbf{V}$	L/T	0–100 cm/s
3. Transport:			
Viscosity	μ	$ML^{-1}T^{-1}$	0.010 g/(cm)(s)
Conductivity	k	$MLT^{-3}\Theta^{-1}$	1.4×10^{-3} cal/(cm)(s)(°C)
Diffusion coefficient	D	L^2/T	1.3×10^{-5} cm²/s

The transport properties listed are the "molecular" coefficients, which are effective only for very small-scale or "laminar" motions. For typical ocean flows, the scale is so large that mean convective transport due to "turbulent" or "eddy" motions is usually much more important.

2-2.2 The differential equations of motion

In what follows, we adopt the usual hydrodynamics premise of an eulerian coordinate system, in which the fluid is considered to be a "field" described by fixed coordinates. Particles move through this field at will, so that all fluid properties are functions of the four independent variables (x,y,z,t). This means that the time rate of change of any property P is rather complicated mathematically:

$$\frac{dP}{dt} = \frac{\partial P}{\partial t} + u\frac{\partial P}{\partial x} + v\frac{\partial P}{\partial y} + w\frac{\partial P}{\partial z} \tag{2-2}$$

where (u,v,w) are the components of the velocity vector $\mathbf{V}$. This is preferable to tagging each particle with a set of moving coordinates—the langrangian system popular in solid mechanics. However, Eq. (2-2) is obviously a difficult nonlinear expression which in fact is responsible for most of our mathematical troubles—nonuniqueness, nonlinear boundary conditions, turbulent stresses—in hydrodynamics. For details, see, for example, Neumann and Pierson (1966).

Consider a particle of mass m, volume B, and density $\rho = m/B$. The "dynamic" statement of mass conservation for this particle is

$$\frac{dm}{dt} = \frac{d}{dt}\rho B = \rho\frac{dB}{dt} + B\frac{d\rho}{dt} = 0 \tag{2-3}$$

But, by pure geometrical reasoning, the rate of change of volume of any particle, expressed as a percentage, is equal to the sum of the normal strain rates of the particle:

$$\frac{1}{B}\frac{dB}{dt} = \frac{\partial u}{\partial x} + \frac{\partial v}{\partial y} + \frac{\partial w}{\partial z} = \nabla \cdot \mathbf{V} \tag{2-4}$$

Combining Eqs. (2-3) and (2-4) and eliminating the volume, we have an expression of conservation of mass for an elementary particle:

$$\frac{d\rho}{dt} + \rho\nabla \cdot \mathbf{V} = 0 = \frac{\partial \rho}{\partial t} + \nabla \cdot (\rho\mathbf{V}) \tag{2-5}$$

This is the desired form for the "continuity" relation, because it is expressed solely in terms of local physical properties from our previous list.

In a similar manner, we may derive and write down the desired expressions for conservation of momentum, energy, and salt—all in terms of local properties from our list:

Momentum: $$\rho\left(\frac{dV}{dt} + 2\mathbf{\Omega} \times \mathbf{V}\right) = -\nabla p + \rho \mathbf{g} - \mu\nabla^2\mathbf{V} \tag{2-6}$$

Energy: $$\rho C_v\frac{dT}{dt} = k\nabla^2 T + \mu\left(\frac{\partial u_i}{\partial x_j} + \frac{\partial u_j}{\partial x_i}\right)\frac{\partial u_i}{\partial x_j} \tag{2-7}$$

Salt: $$\frac{dS}{dt} = D\nabla^2 S \tag{2-8}$$

Here, $\mathbf{\Omega}$ is the angular velocity of the earth and $\mathbf{g}$ is the local geopotential gravity vector, both assumed known. Equations (2-5) through (2-8) are the fundamental equations of hydrodynamics. There are only four equations but five variables: ρ, $\mathbf{V}$, p, T, and S. Hence a fifth equation is needed—the thermodynamic state relation for seawater, which may be taken in the following linearized form:

$$\rho = \rho_0[1 + K_p(p - p_0) + K_T(T - T_0) + K_S(S - S_0)] \tag{2-9}$$

where subscript 0 denotes some reference state—the ocean at rest, say. Average values of the "expansion" coefficients K_p, K_T, K_S for seawater are given, for example, in Neumann and Pierson (1966). The solution of the above system of equations must be subject to known conditions at every boundary of the flow:

1. At a fixed solid boundary:
   (*a*) The velocity vanishes (no-slip condition).
   (*b*) The temperature or temperature gradient must be known.
   (*c*) The salt or salt gradient must be known.

2. At the interface of the ocean with the atmosphere:
   (*a*) Equality of normal velocity, normal stress, shear stress, heat flux, salt flux, and mass flux must exist across the interface.
3. At an "inlet" or "exit" boundary of the flow:
   (*a*) Complete distributions of velocity, temperature, pressure, and salinity must be known.

This constitutes the complete mathematical problem. In practice, many simplifying approximations are made to achieve engineering formulas. Of particular interest is the pressure condition at the ocean surface. Neglecting viscous normal stresses, we have:

$$p(\text{surface}) = p_{\text{atm}} - \gamma(r_1^{-1} + r_2^{-1}) \tag{2-10}$$

where r_1 and r_2 are the principal radii of the presumably distorted surface, and γ is the coefficient of surface tension, which for a clean sea surface has a value of about 75 dyn/cm.

2-2.3 The Boussinesq approximation

The most important of the various simplifying assumptions for our rather formidable equations is the so-called Boussinesq approximation, which is stated as follows: Since the density in the ocean varies only a few percent at best, we may regard density as constant in all parts of the fundamental equations *except the buoyant force term* in the momentum equation. The strongest result of this assumption is that the continuity equation (2-5) reduces to:

$$\nabla \cdot \mathbf{V} \cong 0 \qquad \rho \cong \rho_0 \tag{2-11}$$

a relatively simple linear relation. Even the buoyant force term is negligible in, say, wave mechanics and tidal currents, but it must be retained in problems where vertical motion is important, such as floating body dynamics and turbulent mixing in surface layers.

2-2.4 Dimensional analysis and scale effects

Suppose now we nondimensionalize all of our variables, both dependent and independent, with respect to a reference speed V_0, a reference length L, and a characteristic time (L/V_0). Then we find by substitution into Eqs. (2-6), (2-7), and (2-8) that the following characteristic dimensionless groups define the motion:

$$
\begin{aligned}
&a.\ \frac{\rho_0 V_0 L}{\mu} &&= \text{Reynolds number} \\
&b.\ \frac{V_0}{L\Omega} &&= \text{Rossby number} \\
&c.\ \frac{\rho_0 C_v V_0 L}{k} &&= \text{Peclet number} \\
&d.\ \frac{\rho_0 C_v L T_0}{\mu V_0} &&= \text{modified Eckert number} \\
&e.\ \frac{V_0 L}{D} &&= \text{diffusive Reynolds number}
\end{aligned}
\qquad (2\text{-}12)
$$

Now, for typical coastal engineering problems, the velocity scale V_0 is about 50 cm/s and the length scale L about 1 km. With the physical properties of seawater included, all five of these dimensionless parameters are quite large, of order 10^3 to 10^9. Only the Rossby number could possibly be "small" and then only for geophysical scale motions, where L is 1,000 km or more. For coastal problems, then, the large magnitude of these parameters allows us to neglect several terms in the fundamental equations which vary inversely with the parameters. If we do this, the "significant" engineering hydrodynamic terms which remain are as follows:

$$
\begin{aligned}
&a.\ \text{Continuity:} && \nabla \cdot \mathbf{V} \cong 0 \\
&b.\ \text{Momentum:} && \rho \frac{d\mathbf{V}}{dt} \cong -\nabla p + \rho \mathbf{g} \\
&c.\ \text{Energy:} && \frac{dT}{dt} \cong 0 \qquad T = \text{constant} \\
&d.\ \text{Salinity:} && \frac{dS}{dt} \cong 0 \qquad S = \text{constant}
\end{aligned}
\qquad (2\text{-}13)
$$

Further, we can take the density ρ as approximately constant. Obviously, these equations are profoundly simpler than the more exact relations (2-5) through (2-8). They imply that engineering ocean flow problems are well approximated by inviscid, incompressible flow of a liquid at nearly constant temperature and salinity. These simplifications fail only in three ways: (1) in geophysical flows, where the Coriolis term must be retained; (2) near the bottom "boundary layer," where shear and heat conduction are important; and (3) near the free surface under highly agitated (storm) conditions, when velocity, temperature, and salinity gradients are all important.

2-2.5 The potential flow approximation

If we neglect the top and bottom "boundary layers" of the ocean, a simple and valuable mathematical substitution can be used to great advantage in Eqs. (2-13). Taking the curl of Eq. (2-13*b*) and assuming zero vorticity at the boundaries, we find that (curl **V**) must vanish everywhere in the fluid. From vector calculus, this means that the velocity vector **V** must be the gradient of a scalar potential ϕ: if

$$\text{curl } \mathbf{V} = 0$$

then

$$\mathbf{V} = \nabla\phi \tag{2-14}$$

This is the "potential flow" approximation, and ϕ is called the "velocity potential." Introducing ϕ into Eq. (2-13*b*), we obtain the following striking algebraic relation:

$$\frac{p}{\rho} + \frac{\partial\phi}{\partial t} + \tfrac{1}{2}V^2 + gz = f(t) \tag{2-15}$$

This is the celebrated Bernoulli relation for unsteady inviscid flow. We have taken the z coordinate to be positive upward, as in Fig. 2-1 (the usual oceanography coordinate system). The $f(t)$ on the right-hand side is arbitrary and can be absorbed into the potential ϕ if desired.

Introducing ϕ into Eq. (2-13*a*), we obtain the basic differential equation of engineering ocean hydrodynamics:

$$\nabla^2\phi = \frac{\partial^2\phi}{\partial x^2} + \frac{\partial^2\phi}{\partial y^2} + \frac{\partial^2\phi}{\partial z^2} = 0 \tag{2-16}$$

This is the linear Laplace equation, of which many solutions are known, particularly for wave motion, as discussed in the next section. Equation (2-15) now becomes simply a boundary condition to be applied at the free surface of

Fig. 2-1 Coordinate system for two-dimensional waves.

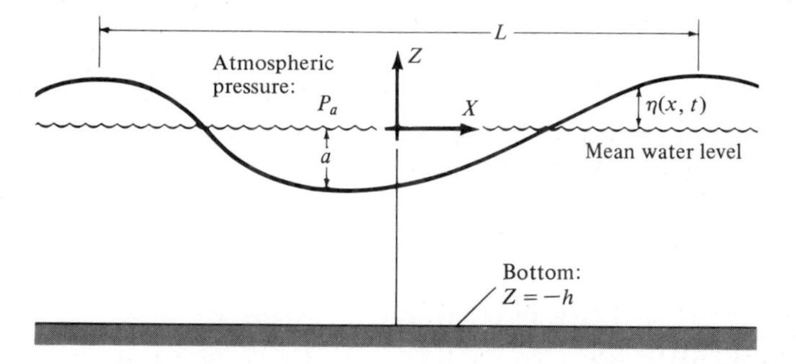

the flow. Also, once ϕ is known, Eq. (2-15) may be used to compute the pressure p at any position in the flow, since V^2 is directly related to ϕ from Eq. (2-14).

2-3 THE MOTION OF SURFACE WAVES

The following will serve as a brief introduction to the mechanics of wave motion in both deep and coastal waters. If the reader desires more details on this subject, there are several excellent treatises on wave motion, notably those of Kinsman (1965), Wiegel (1964), and Phillips (1966).

2-3.1 Small-amplitude two-dimensional waves

A relatively simple linearized solution, which delineates most surface wave effects, was given long ago by G. B. Airy (1842). We assume that the waves have no variation in the y direction and adopt the coordinate system shown in Fig. 2-1. The basic differential equation then becomes:

$$\frac{\partial^2 \phi}{\partial x^2} + \frac{\partial^2 \phi}{\partial z^2} = 0 \tag{2-17}$$

To eliminate nonlinear terms in the boundary conditions at the free surface, it is assumed that the wave amplitude a is much smaller than the wavelength L. The bottom is taken at a uniform depth h below the mean water level. The boundary conditions then become:

a. At the bottom ($z = -h$):

Zero normal velocity: $$\frac{\partial \phi}{\partial z} = 0$$

b. At the free surface ($z = \eta$):

Normal velocity match: $$\frac{\partial \phi}{\partial z} = \frac{\partial \eta}{\partial t} + \cancel{u\frac{\partial \eta}{\partial x}} \tag{2-18}$$

Bernoulli relation: $$\frac{p}{\rho} + \frac{\partial \phi}{\partial t} + \cancel{\tfrac{1}{2}V^2} + g\eta = \frac{p_a}{\rho}$$

Normal stress match: $$p = p_a - \gamma \frac{\partial^2 \eta/\partial x^2}{[1 + \cancel{(\partial \eta/\partial x)^2}]^{3/2}}$$

Here the full expressions have been given, but we have scratched through the terms which are negligible if $a \ll L$. We can then eliminate p and η among the three surface conditions (2-18b), giving a single linear condition on the velocity potential:

At $z = \eta$: $$\rho g \frac{\partial \phi}{\partial z} + \rho \frac{\partial^2 \phi}{\partial t^2} - \gamma \frac{\partial^3 \phi}{\partial z \partial x^2} = 0 \tag{2-19}$$

Now in fact *many* solutions of (2-17) satisfy (2-18) and (2-19). It is at this point that we specify *wave* motion, that is, the free surface moving in some regular periodic fashion. We may specify progress either to the left or to the right:

$$\begin{aligned}
&a.\ \text{Standing wave:} && \eta = a \cos(kx)\cos(\sigma t) \quad k = \frac{2\pi}{L} \quad \sigma = \frac{2\pi}{T} \\
&b.\ \text{Right-running wave:} && \eta = a \cos(kx - \sigma t) \qquad (2\text{-}20) \\
&c.\ \text{Left-running wave:} && \eta = a \cos(kx + \sigma t)
\end{aligned}$$

The parameters (k,σ,L,T) are called the wave number, frequency, wavelength, and period of the waves, respectively. Equations (2-20) describe the shape of the surface only. To complete the picture of a fluid in two-dimensional progressive wave motion, we postulate a variation of ϕ in the z direction also:

$$\text{Right-running wave:} \quad \phi(x, z, t) = f(z)\sin(kx - \sigma t) \tag{2-21}$$

We have chosen the sine rather than the cosine in order that η itself will have a cosine shape. It makes little difference, though. Now substitute Eq. (2-21) into (2-17). The result is

$$\frac{d^2 f}{dz^2} - k^2 f = 0$$

or (2-22)

$$f(z) = A \cosh[k(z+h)] + B \sinh[k(z+h)]$$

and $B = 0$ to satisfy the bottom boundary condition, Eq. (2-18*a*). Using this expression for ϕ, we find that the surface condition, Eq. (2-19), can be satisfied only if

$$\sigma^2 = \left(gk + \frac{\gamma k^3}{\rho}\right)\tanh[k(h+\eta)] \tag{2-23}$$

This is the small-amplitude frequency-wavelength-depth relation for both progressive *and* standing waves. The term $h + \eta$ is commonly approximated simply as h, although it need not be and in fact it hints correctly at a nonlinear effect, namely, that wave crests (positive η) move faster than wave troughs (negative η), causing wave forms to steepen (and break) in shallow water. Now the "phase speed" or local propagation speed of a progressive wave is defined as

$$C = \frac{L}{T} = \frac{\sigma}{k} \tag{2-24}$$

Substituting into Eq. (2-23), we find the following classic formula for small-amplitude or Airy waves:

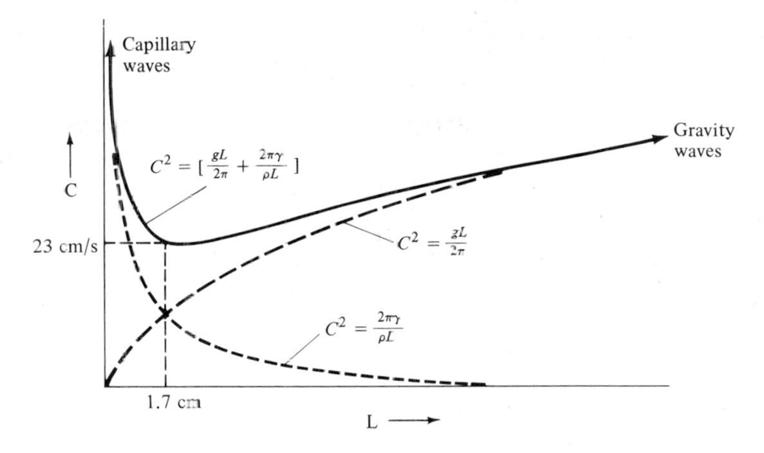

Fig. 2-2 The phase speed of very short surface water waves.

$$C^2 = \left(\frac{g}{k} + \frac{\gamma k}{\rho}\right)\tanh kh \tag{2-25}$$

The formula includes both gravity and surface tension effects. Which is more important? To see this, we assume $\gamma = 75$ dyn/cm and $\rho = 1.025$ gm/cm^3 for clean seawater and plot C versus L from Eq. (2-25) for "deep water," where $\tanh(kh) = 1.0$. The results are shown in Fig. 2-2. We see that the phase speed reaches a minimum where both are equally important, at $L = 1.7$ cm and $C = 23$ cm/s. Below this value, surface tension or "capillary" waves are more important, and above C_{min} "gravity" waves are dominant. Since 1.7 cm is a small wave indeed, we conclude that capillary waves are negligible in engineering calculations of ocean waves, except perhaps for small model studies. The usual formula for wave propagation, then, is simply:

$$\text{Gravity waves:} \quad C^2 = \frac{g}{k}\tanh kh \tag{2-26}$$

This expression is particularly intriguing in that it contains no physical properties of the fluid. That is, surface waves on water, mercury, gasoline, and oil all propagate, to this order of approximation, at the same speed, the only difference being that waves in the more viscous fluids damp out faster. This is a consequence of the potential flow assumption and the fact that gravity phenomena are independent of density.

Equation (2-26) plus the definition $L = CT$ implies that a relation exists between any three of the four parameters (C,L,T,h). Figure 2-3 shows two typical plots of the Airy formula: the (C,T,h) relation and the (T,L,h) relation.

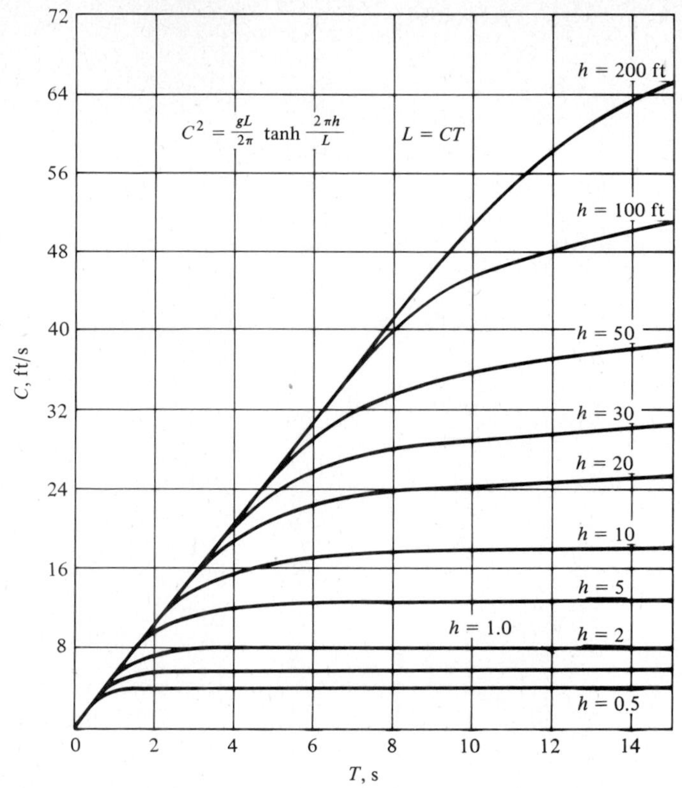

Relationship of wave velocity, period, and depth

(a)

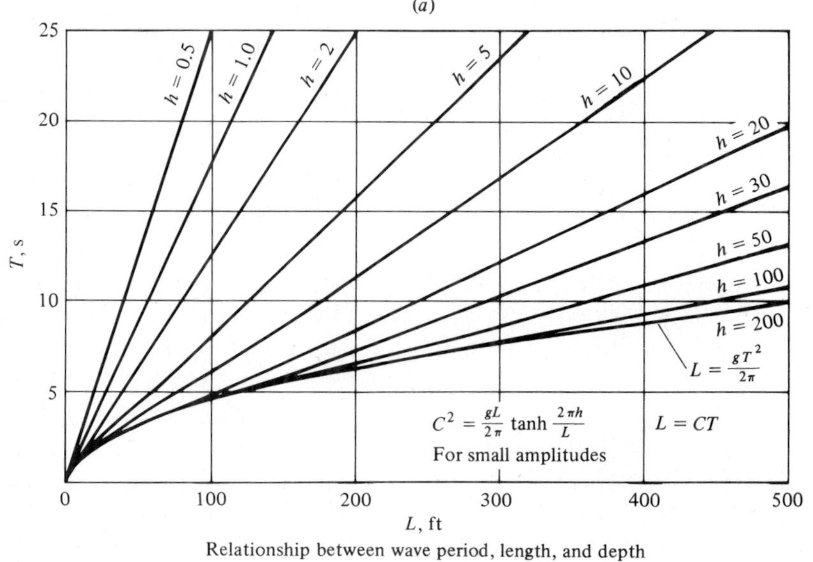

Relationship between wave period, length, and depth

(b)

Fig. 2-3 Characteristics of small-amplitude gravity waves, Eq. (2-26).

From these we see that, generally speaking, wave phase speed increases with increasing depth, wavelength, and period. Two special cases are of interest, "shallow" water and "deep" water:

$$\begin{aligned} &a.\ \text{Shallow}\quad (kh \leqslant 0.25): \quad C^2 \doteq gh \\ &b.\ \text{Deep}\quad (kh \geqslant 3): \quad C^2 \doteq \frac{g}{k} = \frac{gL}{2\pi} \end{aligned} \tag{2-27}$$

Thus shallow-water waves all move at the same speed independent of wavelength, and deep waves are influenced entirely by wavelength and not by depth. In between ($0.25 < kh < 3$) we have "intermediate" waves, for which the full relation (2-26) must be used.

The above wave theory is, strictly speaking, valid only for small-amplitude inviscid waves, uniform in character, infinite in extent, propagating in a constant-depth fluid. In practice, of course, oceans waves occur in "packets" of various frequencies and lengths and travel through waters of varying depth. Yet our theory is a surprisingly good approximation to the behavior of any particular "component" of a packet of ocean waves. This is primarily due to the fact that waves of different periods interact with each other only very weakly.

It should also be noted that waves may arise in any fluid where density differences occur. Here we have discussed "surface" waves, where the density changes abruptly from seawater to nearly zero (the atmosphere). But seawater density differences also occur *below* the surface, particularly in the thermocline. At these deep density interfaces, "internal waves" may arise and follow theoretical relations very similar to the above; see, for example, the text by Phillips (1966). Internal waves, being invisible from the surface, are difficult to monitor and study. They are often suspected of being the villains behind instrumentation errors in physical oceanography experiments.

2-3.2 Orbital motion

Propagation speed is not the only wave characteristic which can be studied by the Airy theory. So far we have found that the velocity potential of a single component wave train is given by

$$\phi(x, z, t) = A \cosh\,[k(z + h)]\,\sin\,(kx - \sigma t) \tag{2-28}$$

Like all linearized theories, the solution is independent of the magnitude of the constant coefficient A. Yet we may relate A to the wave amplitude a (see Fig. 2-1) through the velocity condition at the surface, Eq. (2-18b). Integrating, we have

$$\eta(x, t) = a \cos\,(kx - \sigma t) = \int \frac{\partial \phi}{\partial z}\,dt \tag{2-29}$$

Substituting for ϕ from Eq. (2-28) and carrying out the integration, we find that $A = aC/\sinh(kh)$. Thus our desired expression for the velocity potential is:

$$\phi(x, z, t) = aC \frac{\cosh [k(z + h)]}{\sinh (kh)} \sin (kx - \sigma t) \tag{2-30}$$

From this relation, we may calculate the local velocities:

$$\begin{aligned} u &= \frac{\partial \phi}{\partial x} = aCk \frac{\cosh [k(z + h)]}{\sinh (kh)} \cos (kx - \sigma t) \\ w &= \frac{\partial \phi}{\partial z} = aCk \frac{\sinh [k(z + h)]}{\sinh (kh)} \sin (kx - \sigma t) \end{aligned} \tag{2-31}$$

As we shall see in Sec. 2-5, these oscillating velocities are the primary cause of wave forces on structures in the ocean. Both components are proportional to $ak = 2\pi a/L$ and thus verify our original assumption that their squares are negligible if $a \ll L$. From the velocities, we may compute the local particle trajectories:

$$\begin{aligned} X &= \int u \, dt = -a \frac{\cosh [k(z + h)]}{\sinh (kh)} \sin (kx - \sigma t) \\ Z &= \int w \, dt = +a \frac{\sinh [k(z + h)]}{\sinh (kh)} \cos (kx - \sigma t) \end{aligned} \tag{2-32}$$

Thus the particles also oscillate and form, to this order of approximation, a closed curve, as the following rearrangement shows:

$$\frac{X^2}{X_0^2} + \frac{Z^2}{Z_0^2} = 1 \tag{2-33}$$

where

$$X_0 = a \frac{\cosh [k(z + h)]}{\sinh (kh)} \qquad \text{and} \qquad Z_0 = a \frac{\sinh [k(z + h)]}{\sinh (kh)}$$

This relation shows that particles in simple harmonic wave motion move in closed elliptical curves whose vertical or minor axis equals $2a$ or the wave height at the surface and zero at the bottom and whose horizontal (major) axis is greater than or equal to $2a$ at $Z = \eta$. In very deep water ($kh > 3$), the particle "orbits" become circles, with $X_0 = Z_0 = ae^{kh}$. Some examples are shown in Fig. 2-4, where we see that the wave orbits become flatter ellipses as the depth becomes more shallow. Also, in intermediate or shallow water, the velocity parallel to the bottom does not vanish in this inviscid approximation. Actually, it *must* vanish owing to viscosity; hence a bottom "boundary layer" is set up which tends to damp the wave amplitude in shoaling waters. The second primary cause of wave damping is the viscous straining within the orbits themselves. That is, for orbital motion, the dissipation or viscous term of Eq. (2-7) does not

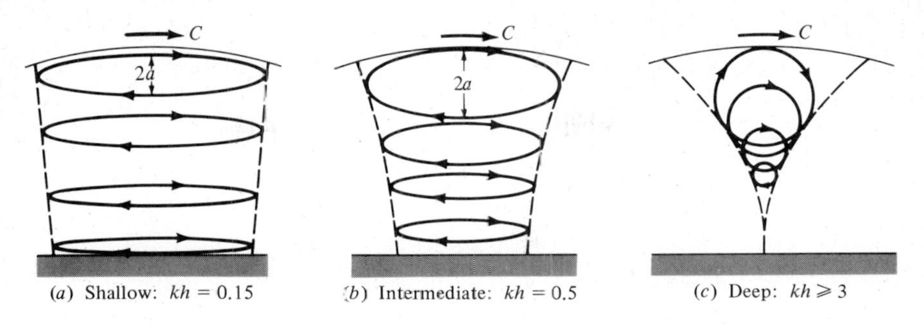

(*a*) Shallow: $kh = 0.15$ (*b*) Intermediate: $kh = 0.5$ (*c*) Deep: $kh \geqslant 3$

Fig. 2-4 Illustration of wave orbits in different depths.

vanish, so that even deepwater waves damp somewhat as they propagate across open ocean.

2-3.3 Wave energy, wave power, wave damping, and group velocity

The total energy contained within an Airy wave of length L is the sum of kinetic and potential energy:

$$E = E_k + E_p = \int_0^L \int_{-h}^0 \tfrac{1}{2}\rho(u^2 + w^2)dx\,dz + \int_0^L \int_0^\eta \rho gz\,dx\,dz \qquad (2\text{-}34)$$

The latter is the "net" potential energy, subtracting off the energy of the ocean at rest. Introducing u, w, and η from the previous theory, we find that both integrals equal $(\rho g a^2 L/4)$, which we might have expected because the motions are oscillatory. Thus the total wave energy is

$$E\,(\text{one wavelength}) = \frac{L\rho g a^2}{2} = \frac{L\rho g H^2}{8} \qquad (2\text{-}35)$$

Waves transmit this energy at a rate commensurate with their so-called "group velocity":

$$C_g = \frac{d\sigma}{dk} = C + k\frac{dC}{dk} = \frac{C}{2}\left[1 + \frac{2kh}{\sinh(2kh)}\right] \qquad (2\text{-}36)$$

This expression is plotted in Fig. 2-5, where we see that $C_g = C$ in shallow water and equals $C/2$ in deep water. Gravity waves are "dispersive"; that is, dC/dk is negative and the long waves move out ahead and the short waves fall behind. From Fig. 2-1 we see that, conversely, capillary waves are nondispersive. To verify that the group velocity is associated with wave energy transmission, we may compute the power transmitted across a vertical plane by a wave of length L:

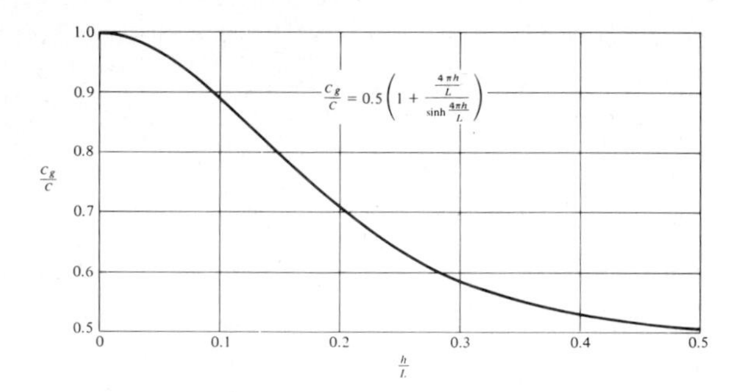

Fig. 2-5 The relation between (C_g/C) and (h/L), Eq. (2-36).

$$P = \int_0^L \int_{-h}^0 (p + \rho g z) u \, dz \, dx = \frac{C_g E}{L} \tag{2-37}$$

where C_g and E have the same meaning as in Eqs. (2-35) and (2-36). This is the power required to maintain a long train of waves of one period; the formula cannot be used to judge how "fast" energy travels with a finite packet of waves. With these formulas for energy and velocity, we may compute the damping of wave amplitude due to laminar orbit straining and bottom friction:

a. Orbit straining (Lamb 1932): $\frac{a}{a_0} = e^{-b_1 t} \qquad b_1 = 2nk^2$

b. Bottom friction (Keulegan 1959): $\frac{a}{a_0} = e^{-b_2 C_g t}$ (2-38)

$$b_2 = \left(\frac{nT}{\pi}\right)^{1/2} \frac{k}{L} \frac{(1 + \pi - 2kh)}{\sinh(kh)}$$

Here a_0 refers to the initial amplitude (at $t = 0$) and $n = \mu/\rho$. These formulas were derived for our constant-depth model of waves but may be applied to variable-depth situations in incremental form:

Wave damping: $$\frac{da}{a} = -dt(b_1 + b_2 C_g) \tag{2-39}$$

Other less important sources of wave damping are (1) turbulence (for very long waves); (2) bottom permeability; and (3) sidewall friction (when waves move into narrow channels).

2-3.4 Shoaling waters—transformation, refraction, and diffraction

What happens when a wave train moves into shallow water? The exact theory of wave motion with a sloping bottom is extremely complex—see the text by Stoker (1957)—and not yet suitable for engineering application. The usual engineering assumption follows a 60-year-old suggestion by Lord Rayleigh that waves move into shoaling waters with their energy neither damped nor reflected. Hence the power transmission at any intermediate depth is set equal to the deepwater value (subscript 0):

$$EC_g W \cong E_0 C_{g0} W_0$$

$$\frac{H}{H_0} = K_s = \left(\frac{2\cosh^2 kh}{2kh + \sinh 2kh}\right)^{1/2} \tag{2-40}$$

Here W refers to the "width" between wave "orthogonals," that is, the distance between lines drawn normal to the wave crests (see Fig. 2-6). It is possible for the wave orthogonals to diverge or converge if the water depth varies laterally in the direction of the wave crests. This is because decreasing depth tends to slow down the wave phase speed and gives the impression that waves are turning toward especially shallow portions of a coast. Some of these effects are shown in Fig. 2-6, which illustrates wave divergence, wave convergence, and the tendency of oblique waves to turn parallel to a uniformly sloping coast. These are qualitative effects; to predict the actual shapes of the orthogonals and wave crests, one must have a formula for incremental wave turning. Since we have assumed in Eq. (2-40) that local phase speed is uniquely determined by local depth and wavelength through Airy theory, it follows that Snell's law from optics is nearly valid:

$$\frac{\sin\theta_1}{\sin\theta_2} = \frac{C_1}{C_2} \tag{2-41}$$

Fig. 2-6 Shallow-water effects: (*a*) waves diverge from a submarine valley; (*b*) waves converge toward a submarine ridge; (*c*) oblique waves tend to turn parallel to a uniformly sloping coast.

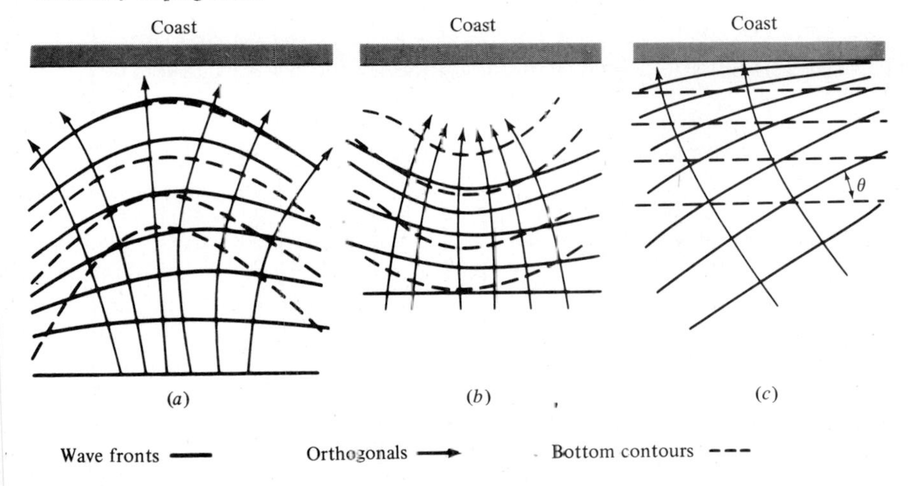

where θ is the angle between the wave crests and the local bottom contours (see Fig. 2-6*c*). By combining Eqs. (2-40) and (2-41) and the Airy theory for phase speeds, one can construct complete theoretical patterns of coastal wave action for a given deepwater wave train. Early wave patterns of this type were computed graphically–Johnson, O'Brien, and Isaacs (1948)–and more recently the calculations have been programmed for high-speed computers; see, for example, Karlsson (1969) or Orr and Herbich (1969). Since the patterns must be repeated for each deepwater period and direction to paint a complete picture of coastal wave action, the digital computer solutions are recommended. Also, the interpretation of computed refraction patterns requires considerable engineering judgment, since crossed orthogonals ("caustics") and other pathological results often appear.

A second type of coastal wave effect is the phenomenon of "diffraction," which occurs when waves are partially blocked by an obstacle such as a portion of a breakwater. In this case, the waves are partially reflected and partially turned inside the obstacle by the "ends" of the breakwater, which serve as "sources" for generation of new and smaller waves. The theory of diffraction from acoustics and optics is applicable here to certain idealized breakwater problems, and examples have been computed by Penny and Price (1952) and by Johnson (1952). The chief problems studied are "ends" and "gaps" in breakwaters.

2-3.5 Nonlinear effects–finite-amplitude waves

The Airy theory fails when either (1) the amplitude is not small compared with the wavelength or (2) the amplitude is not small compared with the depth. Both cases violate our linearized free surface conditions, Eqs. (2-18*b*). Because different parameters are involved, two different theories have arisen to cover these two cases:

a. Deep water or "Stokes" theory: Series expansion in powers of $\frac{a}{L}$

b. Shallow water or "cnoidal" theory: Series expansion in powers of $\frac{a}{h}$

(2-42)

An excellent presentation of both these types of nonlinear theories is given in the text by Wiegel (1964). The chief results of the finite wave theory are several–first, that the phase speed is somewhat larger than an Airy wave:

a. Stokes' waves (Hunt 1953):

$$C^2 = \frac{g}{k} \tanh(kh) \left\{ 1 + \left(\frac{\pi a}{2L}\right)^2 \left[\frac{8 + \cosh(4kh)}{8 \sinh^4(kh)}\right] + \cdots + \mathcal{O}\left(\frac{a}{L}\right)^4 \right\} \qquad (2\text{-}43)$$

b. Solitary waves (Keulegan and Patterson 1940):

$$C^2 = gh\left(1 + \frac{2a}{h}\right)$$

The term in square brackets [ ] in the Stokes wave theory approaches unity in deep water. We see that Stokes' waves progress up to 5 percent faster and solitary waves up to 35 percent faster ($2a/h = 0.78$) than Airy waves. The second important result is that both theories give a criterion for breaking of a wave:

a. Stokes' waves (Michell 1893): $$\left(\frac{H}{L}\right)_{max} = 0.142 \tanh (kh)$$

$$\tag{2-44}$$

b. Solitary waves (McCowan 1891): $$\left(\frac{H}{h}\right)_{max} = 0.78$$

Finally, a striking phenomenon predicted by the nonlinear theories is the presence of a finite particle motion or "mass transport":

Stokes (1880): $$U_0 = \frac{\pi^2 H^2}{TL}\left[\frac{\cosh (2kh)}{2 \sinh^2 (kh)}\right] \tag{2-45}$$

Again the square bracket term is unity in deep water. The solitary wave also predicts a net particle motion in the direction of propagation, but no steady velocity, since it is only a single wave rather than a train of waves. This Stokes "drift" is a good approximation in deep water but is really quite small, i.e., $U_0 =$ 1 in/s for a 10-f wave of 15-s period. In coastal waters, boundary effects modify the formula greatly, and wind drift is usually a more significant cause of particle motions. See, for example, Teeson et al. (1970).

The two wave-breaking indices, Eqs. (2-44), are very important for design and placement of coastal structures, since breaking waves create very high impulsive forces (Kamel 1968). One uses the transformation theory, Eq. (2-40), to compute local H and L for a wave moving inshore, and breaking is predicted by whichever of the two criteria, (2-44*a*) or (2-44*b*), occurs first. The subsequent behavior of a breaking wave is dependent on the local depth and bottom slope, with three possibilities: "plunging," "spilling," or "surging" breakers. These are discussed in detail by Wiegel (1964), Galvin (1968), and Canfield and Street (1969).

2-4 WIND GENERATION OF WAVES

The Airy wave theory just discussed is ready for service if one knows the period and length of the local waves. But this is no simple matter: waves naturally occur in a continuous spectrum of frequencies, wavelengths, and heights. Figure 2-7 shows a schematic picture of the frequency distribution ("power spectrum") of wave energy in the real ocean, after a suggestion of Professor W. H. Munk. We see that, besides "gravity" and "capillary" waves, several other wave types exist,

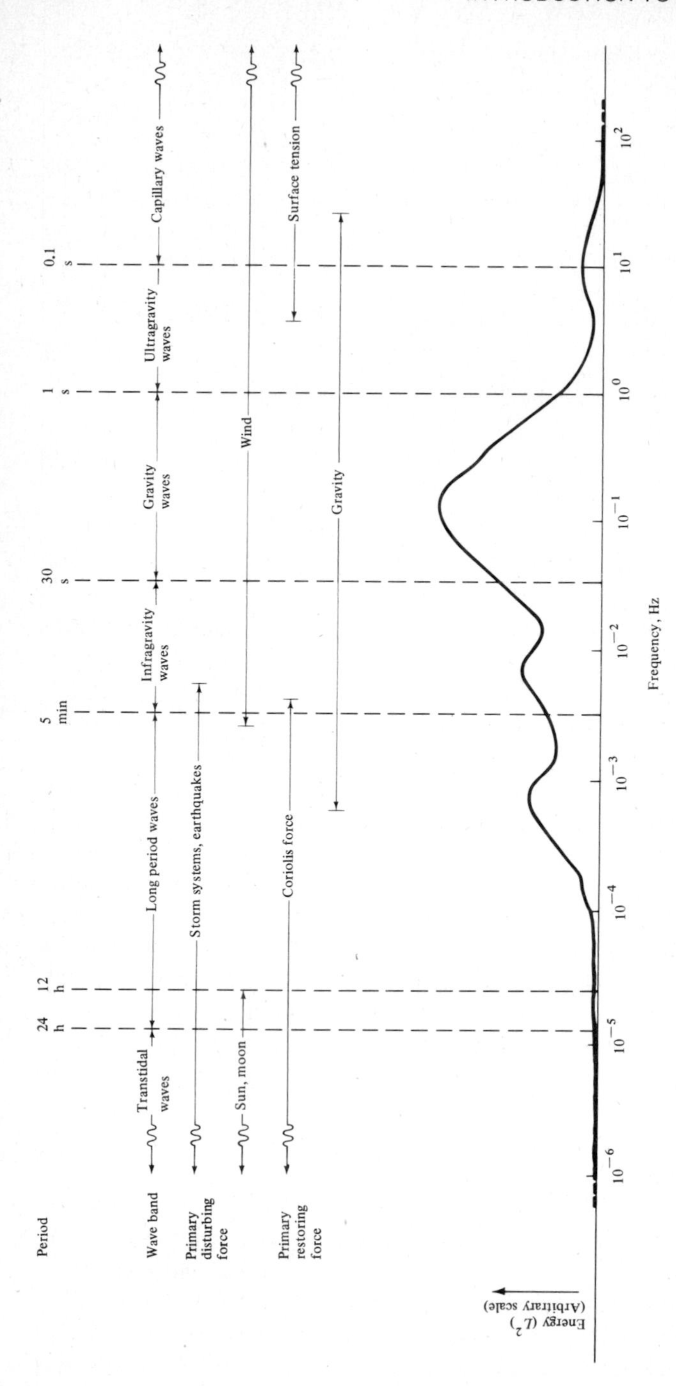

Fig. 2-7 Schematic drawing of the energy versus frequency distribution of naturally occurring ocean waves, after W. H. Munk.

with nearly discrete peaks at the diurnal (24-h) and semidiurnal (12-h) tides. The very-low-frequency waves ("long" waves) are the result of storms, earthquakes, and planetary motions. The high-frequency (0.1- to 10-Hz) waves are generated by wind acting on the air-sea interface. By far the majority of wave energy is contained in this range; hence, our practical problem is to predict the growth and subsequent behavior of gravity waves under wind excitation. Generally speaking, the larger the wind velocity and the longer its fetch and duration, the higher and longer will be the waves generated. Waves under direct wind action are called "sea" and are agitated and peaked in appearance. Waves running free of the storm are smoother, tend to decay with travel, and are called "swell." There are prediction techniques for generation of sea and also for decay of swell. Since gravity waves are dispersive (see Fig. 2-2), the long waves will move out ahead and the short waves will fall behind. There is constant group activity, and the spectrum of a packet of wind-generated waves is constantly changing during its run across the ocean.

There are two types of wave prediction methods: (1) analysis of gross average values—the "significant wave" technique; and (2) analysis of statistical distributions—the "wave spectra" methods. The latter have not yet become accepted as practical engineering tools but are being constantly perfected (Barnett 1968).

Still a third area of study might be called the "scientific" analysis of the precise mechanism of wind wave generation. How does an otherwise smooth ocean erupt into wind waves? Early work (Lamb 1932) attributed the initial growth to a Kelvin-Helmholtz instability, but the numerical values are unrealistic (60-kn wind speeds) unless the surface has extreme tension, i.e., a dense slick. The presently accepted mechanism of wave initiation is due to Phillips (1957), who postulated that atmospheric turbulent pressure fluctuations momentarily dapple the surface to allow the wind to get a grip, so to speak. Subsequent growth of the initiated waves is predicted by a wind-pressure–wave-form mechanism theorized by Miles (1957, 1960). Neither theory leads to tight engineering numbers, but together they suggest a plausible mechanism to build upon.

2-4.1 Gross prediction by dimensional analysis

The significant wave technique appeared in World War II with the work of Sverdrup and Munk (1947), which later was smoothed and reinterpreted by Bretschneider (1959). Thus we generally term this approach the SMB method. Many thousands of wave measurements, summarized by Longuet-Higgins (1952), had to be studied to obtain their gross average characteristics. It was found that the human eye tends to notice and average only the highest and longest one-third of the waves. A "significant wave," then, is by definition the average of this highest one-third. The significant wave height is denoted by $H_{1/3}$ and the significant period by $T_{1/3}$.

Figure 2-8 illustrates a view from above of a schematic wave generation area. We postulate that the average wave height and period at the end of the generation area (point A in Fig. 2-8) should depend on the fetch, the width, the wind velocity and duration, the water depth, and, of course, gravity. Thus we have

$$H_{1/3} \quad \text{or} \quad T_{1/3} = f(F, W, U, t, h, g) \tag{2-46}$$

By dimensional analysis (Bretschneider 1959), we may reduce this to a dimensionless relation involving a minimum number of parameters:

$$\frac{gH_{1/3}}{U^2} \quad \text{or} \quad \frac{gT_{1/3}}{U} = f\left(\frac{gF}{U^2}, \frac{gt}{U}, \frac{W}{F}, \frac{gh}{U^2}\right) \tag{2-47}$$

If the wind field is narrow (W less than F), as in a channel or bay, the wind is not as effective and wave heights will be less. Measurements by Saville (1954) indicate an effective height of 65 percent, 88 percent, and 98 percent for $W/F =$ 0.5, 1.0, and 1.5, respectively. In the open ocean, W/F is almost always greater than unity, indicating negligible width effect.

Further, if the depth is such that gh/U^2 is greater than about 3.0 (equivalent to h = 240 ft for a 30-kn wind), the effect of bottom friction is negligible. This leaves us with a "deepwater" functional relation for wave generation:

$W/F \geqslant 1.0, gh/U^2 \geqslant 3.0$:

$$\frac{gH_{1/3}}{U^2} \quad \text{and} \quad \frac{gT_{1/3}}{U} = f\left(\frac{gF}{U^2}, \frac{gt}{U}\right) \tag{2-48}$$

After analyzing numerous experiments from various parts of the ocean, Bretschneider (1959) recommended numerical values for the functional relations which are shown in Table 2-1. To use this table, one computes gF/U^2 and gt/U from the known wind conditions and examines the table; whichever of these two occurs earlier in the table is the correct line for reading $H_{1/3}$ and $T_{1/3}$. If gF/U^2

Fig. 2-8 Schematic of idealized wave generation parameters, viewed from above.

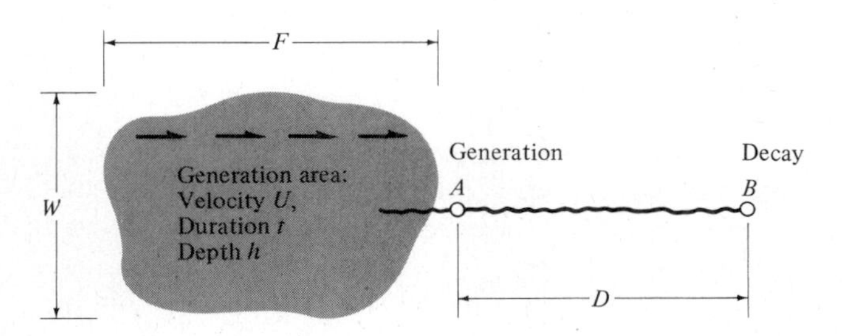

Comparison of
Building Seas

Table 2-1 Deepwater wave generation parameters, after Bretschneider (1959)

gF/U^2	gT/U	$gH_{1/3}/U^2$	$gT_{1/3}/2\pi U$
0.01	0.63	0.000574	0.0247
0.02	1.14	0.000611	0.0258
0.06	2.92	0.000867	0.0316
0.1	4.50	0.00105	0.0353
0.2	8.00	0.00143	0.0425
0.6	20.3	0.00235	0.0591
1.0	31.0	0.00301	0.0695
2.0	54.0	0.00430	0.0869
6.0	129.	0.00743	0.124
10.	192.	0.00951	0.147
20.	306.	0.0129	0.179
60.	654.	0.0208	0.240
100.	920.	0.0255	0.279
200.	1,520.	0.0337	0.337
600.	3,300.	0.0522	0.453
1,000.	4,800.	0.0641	0.519
2,000.	8,000.	0.0841	0.618
6,000.	18,960.	0.130	0.816
10,000.	28,100.	0.157	0.924
20,000.	48,200.	0.195	1.10
60,000.	112,800.	0.253	1.39
100,000.	168,000.	0.270	1.54
200,000.	286,000.	0.279	1.74
600,000.	702,000.	0.282	1.95

occurs first, the waves are said to be "fetch-limited"; if gt/U comes first, they are "duration-limited." In other words, either the fetch or the duration is too short for fully effective waves to form. If both gF/U^2 and gt/U should exceed the values in the table—which is unlikely—we have a "fully arisen sea," for which the last two columnar values of 0.282 and 1.95 are assumed to remain constant.

As an example, suppose the wind velocity is 35 kn (59 ft/s), the fetch is 125 mi, and the duration 4 h. Then gF/U^2 equals 6,000 and gt/U equals 8,000. The latter value occurs first in Table 2-1; hence the waves are duration-limited, and we read $gH/U^2 = 0.0841$ and $gT/2\pi U = 0.618$. Substituting for g and U, we compute a significant height of 9 ft and a significant period of 7 s. Had the wind blown longer than 9.7 h, for which gt/U equals 18,960, this same example would

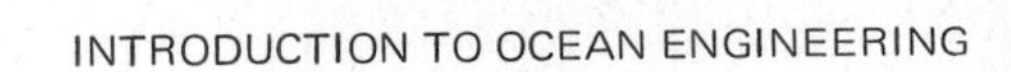

be fetch-limited. To the writer's knowledge, this simple procedure still affords the best present general estimate of wind-generated waves.

In shallow water, we must include the additional parameter gh/U^2. This is shown in Fig. 2-9 for a fetch-limited wind. We enter this chart at gh/U^2 and read gH/U^2 for the appropriate value of gF/U^2, the values of which at the far right side exactly coincide with the deepwater prediction, Table 2-1. Suppose for the example above that the average depth in the generating area was 20 ft. Then $gh/U^2 = 0.18$, and we find from Fig. 2-9 that the wind is severely depth-limited and we read $gH/U^2 \doteq 0.047$. For the given wind velocity of 59 ft/s, this corresponds to a wave height $H = 5$ ft, compared to the deepwater value of 9 ft. Figure 2-9 may be used for duration-limited winds also if we first determine the correct value of gF/U^2 from Table 2-1 which matches the dimensionless duration gt/U.

2-4.2 Statistical estimates of wind wave generation

The significant-wave concept is a gross parameter and does not indicate the spectral distribution of waves in a real sea. This task has been the object of extensive study in the past 20 years, and the early work is summarized by Longuet-Higgins (1952). The first practical technique was assembled by Pierson, Neumann, and James (1955), but this PNJ method, which has been included in a recent design handbook by the Corps of Engineers (U.S. Army 1961), is now rather dated. The state of the art in ocean wave spectra was reviewed in 1963 by the National Academy of Sciences (1963). Statistical predictions must take account of nonlinear interactions between the various wave components. These interactions have been reviewed by Lighthill et al. (1967). Modern theories take advantage of the digital computer, and several approaches are now under intense study: at New York University (Pierson et al. 1966), at the U.S. Navy (Moskowitz 1967), and in France (Fons 1966). Review articles on these techniques have been given by Barnett (1968) and by Bunting (1970).

In the meantime, it is desirable to have some estimate of the statistics of the sea surface, particularly the spectral density or "power spectrum" of surface elevation, $S_\eta(\omega)$. One estimate of this spectrum is that of the PNJ method (Pierson et al. 1955), but a surprisingly simple and reliable estimate is the formula proposed by Bretschneider (1959):

$$S_\eta(\omega) \doteq 1.35 \frac{H^2}{\omega}\left(\frac{2\pi}{\omega T}\right)^4 e^{-0.675(2\pi/\omega T)^4} \tag{2-49}$$

where H and T are significant wave height and frequency from, say, Table 2-1 or Fig. 2-9 and ω is the frequency. Some parametric plots of Eq. (2-49) for unit wave height and various periods are shown in Fig. 2-10. In linear analyses, Eq. (2-49) may be substituted directly into the usual linear spectrum formulas to calculate root-mean-square response of ocean systems to wave excitation. For

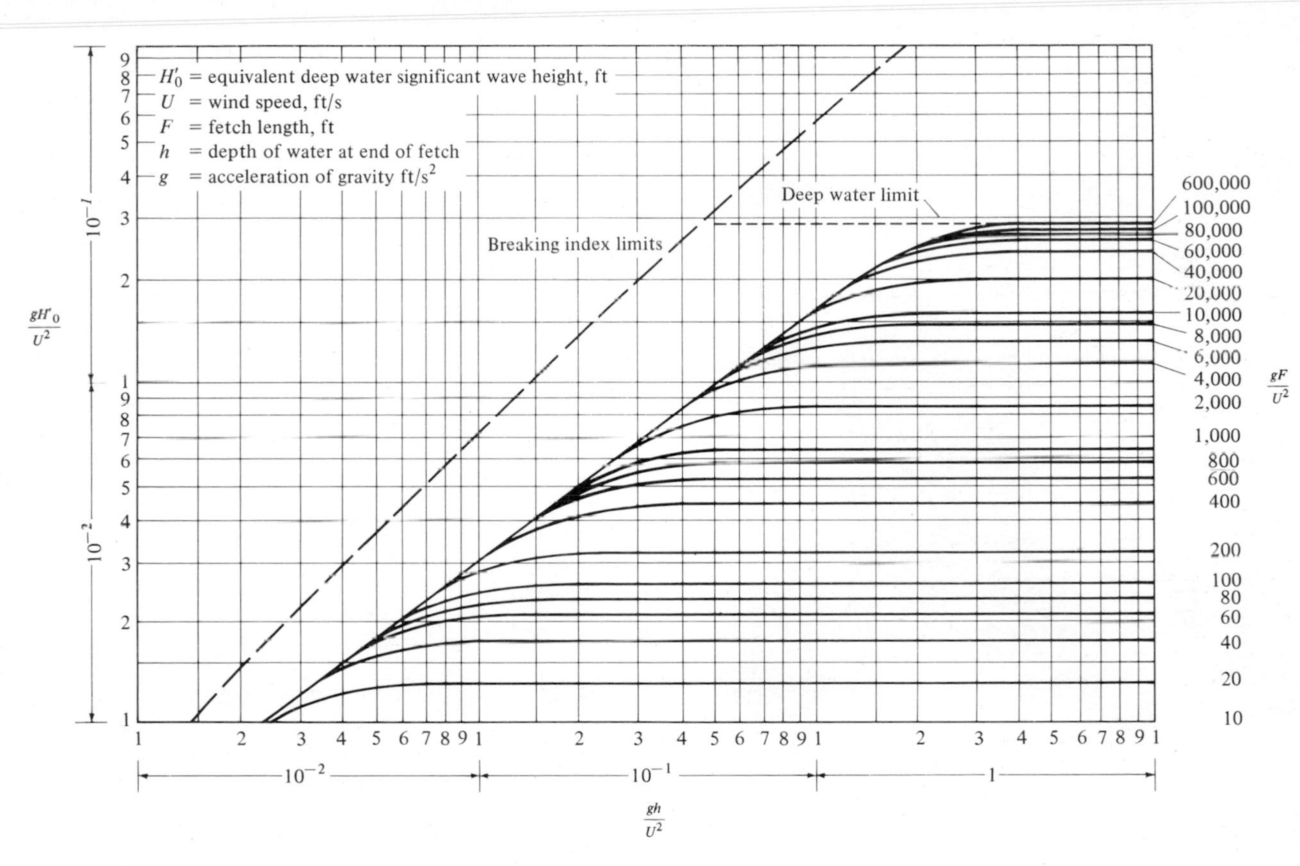

Fig. 2-9 Wave generation in shallow water for fetch-limited case [after Bretschneider (1959)].

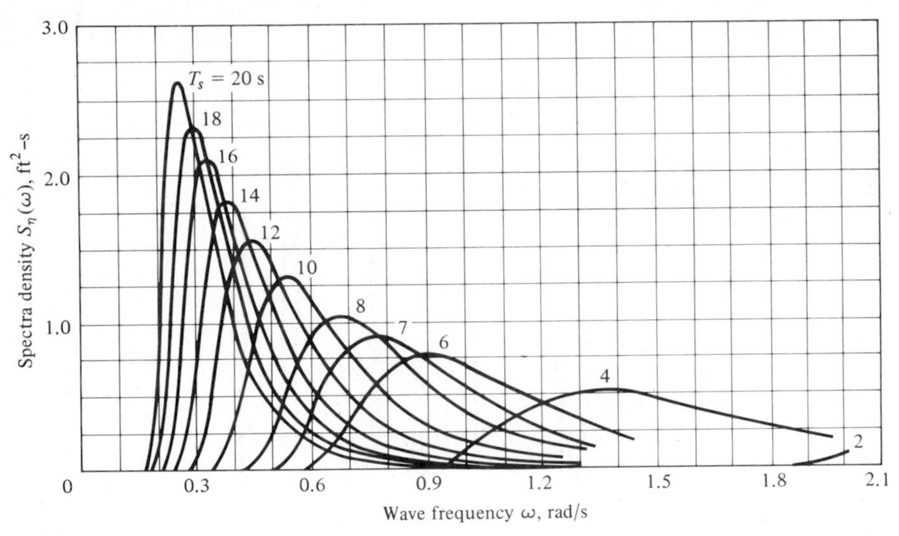

Fig. 2-10 Parametric plots of the Bretschneider spectral density Eq. (2-49), for a unit significant wave height.

nonlinear analyses, Borgman (1969) has shown how the spectral density function may be formulated into a random sea state time series which can be used as a forcing function for nonlinear systems. Note that the spectral density in Eq. (2-49) is consistent with the fact that, except for severe storms, wave heights in the ocean have approximately a Rayleigh-type probability distribution:

$$p(H) = \frac{2H}{Q} e^{-H^2/Q}$$

where

$$Q = \int_0^\infty S_\eta(\omega)\, d\omega \tag{2-50}$$

Similarly, in deep water at least, the square of the wave period also follows a Rayleigh distribution, $p(T^2) \sim T^2 \exp(-T^4)$. For more details, consult the paper by Bretschneider (1959). This paper also contains data on decay of wind waves. These data show extreme scatter and for some reason have not been nondimensionalized.

2-5 WAVE INTERACTION WITH STRUCTURES

The computation of forces due to wave motion past a structure is a relatively simple matter if one is willing to accept two engineering approximations. Fluid forces are of three types: pressure, viscous, and inertia forces. Pressure and viscous forces are commonly measured in steady flow experiments, such as wind or water tunnels, where inertia forces are absent because the fluid is unaccelerated relative to the structure. The measured force is not split into "pressure" plus "viscous"–which would be a difficult breakdown–but is simply resolved into components: the "drag" parallel to the flow velocity and the "lift" normal to the flow. On the other hand, inertia forces arise from unsteady flow. The fluid must divide and accelerate to pass around the obstacle, and the momentum imparted to the particles reflects in an equivalent reaction force on the body. This force is proportional to the relative fluid acceleration and hence the proportionality constant has mass units and is commonly called the "virtual" or "hydrodynamic" mass of the body. Some experimental data are available for virtual mass, but commonly the coefficients are computed from potential theory (Lamb 1932).

For engineering purposes, then, we consider drag, lift, and inertia forces on the body, as shown in Fig. 2-11. By definition,

a. Pressure plus viscous forces:

1. Drag: $F_D = C_D \frac{1}{2} \rho V^2 A$

2. Lift: $F_L = C_L \frac{1}{2} \rho V^2 A$ (2-51)

b. Inertia force:

$$F_I = M_0 \frac{dV}{dt} = C_M \rho B \frac{dV}{dt}$$

Here A and B are a reference area and reference volume of the body, respectively, and ρ is the *fluid* density–not the body density, which is of no importance except in computing the buoyant force and acceleration of the body. These forces do not necessarily pass through the centroid of the body. The inertia force, from potential theory, passes through the centroid computed for a body of uniform density, but the drag and lift are primarily influenced by body shape (flaps, wings, etc.) and do not pass through the centroid unless both

Fig. 2-11 Sketch of fluid forces on a fixed solid body.

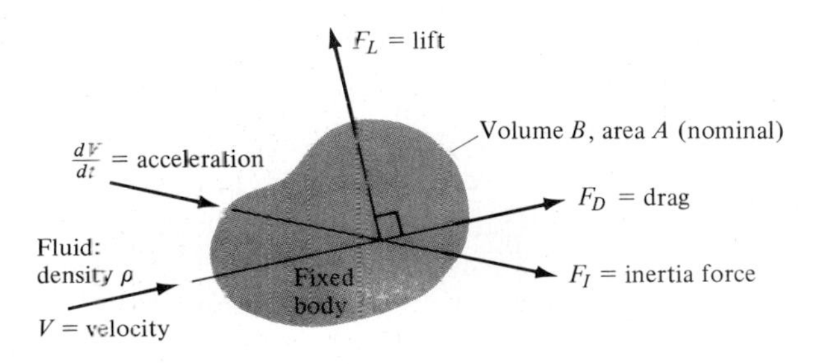

the body and the flow field are symmetrical, e.g., a uniform current about a sphere.

The three quantities C_D, C_L, and C_M are dimensionless and are called the drag, lift, and inertia coefficients, respectively. They vary with body shape even in the simplest theory and, in a practical ocean problem, there are effects due to Reynolds number, Froude number, body surface roughness, vortex shedding, wave frequency (Strouhal number), body diffraction, and the proximity of other bodies. Further, even in careful wave tank experiments with simple shapes, the scatter in drag and inertia data is painful to behold (Wiegel 1964, pp. 258 and 262). Also, strictly speaking, the inertia coefficient C_M is a tensor quantity.

In the face of this confused experimental situation, it is customary to make engineering approximations: (1) C_D is constant; (2) C_M is a scalar and approximately constant; and (3) if the body has lifting characteristics, C_L is proportional to the angle of attack, i.e., the angle between the stream velocity and the chord line of the body. The two most common shapes in ocean structures are spheres and cylinders:

$$\begin{aligned} &a.\ \text{Sphere:} \quad C_D \doteq 0.7 \quad C_M \doteq 1.5 \\ &b.\ \text{Cylinder:} \quad C_D \doteq 1.2 \quad C_M \doteq 2.0 \end{aligned} \tag{2-52}$$

These are referred to as the frontal area and the body volume. An extensive table of average mass coefficients for various body shapes is given by Patton (1965).

Consider a small submerged body whose diameter is much less than the water depth or the length of waves passing by. If the waves have period T and height H and the body is nonlifting, the total horizontal force on the body can be computed from Airy theory based on the horizontal velocity u from Eq. (2-31) and its time derivative. If the body is submerged at depth z, the total oscillating force is, from Eqs. (2-52),

$$\begin{aligned} F_h = &-2\pi^2 C_M \rho B \frac{H}{T^2} \frac{\cosh k(z+h)}{\sinh (kh)} \sin (\sigma t) \\ &+ \tfrac{1}{2}\pi^2 C_D \rho A \frac{H^2}{T^2} \left[\frac{\cosh k(z+h)}{\sinh (kh)}\right]^2 |\cos (\sigma t)| \cos (\sigma t) \end{aligned} \tag{2-53}$$

Since both terms vary inversely with T^2, we see that long period waves are no problem. The maximum force occurs neither at the wave crest nor in the trough but leads the crest by an amount which depends on the ratio B/AH, i.e., the body diameter to wave height ratio. If this ratio is small–small diameter, high waves–the drag force is the most important term, and vice versa. If the body extends over a considerable depth, such as a long piling, an engineering estimate is to integrate Eq. (2-53) with respect to z, considering B and A as the volume and area of a two-dimensional slice of height dz. For example, for a vertical piling of diameter D, $dA = D\,dz$ and $dB = (\pi D^2/4)\,dz$. If the piling extends from the bottom to the surface, the total horizontal force on the piling is:

$$F_h(\text{vertical piling}) = -\tfrac{1}{4}\pi^2 C_M \rho \frac{D^2 HL}{T^2} \sin(\sigma t)$$

$$+ \pi C_D \rho \frac{DH^2 L}{T^2} \left[\frac{2kh + \sinh(2kh)}{16 \sinh^2(kh)} \right] |\cos(\sigma t)| \cos(\sigma t) \quad (2\text{-}54)$$

and again we see that the ratio D/H determines the relative importance of inertia compared with drag. Remember this is hardly an exact solution for the piling force but instead is an incremental approximation based on simple drag and inertia and wave theories.

Equation (2-54) is the force due to a single wave component (H,L,T). Since we have already seen that the seaway contains a spectrum of heights and periods, Eq. (2-54) is only appropriate in the "significant wave" sense or else should be applied to a statistical theory. Borgman (1969) has shown how to simulate the seaway statistically, and Brown and Borgman (1967) give tables of statistical wave forces and estimates of drag and inertia coefficients. Grace and Casciano (1969) apply these concepts to subsurface spheres, and Kamel (1968) considers forces on seawalls and breakwaters. Plate and North (1969) consider the modeling of structures for experimental studies, and Foster (1970) has reviewed nonlinear effects in ocean structure analysis. For further details on ocean structures, consult Chap. 9; for further details on the resistance of bodies in fluids, see Chap. 4.

2-6 HYDRODYNAMICS OF ESTUARIES

We found earlier in this chapter that waves and currents can be computed to good accuracy without considering variations in temperature or salinity. The same is true of flow conditions in a tidal estuary, although there are significant added motions if the density and temperature are strongly stratified. The opposite is not at all true: temperature and salinity and other thermodynamic properties, such as dissolved oxygen, are strongly dependent on flow conditions because of "convective" effects. The computation of currents and tides in an estuary is an established procedure (Leendertse 1969), but prediction techniques for thermal, salt, and oxygen diffusion are still being developed.

2-6.1 Equations of mean turbulent motion

All coastal flow problems are of such a large scale that they are naturally "turbulent"; i.e., the flow properties continually fluctuate in time with a spectrum of frequencies in the range of 1 to 1,000 kHz. These fluctuations are superimposed on tidal variations which have much longer periods, of the order of hours. Therefore, following a classic idea by Osborne Reynolds in 1895, we may "average" the flow properties over a 1- or 2-min time period which is "long" compared to turbulent fluctuations, yet "short" compared to tidal variations. We denote the average value of any property q by $\bar{q}$ and its turbulent fluctuation by q'. Thus we superimpose as follows:

$$q = \bar{q} + q'$$

where

$$\bar{q} = \frac{1}{\Delta t} \int_0^{\Delta t} q \, dt \tag{2-55}$$

By definition, $\overline{q'}$ vanishes. The averaging period Δt resolves the turbulence but still allows $\bar{q}$ to contain tidal variations. Suppose now we define $u = \bar{u} + u'$, $T = \bar{T} + T'$, etc., substitute into our basic equations of motion, Eqs. (2-5) to (2-8), and carry out the averaging process on each entire equation. The results are the Reynolds equations for mean turbulent motion:

a. Continuity:

$$\frac{\partial \bar{u}}{\partial x} + \frac{\partial \bar{v}}{\partial y} + \frac{\partial \bar{w}}{\partial z} = 0$$

b. X momentum (neglecting Coriolis and gravity effects):

$$\frac{d\bar{u}}{dt} = -\frac{1}{\rho}\frac{\partial \bar{p}}{\partial x} + \frac{\partial}{\partial x}\left(n \frac{\partial \bar{u}}{\partial x} - \overline{u'u'}\right) + \frac{\partial}{\partial y}\left(n \frac{\partial \bar{u}}{\partial y} - \overline{u'v'}\right) + \frac{\partial}{\partial z}\left(n \frac{\partial \bar{u}}{\partial z} - \overline{u'w'}\right)$$

c. Energy (neglecting viscous dissipation effects):

$$\frac{d\bar{T}}{dt} = \frac{\partial}{\partial x}\left(\alpha \frac{\partial \bar{T}}{\partial x} - \overline{u'T'}\right) + \frac{\partial}{\partial y}\left(\alpha \frac{\partial \bar{T}}{\partial y} - \overline{v'T'}\right) + \frac{\partial}{\partial z}\left(\alpha \frac{\partial \bar{T}}{\partial z} - \overline{w'T'}\right) \tag{2-56}$$

d. Salt diffusion (or any other conservative species S):

$$\frac{d\bar{S}}{dt} = \frac{\partial}{\partial x}\left(D \frac{\partial \bar{S}}{\partial x} - \overline{u'S'}\right) + \frac{\partial}{\partial y}\left(D \frac{\partial \bar{S}}{\partial y} - \overline{v'S'}\right) + \frac{\partial}{\partial z}\left(D \frac{\partial \bar{S}}{\partial z} - \overline{w'S'}\right)$$

where $\alpha = k/\rho c_v$ is the thermal diffusivity of the fluid. We see that mean continuity is unaffected by fluctuating terms and that each of the other three equations has encumbered three new (and unknown) convective terms involving fluctuating products: terms such as $\overline{u'v'}$ (a turbulent "shear"), $\overline{u'T'}$ (turbulent heat "flux"), and $\overline{u'S'}$ (turbulent "salt flux"). Note the striking similarity between all three equations, particularly if pressure gradients are negligible, as is often the case. It is the task of the analyst to model these turbulent fluxes as a function of the geometric, kinematic, and thermodynamic scales of the motion. One tentative idea is to model each flux term after its molecular neighbor in the same parentheses. For example, we may rewrite the salt diffusion equation in the form

$$\frac{d\bar{S}}{dt} = \frac{\partial}{\partial x}\left(D_x \frac{\partial \bar{S}}{\partial x}\right) + \frac{\partial}{\partial y}\left(D_y \frac{\partial \bar{S}}{\partial y}\right) + \frac{\partial}{\partial z}\left(D_z \frac{\partial \bar{S}}{\partial z}\right) \tag{2-57}$$

where, for example, $D_x = D - \overline{u'S'}/(\partial \bar{S}/\partial x)$ is a composite of both molecular diffusion (which is usually negligible) and turbulent convective flux. This redefinition solves no problems, but one then attempts to correlate (D_x, D_y, D_z) empirically with the velocity and length scales of the motion. In a narrow estuary, the streamwise component D_x may be dominant (Glenne and Selleck 1969); in the open ocean, the lateral diffusion D_y may dominate (Orlob 1959); and near the bottom or in very shallow water, the vertical diffusion D_z may be most important (Vager and Kagan 1969). Little "fundamental" work has been done on these turbulent coefficients; one often simply plots measured D versus stream velocity or channel depth and width. The scatter may be huge.

2-6.2 Tidal currents and heights

One may notice that Eqs. (2-56*a* and *b*) are essentially uncoupled from temperature and salinity. Even without knowing the structure of the turbulent fluxes $\overline{u'v'}$, etc., one may solve them for mean flow parameters by integrating the equations from the bottom to the surface. Define surface height $\eta(x,y,t)$ and mean flow rates:

$$Q_x = \int_{-h}^{\eta} u\,dz \qquad Q_y = \int_{-h}^{\eta} v\,dz \tag{2-58}$$

Also, in an estuary, the pressure is nearly hydrostatic, so that $dp = \rho g\,d\eta$. Therefore the vertically integrated equations of continuity and momentum become, approximately:

$$\begin{aligned}
&\frac{\partial Q_x}{\partial x} + \frac{\partial Q_y}{\partial y} + \frac{\partial \eta}{\partial t} = 0 \\
&\frac{\partial Q_x}{\partial t} + \frac{Q_x}{h+\eta}\frac{\partial Q_x}{\partial x} + \frac{Q_y}{h+\eta}\frac{\partial Q_x}{\partial y} + g\frac{\partial \eta}{\partial x} = [\tau_{xz}(\eta) - \tau_{xz}(-h)]\frac{1}{\rho} \\
&\frac{\partial Q_y}{\partial t} + \frac{Q_x}{h+\eta}\frac{\partial Q_y}{\partial x} + \frac{Q_y}{h+\eta}\frac{\partial Q_y}{\partial y} + g\frac{\partial \eta}{\partial y} = [\tau_{yz}(\eta) - \tau_{yz}(-h)]\frac{1}{\rho}
\end{aligned} \tag{2-59}$$

where $h = h(x,y)$ denotes the variable bottom contour. The surface and bottom stresses on the right-hand side are estimated from wind and bottom drag coefficients, respectively. The equations are nonlinear but may easily be programmed for a digital computer, a notably successful example being the analysis of Leendertse (1969). Alternately, one can linearize the convection and stress terms and obtain analytic solutions for simple geometries, some examples of which are given in the text by Ippen (1966). Such two-dimensional "long-wave" computer models are now being developed for many estuaries throughout the world.

2-6.3 Turbulent diffusion in an estuary

In principle, the flow rates Q_x and Q_y from Eqs. (2-59) may be introduced into the temperature and energy equations (2-58*c* and *d*) to determine vertically averaged temperature and salinity (or other species) in an estuary. The temperature equation would require knowledge of surface and bottom heat flux rates, but the salinity relation would not, since presumably no salt is entering either surface. Computer salinity models are now being developed (Leendertse 1969), and work is proceeding on measurement and analysis of thermal diffusion (Wada 1969) and also dissolved oxygen (Juliano 1969). Much of the work has been limited to one-dimensional models of longitudinal (streamwise) diffusion. If $U(x)$ is the streamwise velocity, we have, approximately:

$$\frac{\partial S}{\partial t} + U \frac{\partial S}{\partial x} \cong \frac{\partial}{\partial x}\left(D_x \frac{\partial S}{\partial x}\right) \tag{2-60}$$

which may be solved easily by breaking the stream up into longitudinal slices. Even so, the estimation of the longitudinal diffusion coefficient $D_x(x,t)$ has not been resolved accurately.

The idea of "vertically averaged" is useful only in well-mixed estuaries, where the tidal prism is large compared to the freshwater flow. Many estuaries have large freshwater flow and are therefore stratified, the heavier saline water being at the bottom. This problem of "salinity intrusion" in an estuary has been extensively studied (Burgh 1968, Dyer and Behnke 1968). Such bottom seawater intrusion is often the cause of heavy sediment shoaling (Harleman and Ippen 1969). Work continues on analysis of diffusion and entrainment in such two-layer flows (Carstens 1970) and the attendant stability effects on turbulent mixing (Bowden 1967). The writer believes that computer modeling of tides, temperature, salinity, dissolved oxygen, etc., will become commonplace in the next decade.

There is also the nonmathematical approach of building an actual physical model of an estuary, including mechanical tide generators and dye studies for diffusion. Such models continue to be studied (Cowley 1969) but have certain drawbacks, notably (1) that the vertical scale must be distorted 100:1 to avoid capillary effects; and (2) that a small model is subject to molecular diffusion, which would be unimportant in the real estuary.

More details on estuarine pollution and its many parameters are contained in Chap. 10. The present chapter is intended to serve only as an introduction to coastal hydrodynamics.

PROBLEMS

2-1. Derive an equation from the material in Sec. 2-3 for the pressure at any depth Z beneath a wave traveling overhead. Note that this equation is crucial to the design of a subsurface pressure transducer for wave measurement. *Hints*: The function $f(t)$ in Eq. (2-15) can be obtained noting that in still water, P is P_a and the other terms are zero. Differentiating Eq. (2-30) will give $(\partial\phi/\partial t)$ which, with Eq. (2-29), will give a usable expression for $\rho(\partial\phi/\partial t)$ in Eq. (2-15), providing we use Eq. (2-26) to obtain an expression σc. We then assume the kinetic energy term $\frac{1}{2}\rho V^2$ is small in Eq. (2-15) to obtain the final result.

2-2. A pressure transducer placed at the bottom at 20 ft will register a gauge pressure of 1,280 lb/ft^2. If a harmonic wave of amplitude $a = 1$ ft and length $L = 100$ ft passes over this tranducer, use the equation obtained in Prob. 2-1 to find the total pressure when the wave is directly overhead and the percentage increase in pressure over the no-wave condition.

2-3. A two-dimensional (linearized wave is found to have a phase speed of $c = 25$ ft/s when traveling in water of uniform depth of $h = 20$ ft. What is its length?

2-4. A deepwater wave has a length of $L = 1{,}000$ ft and an amplitude of $a = 5$ ft. Assuming a gradual transformation of depth with no change in width, at what depth will the amplitude have risen to 7 ft?

2-5. A 100-ft-long wave of height 6 ft and period 6 s moves toward a coast from deep water:

(*a*) What is the speed at which the wave approaches the coast?

(*b*) What is the total energy of the wave in ft/(lb$_f$)/(ft) of wave breadth?

(*c*) What horsepower could 1 ft of this wave generate if conversion were total?

(*d*) At what approximate depth will the wave begin to change form owing to bottom friction?

(*e*) What will be the length, period, and height of this wave in a depth of 8 ft? Will it break?

2-6. A horizontal cylinder 10 ft long and 3 ft in diameter is anchored submerged in deep water where waves of length 200 ft and height 10 ft are passing. Considering the maximum impulsive forces only:

(*a*) What is the ratio of maximum inertia force to maximum drag force?

(*b*) If this shape were anchored just beneath the surface in 50-ft-deep water with the same waves running past as above, would the maximum inertia force from the waves be more or less than the deepwater case?

(*c*) By how much would the drag on the cylinder be reduced if it is anchored at 25 ft rather than just under the surface?

2-7. 80 mi/h winds blow across Narragansett Bay for 8 h, a distance of 2,000 yd. Predict the wave heights at a pier in 20 ft of water. Will these waves break? If not, predict where (at what depth) the break will occur.

2-8. A 2-ft spherical buoy is anchored in deep water at a depth 4 ft below mean water level. A wave of length 150 ft and height 6 ft passes by. What will be the maximum force on the buoy owing to drag and inertia forces?

2-9. A half-circular bay 1,000 ft in diameter has a water depth of 50 ft at its center sloping uniformly upward to the beach. A 10-foot-high wave with a period of 6 s enters this bay. Sketch where the break will occur on a scale diagram of the bay.

2-10. A whole gale (70 mi/h) blows for 7 h from the south across Buzzards Bay (Mass.) a distance of 20 mi. What will be the wave height on the north side if the bay averages 50 ft deep?

REFERENCES

Airy, G. B. (1842): Tides and Waves, *Encyclopedia Metropolitana*, vol. 5, pp. 241–396.

Barnett, T. P. (1968): On the Generation, Dissipation, and Prediction of Ocean Wind Waves, *J. Geophys. Res.*, vol. 73, pp. 513–529.

Borgman, L. E. (1969): Ocean Wave Simulation for Engineering Design, *ASCE Waterways Harbors Div. J.*, vol. 95, pp. 557–583.

Bowden, K. F. (1967): Stability Effects on Turbulent Mixing in Tidal Currents, *Phys. Fluids*, vol. 10, no. 9, (II), pp. 278–280.

Bretschneider, C. L. (1959): "Wave Variability and Wave Spectra for Wind Generated Gravity Waves," U.S. Army Corps of Engineers, Beach Erosion Board, Technical Memo no. 118.

Brown, L. J., and L. E. Borgman (1967): "Tables of the Statistical Distribution of Ocean Wave Forces and Methods of Estimating Drag and Mass Coefficients," U.S. Army Coastal Engineering Research Center, Technical Memo no. 24.

Bunting, D. C. (1970): Evaluating Forecasts of Ocean Wave Spectra, *J. Geophys. Res.*, vol. 75, pp. 4131–4143.

Burgh, P. van der (1968): Prediction of the Extent of Saltwater Intrusion into Estuaries and Seas, *J. Hydraulic Res.*, Delft, vol. 6, no. 4, pp. 267–288.

Canfield, F. E., and R. L. Street (1969): The Effect of Bottom Configuration on the Deformation, Breaking, and Run-up of Solitary Waves, *Proc. 11th Conf. Coastal Eng.*, vol. I, pp. 173–189.

Carstens, T. (1970): Turbulent Diffusion and Entrainment in Two-Layer Flow, *ASCE Waterways Harbors J.*, vol. 96, pp. 97–104.

Cowley, J. E. (1969): Tidal Model of Northumberland Strait, *ASCE Hydraulics Div. J.*, vol. 95, pp. 827–838.

Dyer, K. L., and J. J. Behnke (1968): Sea Water Intrusion into a Fresh Water Forebay Due to Wave Action, *J. Hydrol.*, vol. VI, no. 1, pp. 95–101.

Fons, C. (1966): Prévision de la Houle, la Méthode des Densités Spectro-angulaires Non (DSA-5), *Cahiers Oceanog., Bull. Inform.*, vol. 18, part 1, pp. 15–33.

Foster, E. T. (1970): Model for Nonlinear Dynamics of Offshore Towers, *ASCE Eng. Mech. Div. J.*, vol. 96, pp. 41–67.

Galvin, C. J. (1968): Breaker Type Classification on Three Laboratory Beaches, *J. Geophys. Res.*, vol. 73, pp. 3651–3659.

Glenne, B., and R. E. Selleck (1969): Longitudinal Estuarine Diffusion in San Francisco Bay, California, *Water Res.*, vol. 3, no. 1, pp. 1–20.

Grace, R. A., and F. M. Casciano (1969): Ocean Wave Forces on a Subsurface Sphere, *ASCE Waterways Harbors J.*, vol. 95, pp. 291–317.

Harleman, D. R. F., and A. T. Ippen (1969): Salinity Intrusion Effects in Estuary Shoaling, *ASCE Hydraulics Div. J.*, vol. 95, pp. 9–27.

Hunt, J. N. (1953): A Note on Gravity Waves of Finite Amplitude, *Quart. J. Mech. Appl. Math.*, vol. 6, pp. 336–343.

Ippen, A. T. (1966): "Estuary and Coastline Hydrodynamics," McGraw-Hill, New York.

Johnson, J. W. (1952): Generalized Wave Diffraction Diagrams, *Proc. 2d Conf. Coastal Eng.*, pp. 6–23.

——, M. P. O'Brien, and J. D. Isaacs (1948): "Graphical Construction of Wave Refraction Diagrams," U.S. Navy Hydrographic Office, Publication no. 605.

Juliano, D. W. (1969): Reaeration Measurements in an Estuary, *ASCE Sanitary Eng. Div. J.*, vol. 95, pp. 1165–1178.

Kamel, A. M. (1968): "Water Wave Pressures on Seawalls and Breakwaters," U.S. Army Waterways Experiment Station, Vicksburg, Miss., Research Report no. 2-10.

Karlsson, T. (1969): Refraction of Continuous Ocean Wave Spectra, *ASCE Waterways Harbors Div. J.*, vol. 95, pp. 437–448.

Keulegan, G. H. (1959): Energy Dissipation in Standing Waves in Rectangular Basins, *J. Fluid Mech.*, vol. 6, pp. 33–50.

——, and G. W. Patterson (1940): Mathematical Theory of Irrotational Translation Waves, *J. Res. Natl. Bur. Std.*, vol. 24, no. 1, pp. 47–101.

Kinsman, B. (1965): "Wind Waves–Their Generation and Propagation on the Ocean Surface," Prentice-Hall, Englewood Cliffs, N.J.

Lamb, H. (1932): "Hydrodynamics," 6th ed., Cambridge, London (also Dover, New York, 1945).

Leendertse, J. J. (1969): Use of a Computational Model for Two-Dimensional Tidal Flow, *Proc. 11th Conf. Coastal Eng.*, vol. II, pp. 1403–1420.

Lighthill, M. J., et al. (1967): A Discussion on Nonlinear Theory of Wave Propagation in Dispersive Systems, *Proc. Roy. Soc. London*, Ser. A, vol. 299, pp. 1–145.

Longuet-Higgins, M. S. (1952): On the Statistical Distribution of the Heights of Sea Waves, *J. Marine Res.*, vol. 11, pp. 245–266.

McCowan, J. (1891): On the Solitary Wave, *London/Edinburgh/Dublin Phil. Mag. J. Sci.*, vol. 32, pp. 45–58.

Miche, R. (1944): Mouvements Ondulatoires des Mers en Profondeur Constante ou Décroissante, *Annales des Ponts et Chausées*, pp. 25–78, 131–164, 270–292, 369–406.

Michell, J. H. (1893): On the Highest Waves in Water, *Phil. Mag.*, vol. 36, no. 5, pp. 430–435.

Miles, J. W. (1957): On the Generation of Surface Waves by Shear Flows, *J. Fluid Mech.*, vol. 3, part 3, pp. 185–204.

——, (1960): On the Generation of Surface Waves by Turbulent Shear Flows, *J. Fluid Mech.*, vol. 7, part 3, pp. 469–478.

Moskowitz, L. I. (1967): "Evaluation of Spectral Wave Hindcasts Using the Automated Wave Prediction Program of the Naval Oceanographic Office," U.S. Naval Oceanographic Office Informal Report IR-67-78.

National Academy of Sciences (1963): "Ocean Wave Spectra," Prentice-Hall, Englewood Cliffs, N.J.

Neumann, G., and W. J. Pierson (1966): "Principles of Physical Oceanography," Prentice-Hall, Englewood Cliffs, N.J.

Orlob, G. T. (1959): Eddy Diffusion in Homogeneous Turbulence, *ASCE Hydraulics Div. J.*, vol. 85, pp. 75–101.

Orr, T. E., and J. B. Herbich (1969): Numerical Calculation of Wave Refraction by Digital Computer, *Proc. 2d Offshore Technol. Conf.*, vol. II, pp. 533–540.

Patton, K. T. (1965): "Tables of Hydrodynamic Mass Factors for Translational Motion," ASME Paper no. 65-WA/UNT-2.

Penny, W. G., and A. T. Price (1952): The Diffraction Theory of Sea Waves by Breakwaters, and the Shelter Afforded by Breakwaters, *Phil. Trans. Roy. Soc. London*, Ser. A, vol. 244, pp. 236–253.

Phillips, O. M. (1957): On the Generation of Waves by Turbulent Winds, *J. Fluid Mech.*, vol. 2, part 5, pp. 417–445.

——, (1966): "The Dynamics of the Upper Ocean," Cambridge, London.

Pierson, W. J., G. Neumann, and R. W. James (1955): "Practical Methods of Observing and Forecasting Ocean Waves by Means of Wave Spectra and Statistics," U.S. Navy Hydrographic Office, Publication no. 603.

——, L. J. Tick, and L. Baer (1966): Computer-based Procedures for Preparing Global Wave Forecasts and Wind Field Analyses, *Proc. 6th Symp. Naval Hydrodynamics*, pp. 499–532.

Plate, E. J., and J. H. Nath (1969): Modelling of Structures Subjected to Wind Waves, *ASCE Waterways Harbors Div. J.*, vol. 95, pp. 491–511.

Saville, T. (1954): "The Effect of Fetch Width on Wave Generation," U.S. Army Beach Erosion Board, Technical Memo no. 70.

Stoker, J. J. (1957): "Water Waves," Interscience, New York.

Stokes, G. G. (1880): On the Theory of Oscillatory Waves, *Mathematical and Physical Papers*, vol. I, Cambridge, London.

Sverdrup, H. V., and W. Munk (1947): "Wind, Sea, and Swell–Theory of Relations for Forecasting," U.S. Navy Hydrographic Office, Publication no. 601.

Teeson, D., F. M. White, and H. Schenck (1970): Studies of the Simulation of Drifting Oil by Polyethylene Sheets, *Ocean Eng.*, vol. 2, no. 1, pp. 1–11.

U.S. Army (1961): "Shore Protection, Planning and Design," Corps of Engineers, Coastal Engineering Research Center (formerly Beach Erosion Board), Technical Report no. 4.

Vager, B. G., and B. A. Kagan (1969): The Dynamics of the Turbulent Boundary Layer in a Tidal Current, *Akad. Nauk SSSR, Bull. Atmospheric Oceanic Phys.*, vol. 5, no. 2, pp. 88–93 (see also pp. 475–479).

Wada, A. (1969): Studies of Prediction of Recirculation of Cooling Water in a Bay, *Proc. 11th Conf. Coastal Eng.*, vol. II, pp. 1453–1471.

Wiegel, R. L. (1964): "Oceanographical Engineering," Prentice-Hall, Englewood Cliffs, N.J.

3
Submarine Soil Mechanics

Vito Nacci

3-1 INTRODUCTION[1]

The study of the engineering properties of ocean sediments developed as a practical response to the need to understand the behavior of the ocean bottom when subjected to various engineering operations. When considering the feasibility of underwater installations for monitoring, surveillance, and navigation purposes, construction of manned stations, offshore drilling activities, or deep-sea mining, more reliable predictions of sediment behavior are needed for safer and more accurate estimates of the performance of the installations. It is therefore appropriate that the mechanics of sediments be investigated.

The mechanics of particles, soil mechanics, is a branch of applied mechanics which studies the response of particle matter when subject to mechanical and hydraulic loads. The oceanographic application of soil mechanics is complicated by sea pressures, lack of visibility, and the soft sediment condition. About 72 percent of the earth's surface is covered by oceans and the

[1] Contributions to this chapter, particularly from R. H. Helton on marine geology and from C. Katsetos and R. D'Andrea on soil mechanics, are gratefully acknowledged.

mean depth of water is approximately 12,000 ft. The practicing land soil engineer uses the deductive processes employed by geologists to help interpret soil data. The submarine sediment engineer must likewise acquaint himself with basic submarine geology.

3-2 SUBMARINE GEOLOGY

The province of submarine geology begins at the estuary, proceeds through saltwater embayments, continues over the continental margin, descends to the abyssal floor, and rises again over such geomorphic features as the mid-oceanic ridge, seamounts, and islands. This generalized picture is repeated in reverse as one proceeds toward the margin of the other continental areas (Heezen and Menard 1963). The Pacific Ocean is the exception to this picture in that it is surrounded by epicontinental seas, island areas, and trenches. Furthermore, it possesses no mid-oceanic ridge but instead has an expansive East Pacific rise region. The ocean basin is the great receptacle of all continental erosional material primarily transported by wind or water. Of principal interest to the soil mechanician are the sedimentation processes and sediment distribution.

3-2.1 Continental shelf

River-borne sediments consisting of sands, silts, and clays enter the estuary and may pass through into the embayments of the coastal area if the estuary is filled. The coarse particles such as sand settle out of the runoff first, with silts and clays being carried further offshore to form deltaic deposits. The Mississippi Delta is a classic example of this in that it has nearly crossed the breadth of this Gulf Coast continental shelf. Most of the sediment is now carried through the delta and deposits on the continental shelf. These slope sediments rapidly accumulate with resultant slope instability, slumping, and turbidity flows.

The coarser sand left near shore forms the beaches, offshore bars, and barrier islands.

The continental shelf widths vary tremendously around the world—from 200 mi to almost nothing. Relic sediments and paleomicrotopographic features surviving the last sea-level rise at the end of the Pleistocene epoch add to the picture of shelf sediment distribution. New sediment is slowly covering the ancient sediments. Guilcher (1963) reports that deposition on the inner shelf has varied from 12 to 100 cm per century for the past 200 years.

These newly deposited shelf sediments are in equilibrium with their present environment. Emery (1963) interprets the idealized distribution of the chief classes of sediments under equilibrium to be as follows:

1. In the polar regions glacial sediments transported by iceberg rafting predominate.

2. Between the poles, water-contributed detrital sediments predominate where rivers are present. Detrital sediment is derived in a process called detrition, meaning a rubbing or wearing away or erosion process.
3. At the equatorial regions biogenic sediments are present where the planktonic animals having calcareous and siliceous shells settle to the bottom at death and add to the sediment composition. Calcareous sediments predominate on the low-latitude western sides of ocean basins where temperatures are higher, and siliceous sediments predominate on the higher-latitude western boundaries where water temperatures are lower.
4. Out of the equatorial belt, some authigenic sediments are present. These sediments are formed "in place" by crystallization. Chief among these sediments are the phosphorites and glauconites.

Shepard (1963) figuratively summarizes the continental shelf sediment distribution as varied colored splotches likened to the worst imaginable modern art painting. This can be appreciated in the history of exposure of the shelf during the Pleistocene epoch when additional depositional and erosional processes were effective.

3-2.2 Continental rise

The continental rise area, when it extensively exists, is an overall wedge of sediment that is thickest adjacent to the continental block fault zone. This wedge becomes thinner toward its deepwater terminus at the abyssal floor. Heezen (1966) reports that this wedge varies in width from 100 to 1,000 km and that where geosynclinal trenches are absent, it lies at the base of the continental slope. At the base of the continental slope the wedge thickness varies from 1 to 10 km. The theory for the deposition of this wedge is that a geostrophic contour-following bottom current transports the sediment and distributes it on the rise. This current is fastest next to the slope and slowest next to the abyssal floor area. This water velocity profile carries the greatest sediment load at the slope area end, so the greatest deposition supposedly occurs here. This theory explains the characteristic downslope thinning and the accumulative wedge geometry of the continental rise.

The rise sediments, in which a significant fraction consists of particles smaller than 4 μm, have a high water content judging from core analysis and the poor acoustic reflectivity encountered in this area. Where turbidity flows occur, both sand and silt are additionally found.

3-2.3 Continental slope

The worldwide continental slope is the least investigated geomorphic province in the ocean basin. Although research interest in this area is increasing, it will probably be many years before even bathymetric charts are generally available. Shepard (1963) summarizes the available information as follows:

1. From Georges Bank to Cape Hatteras along the eastern continental margin of the United States the continental slope extends from 1,000 fathoms to 1,500 fathoms. There it joins the continental rise, which then continues down to abyssal depths. The slope can be considered to commence at the shelf break whose average worldwide depth is 72 fathoms.
2. For the first 1,000 fathoms of descent the average declivity is 7.5:100. Continental slope declivities usually vary from 1:6 to 1:40, but may even be infinite.
3. The sediments, as reported by surveyors with little geological background, tend to be composed of 60 percent mud, 25 percent sand, 10 percent rock and gravel, and 5 percent shells, tests, and ooze.

The maximum width of the continental slope varies from 30 to 50 mi. Geomorphic features on the slope range from seamounts to terraces and ridges. Some terraces and ridges are erosional features caused by block slides or segments of strata that slide down the slope. These blocks can form ridges or they can be smoothed by sediment to form terraces.

A prime triggering mechanism for block sliding, progressive sedimentary slumping, and resultant turbidity flows is an earthquake epicenter. Barazangi and Dorman of the U.S. Coast and Geodetic Survey have plotted the epicenters of all earthquakes recorded during the 10-year period beginning in 1957. Their plotted data clearly indicate the predominance of tectonism around the Pacific Ocean Basin and along the mid-ocean ridges (Heirtzler 1968). From these data it would appear very foolish to consider the permanent establishment of any structure on the majority of Pacific Ocean continental rise areas without further analysis of the local frequency and probability of earthquake occurrence.

3-2.4 Abyssal floor, oceanic ridges

The greatest part of the ocean basin is composed of the abyssal floor and the mid-ocean ridge provinces. Also present are the rises, abyssal hills, seamounts, and island rises. The ridges are extensive features as exemplified by the Mid-Atlantic Ridge. Ewing (1964) reports that this ridge encompasses one-third of the areal extent of the Atlantic Ocean Basin.

Ewing's seismic analysis of the Mid-Atlantic Ridge region revealed a uniformly rough topographic feature where the red clay and carbonate sediments had collected in sedimentary ponds. These fractions formed cohesive masses that evidently did not flow or slide when they were tilted later by tectonism. On the southern crossings, Ewing found that the sedimentary layer was nearly uniform across the ridge section in contrast to the ponded and sediment-bare regions in the northern crossing from Buenos Aires to Dakar.

Sediments on the abyssal floor are mainly pelagic and biogenic. Nearer the continents, however, with no intervening trenches or island arcs, the sediments also contain turbidite sequences of terrigenous material beds carried down by turbidity flows.

Continuous sedimentation of the abyssal floor is provided by the pelagic sediments which are carried into the area by winds and currents and which then slowly sink to the bottom. The settling rates of these pelagic sediments are very slow. Arrhenius (1963) estimates that most sediment particles settle in less than 100 years, but particles 0.5 μm or smaller may take 200 to 600 years to reach the bottom.

Heezen and Laughton (1963) describe the abyssal plains and rise regions as follows:

1. The normal sediment is pelagic, but the smoothing of the topography results from turbidity flows which carry coarser sands and silts to these areas.
2. In the rise regions (i.e., the Bermuda Rise) only pelagic oozes are found. These oozes are composed of red clays with at least 30 percent fraction of biogenic material. Because of the solutions of biogenic calcium carbonate at higher pressures and lower temperatures, planktonic foraminifera shells are mainly absent below depths of 5,000 M.

3-2.5 Deep ocean sediment

Shepard (1963) reports that the deep ocean sediment is usually soft, has a median particle diameter of 1 μm, and is low in carbonate and silica content. The carbonates and silicas, however, are the suspected cementing agents, so their absence is notable (Bryant 1967).

Sediments on the abyssal floor and mid-ocean ridge provinces are mainly pelagic and biogenic. Pelagic sediments are carried by winds and ocean currents and slowly sink to the bottom. These sedimentation rates, though minute, add up in the geological time frame to considerable thicknesses. Three layers make up the so-called oceanic layer. The upper layer averages 1 km in thickness in the Atlantic Ocean and 0.5 km in the Pacific and is composed of unconsolidated sediments. Layer 2 is considered to be consolidated under the weight of layer 1. Layer 3 is the basaltic mantle.

The Atlantic Ocean, which is ringed by metamorphic and illite-rich sediments, is predominantly illite while the Pacific with its abundance of volcanic material is rich in montmorillonite. In any case, the physical and mechanical properties of these sediments are largely dependent on the rate of deposition. The density of the suspended grains and the interparticle bonding results from solution and redisposition of silica, calcium carbonate, iron, and manganese. The low density of suspended particles would seem to lead to accretion in a particle-by-particle manner indicating a relatively homogeneous "cardhouse" structure as visualized by Lambe and Whitman (1969).

3-3 THE NATURE OF SEDIMENTS

A sediment sample consists of particles, liquid, and possibly gas. The individual particles are primarily carbonates or silicates. They vary in size from 10 Å up to rocks several meters in thickness. Coarse-grained sediments are classified as (1) cobbles, particles larger than 3 in, (2) gravel, from 3 in to 4.76 mm, and (3) sand, from 4.76 mm to 0.06 mm.

In considering the behavior of sediments, the density of packing of the particles is of primary significance and, in this connection, the following definitions are necessary.

Porosity n is the ratio of voids to the total volume of the sediments. Void ratio e is defined as the ratio of voids to the total volume of sediment solid particles. A useful relation between porosity and void ratio is

$$n = \frac{e}{1 + e} \tag{3-1}$$

which we obtain noting that the total sediment volume is the sum of the void volume and the solid volume.

The void volume may be filled with water or water and gas. The degree of saturation S indicates the percentage of the void volume which is filled with water. For a completely saturated sediment, $S = 100\%$.

If G is the specific gravity of the particles and γ_ω the density of water, then the total unit weight of the sediment γ_t is given by

$$\gamma_T = \frac{(G + Se)\gamma_\omega}{(1 + e)} \tag{3-2}$$

while the submerged unit weight γ_B is given by

$$\gamma_B = \frac{(G - 1)\gamma_\omega}{(1 + e)} \tag{3-3}$$

The water content w of the sediment is defined as the weight of water per unit weight of solid particles. An identity that relates these several basic parameters is

$$Gw = Se \tag{3-4}$$

Note that if a relatively undisturbed sample of soil has its weight and volume measured, we can immediately compute its value of γ_T. If we now dry the soil in an oven and weigh it again, we can obtain the water content w. For the usual case of $S = 100\%$, we have two equations (3-2 and 3-4) with two unknowns (e and G). Once these are found, n and γ_B can be computed.

3-3.1 Mineralogy

The spatial arrangement of the solid clay particles together with the associated electrical surface forces is called soil structure. The surface forces can be repulsive or attractive.

If the attractive force predominates, the particle arrangement tends to be nonoriented or flocculent structures. Clays having greater repulsive forces are said to have an oriented or dispersed type of structure.

Clays deposited in an environment containing an excess of cations would be expected to possess a flocculent structure; such is the case with many marine deposits. On the other hand, many freshwater deposits possess a dispersed structure. Figure 3-1 illustrates particle arrangements in marine soil sediments.

Soil behavior is related significantly to soil structures. In very fine-grained clays, montmorillonite, and fine-grained illite, which have large areas of particle surface available to interact with interstitial water and dissolved ion, sediments of extremely high porosity—0.70 to 0.90—may occur. The presence of organic material as well as carbonates and oxides together with an extremely slow rate of clay deposition allows for the development of a slightly cemented but fragile

Fig. 3-1 Sediment structures: (*a*) salt flocculation, (*b*) nonsalt flocculation, (*c*) dispersion.

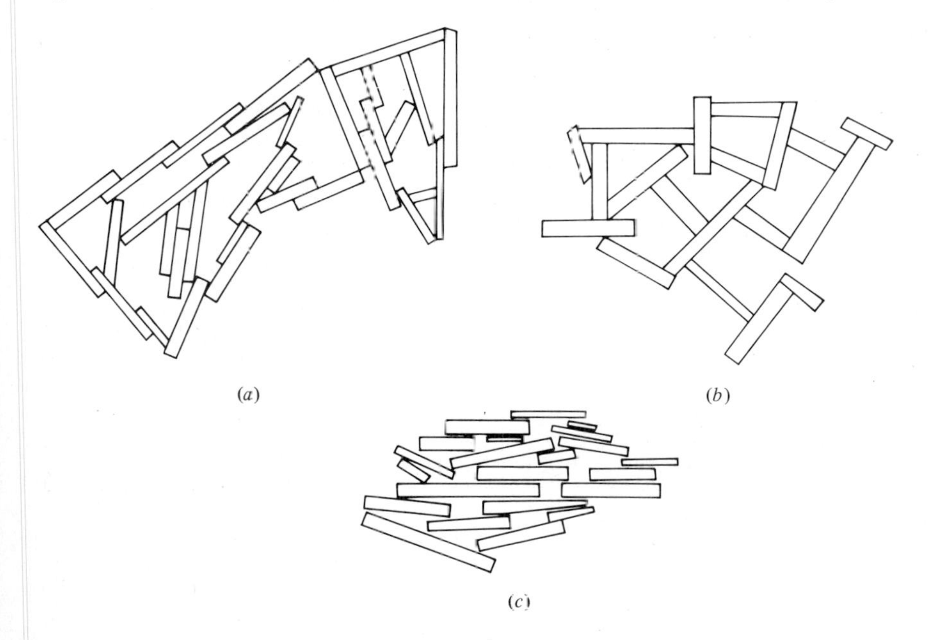

network of minerals. Current evidence indicates that deep-sea clays may resist a natural tendency to increase in density with depth owing to their flocculent structure (Nacci and Huston 1969).

Table 3-1 gives some representative physical properties of sea floor sediments.

3-3.2 Sensitivity

The ratio of undisturbed to disturbed or remolded strengths at natural water content is defined as sensitivity. Sensitivities of from 4 to 8 are typical of normally consolidated clays and may exceed 100 for extra-sensitive or quick clays.

The behavior of quick clays was studied in detail by Bjerrum (1957) in the 1950s. He theorized that clay particles were deposited in a saltwater environment causing a flocculent structure of high porosity. At some later geological age, either the sediment was uplifted or the sea level dropped so that the interstitial salt water was gradually leached and replaced with freshwater. The resulting change in interparticle forces altered the equilibrium conditions and created instability.

The sensitivity of deep ocean clays can hardly be due to the removal of salt water. Instead the precipitation of chemical compounds at particle points causing a metastable structure is thought to be the cause. Bjerrum (1967) described a study by Kenney on cemented clays. By chemically dissolving the cementing agents, an entirely altered stress-strain soil function was obtained.

3-4 EFFECTIVE STRESS PRINCIPLE

Saturated soils may be considered as being composed of a relatively incompressible fluid and a compressible mineral skeleton. Of the stresses that are applied to these components, there is considerable evidence that the mechanical properties of a soil are controlled solely by the stresses that are in the mineral portion.

Consider a plane crossing through the pore spaces and points of mineral contact of a soil sample. The stresses and areas on this plane are:

σ_T, the total stress on the plane
u, the hydrostatic or pore water pressure
a, the effective contact area between minerals
$\bar{\sigma}$, the average intergranular force per unit area of the plane, the effective stress

Then

$$\sigma_T = \bar{\sigma} + (1 - a)u \tag{3-5}$$

Table 3-1 Some typical physical properties[a]

	Continental shelf									Deep sea					
	Sand			*Sandy silt, silty sand*			*Silt*[b]			*Clay*[b]			*Calcinated ooze*		
Property	*Max.*	*Min.*	*Av.*	*Max.*	*Min.*	*Av.*	*Max.*	*Min.*	*Av.*	*Max.*	*Min.*	*Av.*	*Max.*	*Min.*	*Av.*
Density, saturated, g/cc[c]	2.10	1.80	1.90	1.90	1.50	1.75	1.85	1.25	1.45	1.80	1.18	1.40	1.95	1.25	1.70
Bulk density, average mineral grains, g/cc[d]	–	–	2.65	–	–	2.65	–	–	2.65	2.80	2.40	–	–	–	2.72
Porosity, percent	50	35	45	70	45	55	85	50	73	85	50	77	85	45	60
Sound speed, km/s[e]	1.79	1.54	1.70	1.63	1.48	1.54	1.56	1.47	1.51	1.60	1.47	1.49	2.05	1.48	
Shear strength,[f] cohesion, g/cm²	–	–	–	27	4	12	21	3	11	46	4	17	192	4	87
Sensitivity ratio[f,g]	–	–	–	8	3	5	9	2	5	9	2	5	24	2	7

[a]Laboratory; room temperature and pressure; sediment surface; all have exceptions; 2d intergrade.
[b]Including silty clay and clayey silt.
[c]Density of sea water *in situ* ranges from approximately 1.025 g/cc at surface to 1.05 at greatest depth.
[d]Excluding unusual amounts of heavy minerals.
[e]Sediment "surface."
[f]Vane shear.
[g]"Undisturbed" shear strength divided by remolded, or "disturbed," strength.

Terzaghi (1936) and Bishop and Eldin (1950) indicate that the value of a is close to zero, simplifying Eq. (3-5) to

$$\sigma_T = \bar{\sigma} + u \tag{3-6}$$

This formula is probably more correct for granular soils than for clays, which have important attractive and repulsive forces between particles, but it is nevertheless the most important principle in soil mechanics.

Consider the case of a saturated sediment in a static environment. Figure 3-2 shows a section through the water-sediment column. At point A in the sediment mass:

$$\sigma_T = \frac{(h_1 l^2 \gamma_\omega) + (h_2 l^2 \gamma_t)}{l^2} = h_1 \gamma_\omega + h_2 \gamma_t$$

$$u = (h_1 + h_2)\gamma_\omega = h_1 \gamma_\omega + h_2 \gamma_\omega$$

$$\bar{\sigma} = \sigma_T - u = h_2(\gamma_t - \gamma_\omega) = h_2 \gamma_b \tag{3-7}$$

in which γ_b is the submerged unit weight of the sediment. The interrelation between pore pressures, total stresses, and effective stresses can be further explored. Assume a saturated sediment volume V is subjected to an all-around increase in total stress $\Delta\sigma_T$. The decrease in volume of the soil structure is

$$\Delta V = C_c \times V \times \Delta\bar{\sigma}$$

The decrease in the volume of pore water is

$$\Delta V = C_\omega \times n \times V \times \Delta u$$

where C_c and C_ω are the compressibility of the mineral skeleton and pore water respectively. If the drainage is prevented, these changes must be equal:

$$C_c \times V \times \Delta\bar{\sigma} = C_c \times V \times (\Delta\sigma_T - \Delta u) = C_\omega \times n \times V \times \Delta u$$

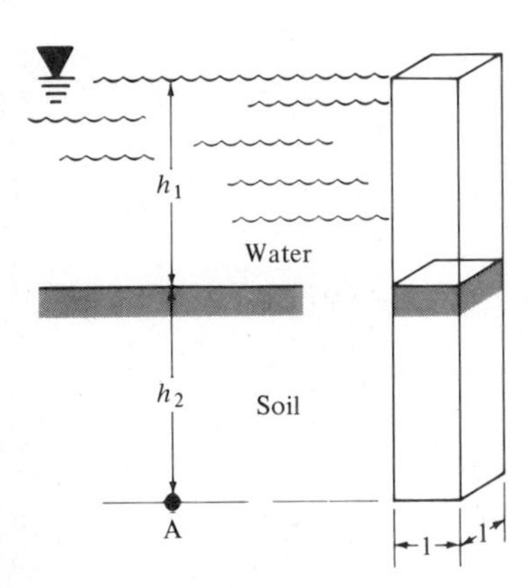

Fig. 3-2 A water-sediment column of unit dimensions.

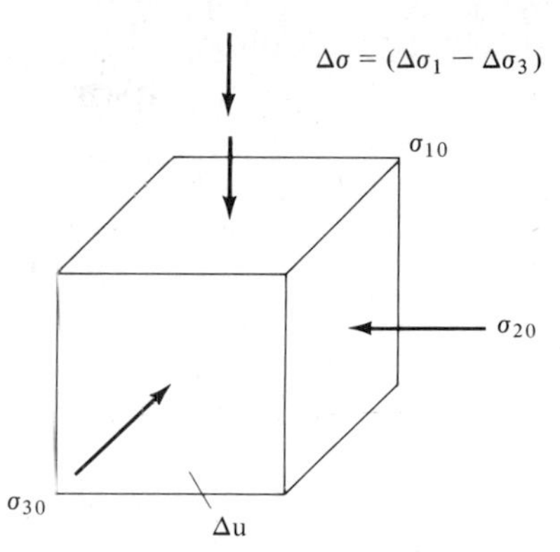

Fig. 3-3 A soil element with all-around stresses subjected to undrained axial loading, producing stress $\Delta\sigma$.

or

$$\Delta u = \frac{1}{1 + (nC_\omega/C_c)}\, \Delta\sigma_T = B \times \Delta\sigma_T$$

where

$$B = \frac{1}{1 + (nC_\omega/C_c)} \tag{3-8}$$

Since the compressibility of water C_ω is very small, B is close to unity for most sediments. The importance of this is that for saturated sediments, if the drainage is prevented, the effective stress, hence strength and compressibility, is independent of the ambient stress. Another common situation exists when a soil element is subjected to undrained uniaxial loading (Fig. 3-3).

The effective stresses are changed by $\Delta\sigma - \Delta u$ in direction 1 and by Δu in directions 2 and 3. The change in stress produces a volume decrease equal to $C_c V(\Delta\sigma - \Delta u)$ and a volume increase equal to $2C_s V\Delta u$, where C_s is a coefficient analogous to the Poisson ratio in solid mechanics. In the undrained state the total volume change is zero and

$$C_c(\Delta\sigma - \Delta u) = 2C_s\Delta u$$

or

$$\Delta u = \frac{\Delta\sigma}{1 + (2C_s/C_c)} = A_f\Delta\sigma$$

The general case of stress change can be obtained by a combination of the two preceding cases giving (Skempton 1954):

$$\Delta u = B\Delta\sigma_3 + A_f(\Delta\sigma_1 - \Delta\sigma_3) \tag{3-9}$$

The coefficients A_f and B are stress-dependent but are usually considered as constants at shear strength levels. Some laboratory-measured values of A_f at failure are shown in Table 3-2.

Table 3-2 Values of A_f

Material	A_f
Very loose fine sand	2 to 3
Sensitive clay	1.5 to 2.5
Normally consolidated clay	0.7 to 1.3
Lightly overconsolidated clay	0.3 to 0.7
Heavily overconsolidated clay	−0.5 to 0

Source: From Lambe and Whitman (1969).

3-5 CONSOLIDATION

The consolidation or compressibility of saturated soils is almost entirely due to an increase in effective stress with resulting reduction of void volume. In silts and clays the drainage of water from the voids requires considerable time; therefore the rate of consolidation is low. In sands and gravels, however, the rate of drainage, hence consolidation, is high, and in many cases the volume change takes place almost simultaneously with load application.

Consider a sediment sample in equilibrium under an effective initial all-around stress $\bar{\sigma}_0$. If the stress is increased, $\Delta\bar{\sigma}$, then a pore pressure will develop:

$$\Delta u = B \times \Delta\sigma$$

or

$$\Delta\bar{\sigma} = (1 - B)\Delta\sigma$$

The new value of effective stress $\bar{\sigma}_1$ will be

$$\bar{\sigma}_1 = \bar{\sigma}_0 + (1 - B)\Delta\sigma$$

But for fully saturated soils it has been suggested that $B \cong 1$. Therefore immediately upon application of a load to saturated sediments there is no increase in effective stress and an excess pore pressure $\Delta u = \Delta\sigma$ is established.

With time the excess pore pressure will dissipate. At any intermediate time after the application of the load, the excess pore pressure will be Δu_t and the effective stress at time t will be

$$\bar{\sigma} = \bar{\sigma}_0 + \Delta\sigma - \Delta u_t \tag{3-10}$$

Eventually the excess pore pressure will be equal to zero and the entire load will become effective under the pressure:

$$\bar{\sigma} = \bar{\sigma}_0 + \Delta\sigma \tag{3-11}$$

3-5.1 One-dimensional consolidation

Natural consolidation of sediments occurs under geological deposition of overburden and in essentially one dimension since lateral strains are zero. Many engineering problems are also one-dimensional; consequently most laboratory work on consolidation is carried out in an apparatus known as a *consolidometer*, in which a sample is contained rigidly between porus stones and is loaded axially. When the load is applied, the rate of consolidation is observed. When equilibrium has been attained, the load is increased and the consolidation is again observed. This process is repeated for as large a load as necessary.

From a knowledge of the initial void ratio e_0 and the compression of the sample under each load increment, the void ratio e under any pressure is readily computed:

$$\frac{e_0 - e}{1 + e_0} = \frac{\Delta h}{h_0} \tag{3-12}$$

where Δh and h_0 are the compression and initial sample thickness.

Typically the void ratio is plotted versus logarithm of pressure as shown in Fig. 3-4, where the curve *abc* is approximately linear and may be expressed:

$$e = e_0 - C_{ci} \log\left(\frac{\bar{\sigma}}{\bar{\sigma}_0}\right) \tag{3-13}$$

where C_{ci} is the compression index. Note that the value of C_{ci} will depend on the logarithm base used in plotting and computing the slope of the line on Fig. 3-4. C_{ci} values in this chapter will be based on base 10 logarithms.

The compressibility or settlement Δh of a sediment of thickness h_0, in one-dimensional consolidation, when the initial stress is increased to $\sigma_0 + \Delta\sigma$ can be calculated from the expression

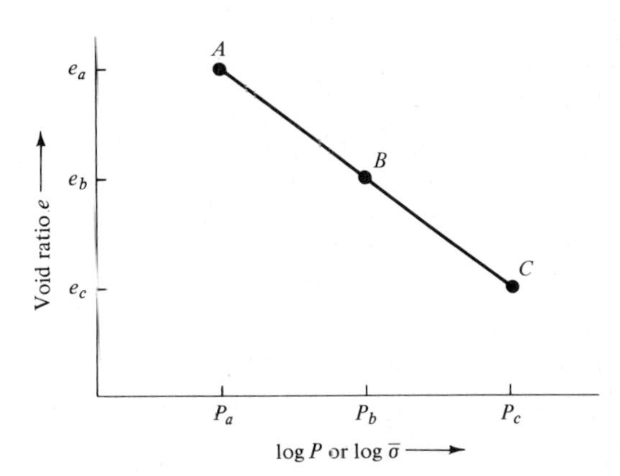

Fig. 3-4 Typical result of a consolidation test with three increases in load *a*, *b*, and *c*.

$$\Delta h = \frac{h_0}{1 + e_0} (e_0 - e) = \frac{h_0}{1 + e_0} \times C_{ci} \times \log \left(\frac{\bar{\sigma}_0 + \Delta\sigma}{\bar{\sigma}_0} \right) \tag{3-14}$$

The fundamental equation of Terzaghi (1943) governing the rate of one-dimensional consolidation is

$$\frac{\partial u}{\partial t} = \frac{k}{\gamma_\omega m_v} \frac{\partial^2 u}{\partial z^2} \tag{3-15}$$

where u is the pore pressure at a distance z from a free-draining surface at a time t after load application, and k and m_v are the permeability and compressibility of the sediment. The ratio of k to m_v is approximately constant; therefore Eq. (3-15) can be written

$$\frac{\partial u}{\partial t} = c_v \frac{\partial^2 u}{\partial z^2} \tag{3-16}$$

where c_v is defined as the coefficient of consolidation. If under the increment of load $\Delta\sigma$ the final settlement of a clay layer of thickness h_0 is Δh and if the settlement that some time t is Δh_t, then it can be shown that

$$\frac{\Delta h_t}{\Delta h} = U = f\left(\frac{c_v t}{H^2}\right) = f(T) \tag{3-17}$$

where U is the degree of consolidation, H is the maximum drainage path of water, and T is a dimensionless time factor. The relation between T and U for most practical problems is given in Table 3-3.

The relation between U and T within the limits of $0 < U < 0.5$ is

$$U = 2\left(\frac{T}{\pi}\right)^{1/2} \tag{3-18}$$

Table 3-3 Solutions of Eq. (3-17)

$U\%$	T
10	0.008
20	0.031
30	0.071
40	0.126
50	0.197
60	0.287
70	0.403
80	0.567
90	0.848

Thus by plotting the results of a consolidation test, U versus t, the value of c_v can be obtained. Note that when the test is performed with only one face of the sample drained, H is equal to the sample thickness h_0. When both sample faces are drained, H will equal $\frac{1}{2}h_0$.

The foregoing discussion on one-dimensional consolidation should be considered as an approximate means of predicting settlements, for actual field consolidation is more likely a three-dimensional compression phenomenon. Lambe and Whitman (1969) discuss settlement under three-dimensional consolidation conditions.

Finally the difference between observed and calculated settlements can also be attributed to elastic deformation and secondary or creep deformation. For most land inorganic clays these deformations can be readily estimated. Such may not be the case for deep-sea, soft, cemented clays.

3-6 SHEAR STRENGTH

The shear strength of saturated sediments is hardly a unique value, being dependent upon failure criteria, method of testing, rate of strain, consolidation pressure, and sample integrity. In order to understand the basic principles that govern the strength of saturated cohesive soils, at the expense of considerable simplification, the following two principles are presented:

Principle I "Strength is uniquely related to the effective stress at failure." The Mohr-Coulomb theory of failure has been found to be successful in defining failure in soils. The theory states that if the shear stress on any plane equals the shear strength of the material, failure occurs. Thus principle I can be expressed

$$s = \bar{c} + \bar{\sigma}_{ff} \tan \bar{\phi} \tag{3-19}$$

where s and $\bar{\sigma}_{ff}$ are the shear stress and effective normal stress on the failure plane at failure and $\bar{c}$ and $\bar{\phi}$ are called the "cohesion" and the "friction angle." The envelope defined by $\bar{c}$ and $\bar{\phi}$ represents the limiting condition for any possible combination of s and $\bar{\phi}$. This strength theory also assumes that the point of tangency of a Mohr circle to the envelope represents conditions on the failure plane. Figure 3-5 illustrates the significance of principle I.

Principle II "Strength is uniquely related to the water content at failure." This principle is illustrated in Fig. 3-6, wherein the log of shear stress is linear when plotted against water content at failure. This principle combines the e log P consolidation linear relation, Eq. (3-13), with principle I. For most insensitive or remolded soils these relations are quite valid and in fact form the backbone of the so-called $\phi = 0$ analysis commonly used in practice. However, for cemented sensitive soils these assumptions may be in error.

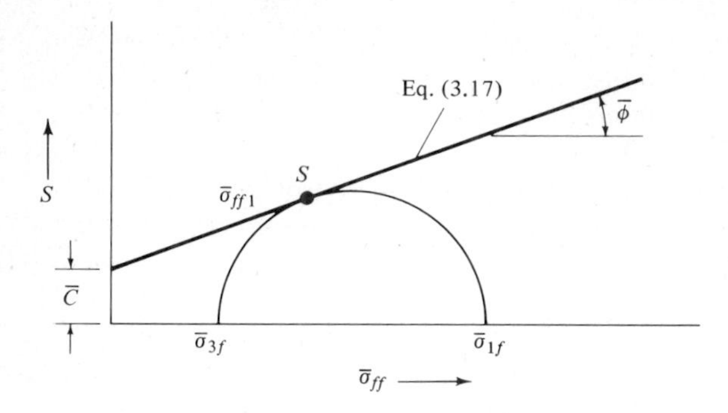

Fig. 3-5 The Mohr-Coulomb circle construction at failure. With an all-around effective stress $\bar{\sigma}_{3f}$, the circle grows as the axial effective stress $\bar{\sigma}_{1f}$ increases until the line of Eq. (3-19) is reached.

3-6.1 Laboratory apparatus for measuring shear strengths

The direct shear test (Fig. 3-7*a*) was the first practical test in soil mechanics. In recent years it has fallen into disfavor owing to high stress gradients that are set up in the sample and the difficulty in controlling drainage. In addition, the directions of the principal stresses rotate during the test, making analysis difficult.

The vane shear test (Aas 1965) is commonly used for sensitive ocean bottom clays, both in the field and in the laboratory, as a means of measuring the undrained shear strength of soils which are too soft to extrude from the coring tube. It suffers from the same limitations as the direct shear test.

The triaxial compression test (Fig. 3-7*b*) is the most widely used test today. Usually the major principal stress σ_1 acts in the vertical axis direction, and the intermediate principal stress σ_2 is equal to the minor principal stress σ_3, which acts horizontally. The apparatus is usually devised to perform at controlled rates of strain and can control drainage during consolidation and shear. There are three cases generally considered: the unconsolidated-undrained (UU), the consolidated-drained (CD), and the consolidated-undrained (CU) types of test.

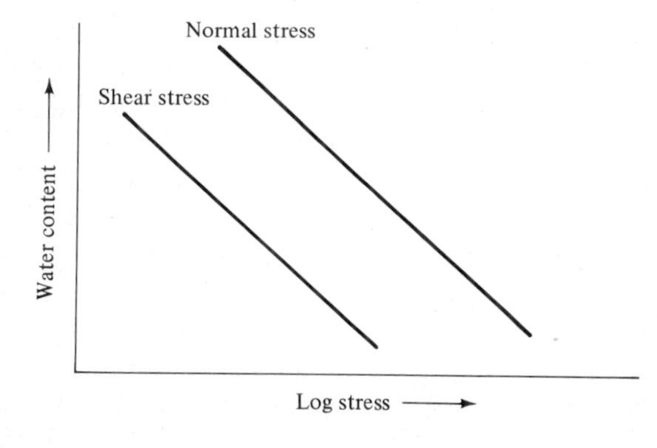

Fig. 3-6 The stress-vs.-water content relations implied by principle II.

In the UU test the cylindrical specimen is placed between the end platens, and no drainage is permitted during the application of ambient pressure or during shear. A simple application of the UU test is the unconfined compression test wherein the ambient pressure σ_3 is zero.

In the CD test the sample is placed on a porous disk which is connected to a burette. This allows consolidation to take place, $\bar{\sigma}_c$, under ambient pressure. Upon completion of consolidation, the sample is sheared slowly, allowing further drainage.

The procedure for the CU test is similar to the CD test except that during shear no drainage is permitted, and usually a null system is used to measure pore pressures.

3-6.2 Stress-strain curves

Before considering the shear strengths determined by different laboratory tests, a discussion of typical stress-strain curves of soils is desirable. In Fig. 3-7*c* test results are shown for sands and clays. Failure for loose sands is typically reached at 15 to 25 percent strain with little change in stress with further deformation. Dense sands are much stronger and failure occurs at about 5 percent strain. With further deformation the stress and density progressively decrease. At large strains, both loose and dense sands have essentially the same strength and density. Normally consolidated clays fail at 3 to 5 percent strain and markedly decrease in strength with further strain. At very large strains, 30 to 50 percent, a strength plateau or residual strength approximately half of the peak value exists (Skempton 1954). Upon remolding a normally consolidated clay at its natural water content, a large drop in strength is usually observed (Fig. 3-7*c*). The ratio of undisturbed to remolded strength is defined as the sensitivity.

3-6.3 Shear strength determination

Mohr's hypothesis and diagrammatic method are much used in soil mechanics in the analysis of shear strength of soils. In the Mohr diagram, Fig. 3-8, shearing stresses are plotted in the direction of the ordinate-vs.-normal stresses as abscissa. Shear stresses are zero on principal planes, so values which correspond to σ_1 and σ_3, the major and minor principal stresses, are plotted along the right of the origin on the abscissa for compression. Through these points a circle is constructed of radius $\frac{1}{2}(\sigma_1 - \sigma_3)$. The center of the circle is on the abscissa at a distance $\frac{1}{2}(\sigma_1 + \sigma_3)$.

The stresses on any plane may be directly derived from the diagram, e.g., plane *AB* in Fig. 3-8.

The undrained strength of a clay tested with measured pore pressures is perhaps the most informative laboratory shear test. It provides strength parameters for long-term stability conditions as well as data for immediate

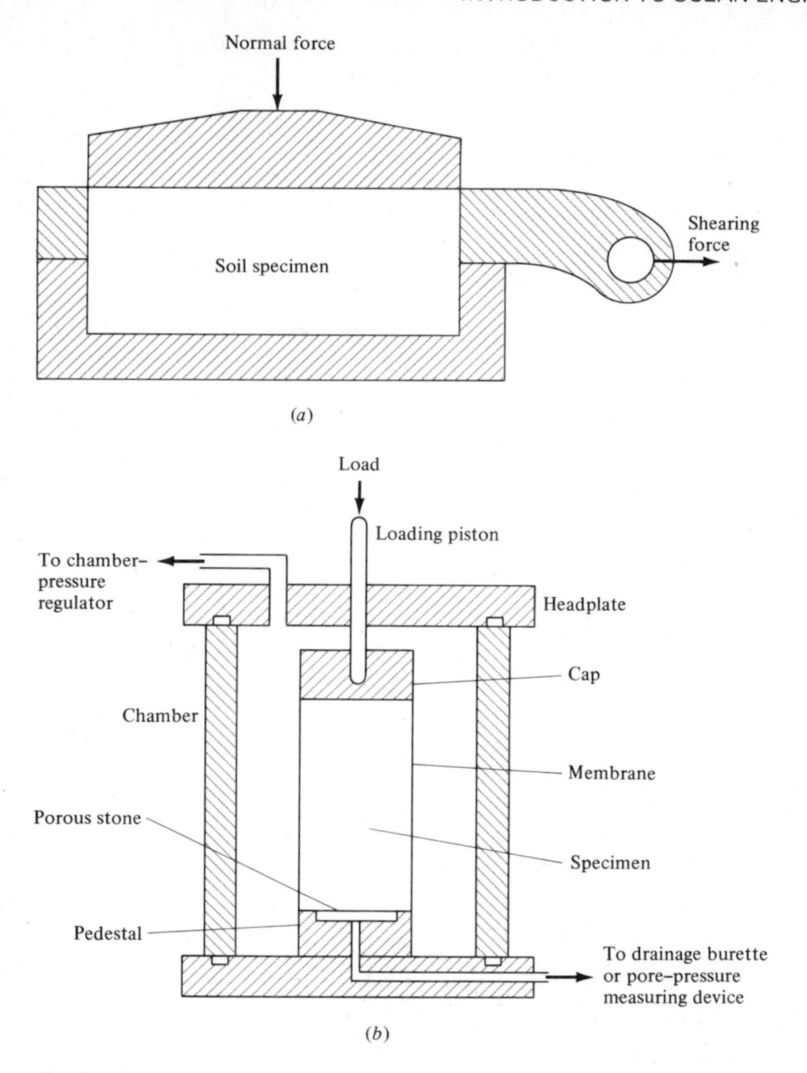

Fig. 3-7 Direct shear (*a*) and triaxial apparatus (*b*).

loadings. Consider a sample of clay consolidated in the triaxial apparatus to all-around stress σ_3. Under conditions of no further drainage but holding ambient pressure constant ($\Delta\sigma_3 = 0$) the axial stress is increased until failure occurs at a *deviator stress* $(\sigma_1 - \sigma_3)_f$. The total principal stresses at failure are

$$\sigma_{1f} = \sigma_3 + (\sigma_1 - \sigma_3)_f$$

$$\sigma_{3f} = \sigma_3 + 0$$

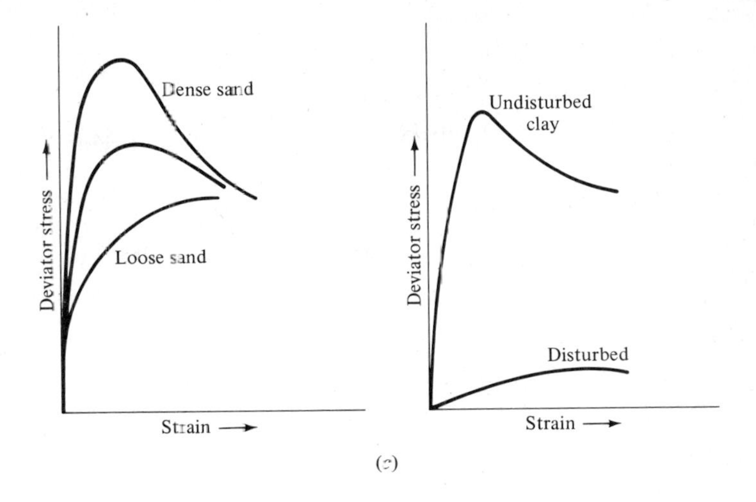

Fig. 3-7 (*cont.*) Typical stress-strain curves (*c*).

and the failure state can be represented by a Mohr circle, Fig. 3-9. However, it is found that the deviator stress induces a pore pressure Δu_f, Eq. (3-9), so that at failure

$$\Delta u_f = \Delta\sigma_3 + A_f(\Delta\sigma_1 - \Delta\sigma_3) \tag{3-20}$$

and the effective stresses at failure are

$$\bar{\sigma}_{1f} = \sigma_3 + (1 - A_f)(\Delta\sigma_1 - \Delta\sigma_3)$$

$$\bar{\sigma}_{3f} = \sigma_3 + A_f(\Delta\sigma_1 - \Delta\sigma_3) \tag{3-21}$$

On Fig. 3-9 are also plotted the results of an unconfined compression test at the same water content. It should be noted that the undrained strength c_u can be readily obtained from this test, but the drained parameters $\bar{c}$ and $\bar{\phi}$ are not obtainable.

The ratio of undrained strength to consolidation pressure for normally consolidated clays has been found to be approximately constant. Since the envelopes to the total and effective stress circles are approximately linear, it follows that

$$c_u = \tfrac{1}{2}(\sigma_1 - \sigma_3) = \frac{\bar{\sigma}_c \sin\bar{\phi}}{1 + (2A_f - 1)\sin\bar{\phi}} \tag{3-22}$$

The consolidation pressure $\bar{\sigma}_c$ at a particular sediment depth D_s is simply found by multiplying the sediment submerged unit weight γ_b times D_s. In problems involving a mass of sediment of substantial thickness, an average consolidation pressure is usually assumed. Selection of the average depth used to compute this average pressure is discussed at the end-chapter application examples.

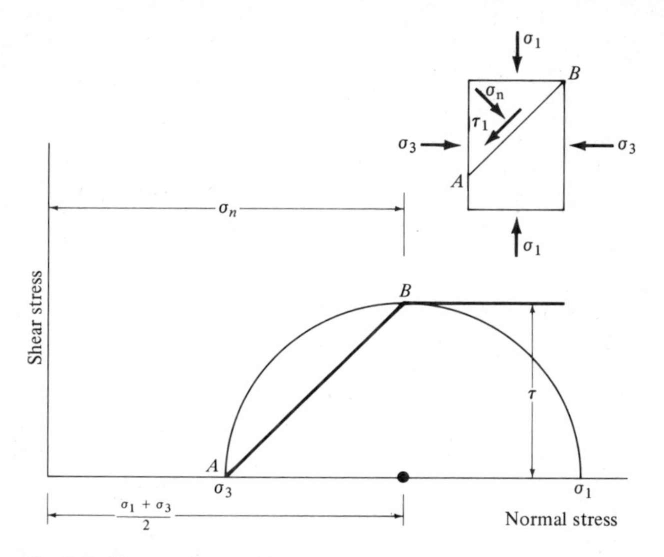

Fig. 3-8 Nomenclature of the Mohr diagram.

3-6.4 Effect of sample disturbance on strength

The goal of laboratory testing is to predict the *in situ* properties of the soil. Ideally the laboratory sample should have the same structure as the field sample. This means that the water content, effective stress, and environment (temperature, pore fluid, etc.) should be the same. Noorany and Seed (1965) point out that even with perfect sampling the *in situ* strength is significantly different from the laboratory strength. While there are means of correcting this variance, actual samples are not perfect since disturbance is created by the

Fig. 3-9 Stress relations at failure in confined and unconfined tests.

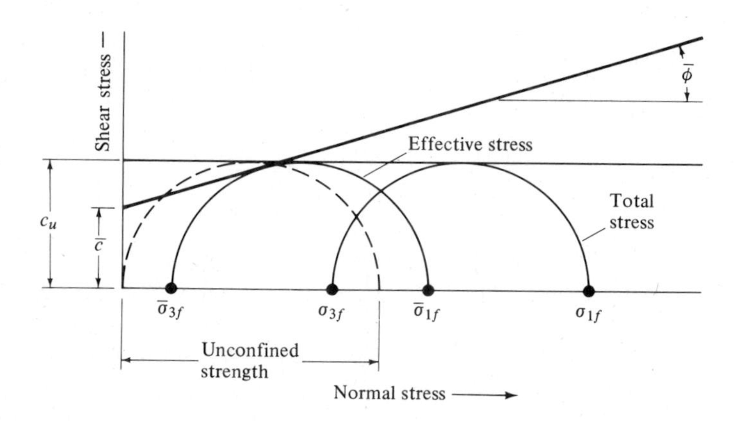

sampling process and subsequent transportation, laboratory storage, and trimming. Much of the effort at soil laboratories is in minimizing such disturbances as well as in evaluating the degree of disturbance.

A factor that may be significant with deep-sea samples is that the volume of pore water expands 1.4 percent per 10,000 ft of elevation. The effect of the expansion on soil structure and strength is as yet unproved although reports of gas bubbles coming out of solution and destroying samples have been made.

3-7 PRACTICAL APPLICATION OF SUBMARINE SOIL MECHANICS

While the application of soil mechanics in water has up to the present time been mainly concerned with nearshore facilities such as piers, breakwaters, and offshore platforms, it is apparent that the future will see expanded activities in offshore drilling, mineral mining, manned and unmanned installations, anchors for buoys, ships, and platforms. The assessment of the properties of ocean sediments, particularly strength and compressibility, will be the most important factor in planning for any of these projects. The preponderance of data available concerns only the top 20 ft of sea floor sediments and is usually confined to grain size, mineralogy, and water content. Where strengths are available, they were usually measured by laboratory vane or unconfined compression and as expected have very high water contents and very low strengths.

While there is good reason to suspect the strength values, grain size and water contents are believed to be correct and the variation of water content with depth is rather interesting.

Contrary to expectations, porosities of deep-sea clays do not consistently decrease with depth. The explanation that an increase in strength with the increase in overburden pressure is responsible for the constant porosity is no doubt true, but may be misleading. Some investigators, including Crawford (1965) and Bjerrum (1967), have noted that when prolonged consolidation is allowed to take place, an increase in shear strength is noted. In the deep ocean, the rate of sediment accumulation is extremely slow, amounting to prolonged consolidation. At the slow rates of deposition, water molecules can orient in the vicinity of mineral contact, and the particles move into an efficient arrangement to form a cohesive bond at contact points. Another type of cohesion is caused by the cementing action of chemical precipitates. The principal difference between the two types of bonds is that the first one is due to electrostatic attractive forces and is "flexible" enough to re-form bonds under shearing strain whereas the second is a "brittle" bond whose strength cannot be reestablished after strain. It can be postulated that during sedimentation due to the ever-increasing overburden load, the clay is subject to a minutely increasing vertical and lateral stress. The slow stress buildup allows for the formation of chemical bonds and efficient structural arrangement that can successfully resist deformation and hence maintain almost constant porosity. It is logical to assume that induced shearing strains from suddenly applied loads, earth shocks, or crude coring operations could permanently destroy much of the undrained shearing strength.

3-8 CORE SAMPLING METHODS AND EQUIPMENT

The main purpose of drilling or coring in the ocean is to obtain a sediment sample. For geological purposes some sample disturbance is tolerable, but for the soil mechanician the degree of disturbance is a major concern. The early core sampling, such as Piggot's gun corer, was mainly for geological purposes. The gun corer was an explosively actuated core barrel and was able to penetrate a 4-ft column of sediment. Present corers replace the explosive actuators with gravity and are either the open-barrel or the piston type.

Open-barrel corers use either a near-bottom tethered free fall or a surface-to-bottom free fall, like the boomerang corer. In the case of the piston corer, a lanyard positions the piston at sediment level while the barrel slides by. The piston aids the sliding of the corer into the barrel by suction, but it also disturbs the core by the variable suction that is introduced. The internal friction between core wall and core created as the sediment enters the core tube limits the length of the core obtained, particularly in the gravity corer; eventually the internal resistance equals the strength of the sediment and causes the tube and sample to act as a plunger. The quality of the sample is influenced by the amount of internal friction and by displacement of sediment during penetration. Sample dimensions based on practical experience were recommended by Hvorslev in 1948 to minimize such disturbances and are still the best guide for sample design. By these standards almost none of the samplers presently in use are acceptable, particularly when plastic core liners are used.

In relatively shallow waters, or where heavily funded, in deeper waters, rotary drilling is used for penetrating the sea floor to great depths. Much of the data for oil companies is obtained in this way. Engineering properties of Gulf of Mexico clays were obtained by drilling in up to 400 ft of water for samples to 397 ft below the mud line. Pilot drilling for the Mohole deep-sea drilling project was carried out in 3,100 ft of water to depths of 1,040 ft below the sea floor. At the Guadalupe site drilling was done in 11,700 ft to a depth of 550 ft into the sediments.

A new project sponsored by the National Science Foundation is currently in progress. The *Glomar Challenger*, a specially designed drilling vessel, is operating in 17,000 ft of water and drilling to depths in excess of 2,500 ft. Few data are available to assess the integrity of the drilled samples.

3-9 IN SITU TESTING OF MARINE SEDIMENTS

Many of the problems associated with sampling are overcome by *in situ* testing. Vane shear tests have been made from the bathyscope *Trieste* and Lockheed's

Deep Quest submersible. The U.S. Naval Civil Engineering Laboratory has an *in situ* device capable of performing vane shear and plate bearing tests and a long-term loading system, LOBSTER, for consolidation properties.

An accelerometer attached to sampling apparatus to give force-displacement data of the sediment has been used.

A deep ocean sediment probe (DOSP) has recently been tested by University of Rhode Island and Naval Underwater Systems Center personnel. The device is a bottom-mounted probe capable of obtaining an undisturbed core as well as *in situ* acoustic, resistivity, nuclear density properties, and temperatures of the sediments. Some of the more practical *in situ* tests are:

Load Test This is one of the simplest and most direct tests of soil properties. Typically a load is applied to a square or circular plate on the undisturbed sediment. The deformation with each increment of load is noted—usually to failure of the soil. The results are then extrapolated for the actual foundation.

Static Cone Penetration Test Load-deformation characteristics of the sediment are obtained by this device as in the plate bearing test. The usefulness of this test is that the soil properties are analyzed over a depth as the cone is forced into the sediment.

Pressure Meter The pressure meter, invented by L. Menard, is basically a cylindrical balloon which fits snugly into a bore hole. When the balloon is inflated, the pressure and volume change are measured, thus producing s stress-strain relation of the sediment.

Conductivity The development of conductivity through dispersive media dates back over a hundred years. Conceivably such parameters as particle orientation and porosity can be inferred from conductivity data.

Acoustic and Seismic The seismic method is based on the measurement of velocity of wave propagation in sediment and rock. The main use has been in subbottom profiling to delineate the sea bottom depth and strata. Some promise is offered that compressional wave velocities may be related to compressibility or strengths of sediments (Del Flache, Bryant, and Cernock 1971).

Radioactive Methods Nuclear methods of determining moisture content and density are standard test apparatus on land compaction studies. High-energy neutrons are used to determine field moisture, and gamma radiation is used to determine field density. Several investigators (Preiss 1966, Keller 1965, and Rose and Roney 1971) have incorporated a nuclear density probe in a sediment profiling device.

3-9.1 Bearing capacities of foundations

When a wall or footing lies flat on the sediment and the load is directed downward, the behavior of the sediment is such that it is pushed out from

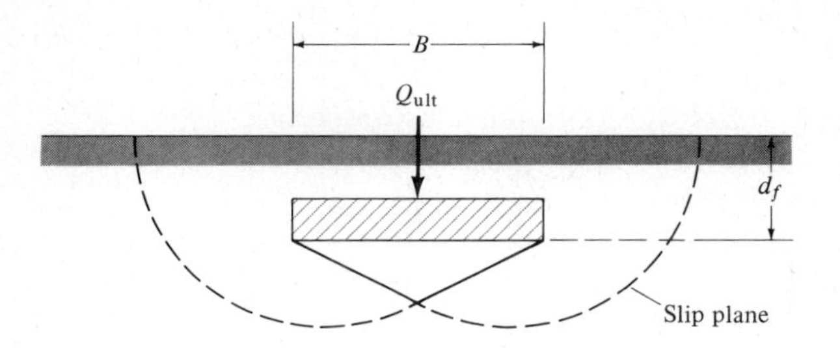

Fig. 3-10 Nomenclature of the footing problem (Example 3-1).

beneath the footing (Fig. 3-10). An approximate solution was derived by Terzaghi (1943):

$$\frac{Q_{\text{ult}}}{A} = q_{\text{ult}} = cNc + 0.5\gamma_b BN_\gamma + \gamma_b d_f N_q \tag{3-23}$$

where B is the width of the footing, d_f is the depth of footing embedment and N_c, N_γ, and N_q are the dimensionless pressure coefficients and are called *bearing-capacity coefficients*. Each coefficient is dependent on $\bar{\phi}$ only (Fig. 3-11). Equation (3-23) applies to "long" footings.

Example 3-1 Determine the ultimate load Q_{ult} on a 4- by 5-ft footing embedded 2 ft into the sediment (Fig. 3-12).

Fig. 3-11 Values of the bearing-capacity coefficients for use in Eq. (3-23) or (3-24).

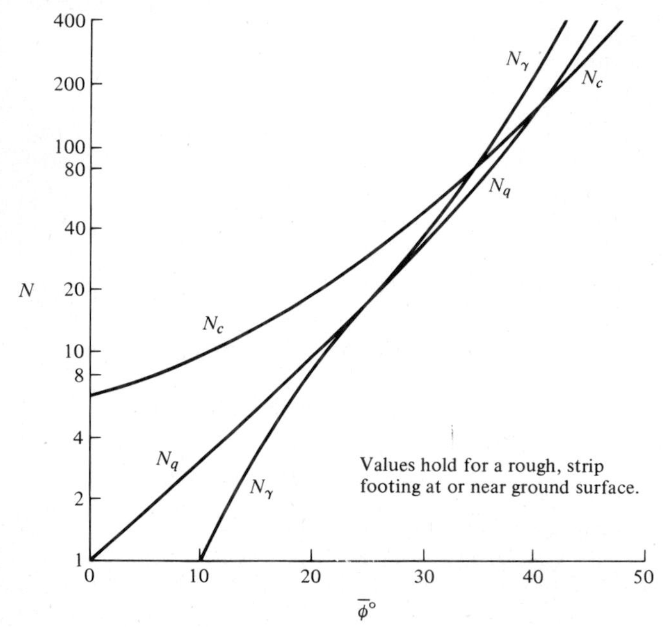

Solution There are two solutions, as there are in most soil engineering problems; one is the "end-of-construction" solution, and the other is the "long-term" case. For the end-of-construction case, the material behaves as a cohesive material; i.e., its strength is independent of total stresses and the strength property of the sediment is C_u, as discussed in Sec. 3-6.3. For a footing problem, the average consolidation pressure is assumed at a depth of $[d_f - (B/2)]$ or, in this case, 4 ft. Then

$$\bar{\sigma}_c = 4\ \text{ft}(104 - 64)\text{lb/ft}^3 = 160.0\ \text{lb/ft}^2$$

and from Eq. (3-22), C_u is 48.0 lb/ft².

Equation (3-23) adapted to square or circular footings is

$$\frac{Q_{\text{ult}}}{A} = 1.3cN_c + 0.3\gamma_b BN_\gamma + \gamma_b d_f N_q \tag{3-24}$$

For the end-of-construction case, c is C_u and $\bar{\phi}$ is zero.

$$\frac{Q_{\text{ult}}}{A} = 1.3 \times 48 \times 5.7 + 0.3 \times 40 \times 4 \times 0 + 40 \times 2 \times 1$$

$$= 435\ \text{lb/ft}^2$$

$$Q_{\text{ult}} = A(435) = 5 \times 4 \times 435 = 8{,}700\ \text{lb}$$

For the long-term case use

$$\bar{\phi} = 22^\circ \qquad \bar{c} = 0 \qquad N_c = 20 \qquad N_\gamma = 6 \qquad N_q = 9$$

$$\frac{Q_{\text{ult}}}{A} = 1{,}008\ \text{lb/ft}^2$$

Therefore

$$Q_{\text{ult}} = 29{,}160\ \text{lb}.$$

Allowing consolidation to occur in the long-term case leads to greater bearing strengths. The load must, of course, be applied gradually.

Fig. 3-12 The Example 3-1 footing system.

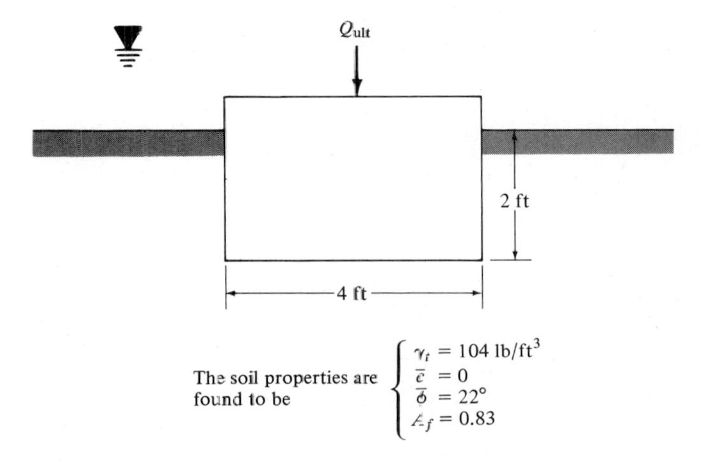

3-9.2 Penetration of objects into ocean bottoms

Many problems are concerned with the penetration of objects into the sediment. Pile driving, anchor penetration, the discovering and reclaiming of lost objects, and the assessment of ocean sediments from dynamic penetration records are some. The object itself may be free-falling or powered, it may have a horizontal component in its trajectory, and it may be embedded to a shallow or deep distance. In any event, the controlling parameters for predicting object penetrations are:

1. The object's impact velocity on the bottom
2. The mass and geometry of the object
3. The impact trajectory
4. The properties of the bottom sediment

Basically the terminal velocity of an object falling through a liquid is given by Stokes' law:

$$\tfrac{1}{2}C_D A \rho_f V^2 = \text{volume}\ (\gamma_0 - \gamma_f) \qquad (3\text{-}25)$$

where C_D = the drag coefficient
A = projected area normal to direction of motion
ρ_f = density of fluid = γ_f/g_c
V = velocity of object
γ_0 = object unit weight
γ_f = density of fluid

The drag force against an object depends on its shape and orientation, the Reynolds number, and the nature of flow. Most books on fluid dynamics will be found useful for further study of these parameters.

The resistance of the sediment to penetration of the object is made up of a static component and a drag component. If we assume that the velocity will quickly moderate, the drag resistance can be eliminated. A simple approach is to equate energy thus:

$$\tfrac{1}{2}mV^2 = Q_{\text{ult}} \cdot x_{\max} \qquad (3\text{-}26)$$

where m is the mass of the object, $x_{\max}$ the penetration, and Q_{ult} the ultimate static bearing capacity as might be found from Eq. (3-23) or (3-24).

Example 3-2 A 3-ft spherical mass, weighing 3,000 lb, is dropped overboard in deep water. C_d for a sphere with a Reynolds number in excess of 80,000 is generally taken to be approximately 0.5. Q_{ult} is estimated for the end-of-construction situation to be 1,325 lb/ft^2. What will be the penetration depth?

Solution The total volume is $\frac{4}{3}\pi(1.5^2)$ ft^3; the displacement is then this figure times 64 lb/ft^3 or 905 lb. The total buoyant weight is (3,000 – 905), or 2,095 lb. From Eq. (3-25)

$$V^2 = \frac{(2)(2{,}095)(64.4)}{(0.5)(\pi 1.5^2)(64)}$$

and V is 34.0 ft/s. A check will reveal that the Reynolds number of the sphere at this velocity is well in excess of 80,000.

Now from Eq. (3-26)

$$x_{\max} = \frac{(3{,}000)(34^2)}{(64.4)(1{,}325)(1.5^2)} = 5.8 \text{ ft}$$

3-9.3 Stress distribution and settlements under foundations

Much of soil mechanics is concerned with the response of soils to the stresses imposed by foundations at ground surface or by piles, tunnels, and trenches. On the one hand, the stress results in soil deformation leading to settlement of structures, and on the other hand, stresses can exceed the sediment strength leading to sediment failure. With the assumptions that the soil is homogeneous and isotropic and that the stress is proportional to strain, the theory of elasticity is used to help to determine stress distribution within the soil mass.

Actually it is very tedious to obtain elastic solutions for a given loading condition; therefore solutions in graphical form are herein presented, Fig. 3-13 being an example.

Following the laboratory determination of field density γ_T and void ratio e_0, the compression index C_{ci} and coefficient of consolidation c_v, the anticipated settlements may be determined from the loading conditions and subsoil data. The following example will illustrate the computation of a settlement problem.

Example 3-3 A 50-ft-diameter cylindrical habitat is placed on a bay bottom. The contact pressure is 400 lb/ft². Determine the ultimate settlement and the settlement after 2 years under the centerline of the structure, given the soil properties shown in Fig. 3-14.

Solution At mid-strata the initial consolidation pressure is

$$\bar{\sigma}_0 = 25(114 - 64) = 1{,}250 \text{ lb/ft}^2$$

The increase in vertical stress (Fig. 3-13) is

$$\frac{X}{R} = 0 \qquad \frac{Z}{R} = \frac{25}{25} = 1$$

so that $\Delta\sigma_z/\Delta p = 0.66$.

$$\Delta\sigma_z = (0.66)(400) = 263 \text{ lb/ft}^2$$

and the ultimate settlement from Eq. (3-14) is

$$\Delta h = \frac{50}{1 + 1.1}(0.8)\log_{10}\frac{1{,}250 + 263}{1{,}250} = 1.74 \text{ ft or } 20.9 \text{ in}$$

After 2 years the theoretical factor

$$T = \frac{C_v t}{H^2} \tag{3-17}$$

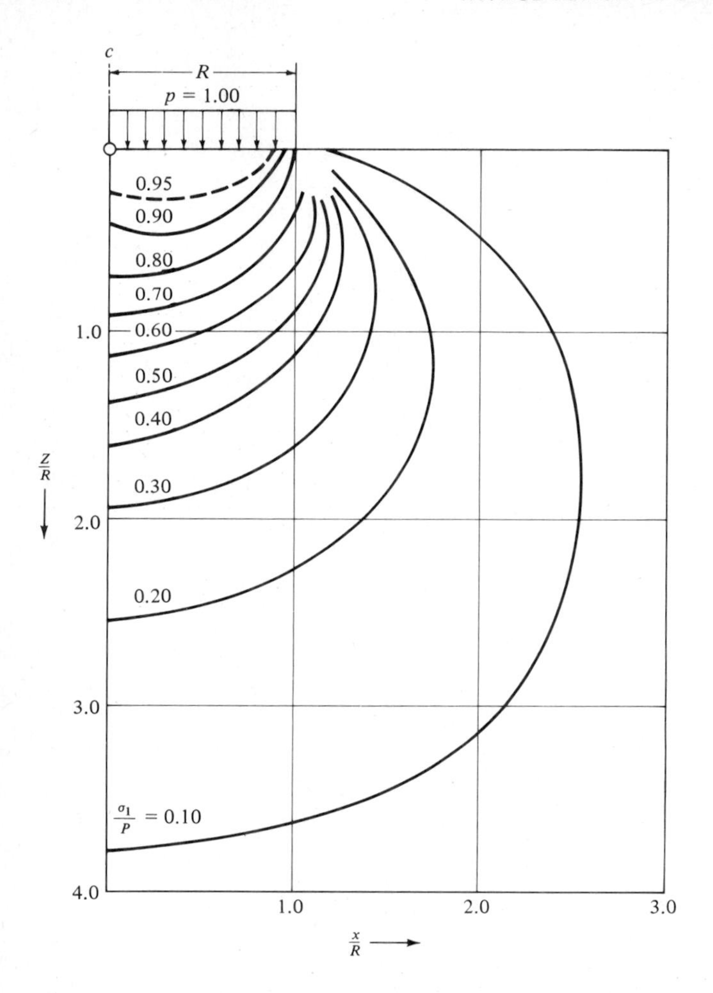

Fig. 3-13 Lines of equal σ_1 for a circular load of radius R on elastic half-space. The contact pressure P and the radius are unity, and Z is the depth under the foundation mass.

gives

$$T = \frac{(20)(2)(12)}{50^2} = 0.192$$

From Table 3-3, the degree of consolidation is found using $U = 49\%$. For $t = 2$ years, $(0.49)(20.9) = 10.3$ in. In this analysis, one-dimensional consolidation has been assumed where in fact settlements are nearer three-dimensional. In addition further distortion occurs even after the pore pressures have dissipated owing to the slippage and rotations of individual particles. This latter phenomenon, called secondary consolidation or creep, is seldom taken into account but may be a more important quantity in sensitive clays.

3-10 SPECIAL STRUCTURES

In addition to the relatively simple structures in this chapter there are many other structures that retain or bear on sediment in one way or another. Although detailed calculations concerning such structures fall outside the scope of this book, a brief description will be given of two such structures.

3-10.1 Double-wall cofferdam

A double-wall cofferdam consists of two rows of sheet piling connected together with tie rods and filled with soil. A more sophisticated type is the cellular cofferdam, in which the tie rods are eliminated and the soil pressure is resisted by the sheet pile interlock tension. Cofferdams usually are temporary structures used to keep a work area dry. The loading on the cofferdam is a combination of the hydrostatic pressure on the outside of the dam and its own weight. To be stable, the cofferdam must resist the tendency to overturn or slide.

3-10.2 Breakwaters, man-made islands

The bases of gravity seawalls, breakwaters, and man-made islands can be viewed as a large footing load on the sediment. The bearing capacity parameters N_c, N_γ, and N_q for both the end-of-construction case and the long-term case and the bearing capacity formula, Eq. (3-23), may be used to form a preliminary judgment on the adequacy of the sediment to resist the superimposed loads.

PROBLEMS

3-1. A saturated sand has a porosity of 35 percent and $G_s = 2.66$. Compute void ratio e, unit weight of the mass γ_t, buoyant unit weight γ_b, and water content w.

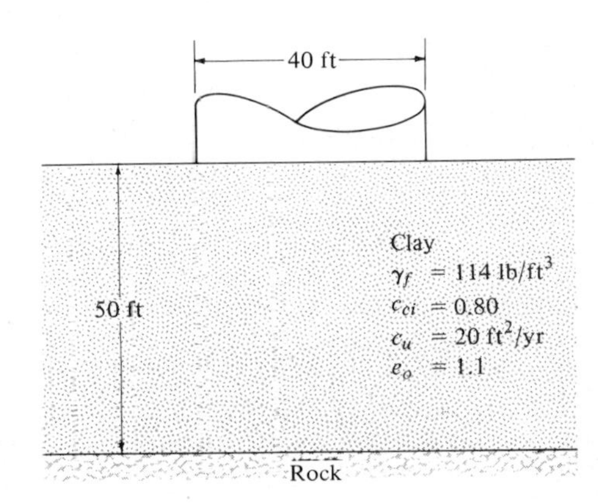

Fig. 3-14 Example 3-3 habitat settlement system.

3-2. A saturated sample of soil has a volume of 1 ft^3 and a weight of 130 lb. The specific gravity of the soil particles is 2.79.

(*a*) Assume that the pore fluid is pure water and determine the water content and the void ratio.

(*b*) Assume next that the pore fluid is salt water with a specific gravity of 1.025. Let the weights of solid matter, pure water, and salt content be designated by w_s, w_w, and w_d, respectively. Determine the void ratio and the values of ratios w_w/w_s, $(w_w + w_d)/w_s$, and $w_w/(w_s + w_d)$.

3-3. A submerged stratum of clay has a thickness of 15 ft. The average water content is 54 percent and the specific gravity of the solid minerals is 2.78. What is the vertical effective stress due to the weight of the clay at the base of the stratum?

3-4. Data from a consolidation test on a clay include the following:

$$e_1 = 1.204 \qquad p_1 = 1.50 \text{ kg/cm}^2$$

$$e_2 = 1.054 \qquad p_2 = 3.00 \text{ kg/cm}^2$$

The sample tested was 2 cm thick under the above conditions. Time required to reach 50 percent consolidation was found to be 10 min. Considering a soil stratum of the same material 30 ft in thickness, determine the time required to reach the same degree of consolidation if (*a*) drained from both top and bottom; (*b*) drained from top only.

3-5. A consolidated undrained triaxial test on a silt soil furnished the following results:

	Test 1, psi	*Test 2, psi*
σ_1	19.0	36.5
σ_3	10.0	19.5
u	7.0	13.5

Plot the graphs representing the stress circles based on total stresses and effective stresses. Determine the strength parameters $\bar{\phi}$, $\bar{c}$, A_f.

3-6. Two triaxial compression tests are performed on a clayey soil. The results are:

	σ_{3f}, *psi*	σ_{1f}, *psi*	u_f, *psi*
Test 1	7	11	4.8
Test 2	17.8	27.4	11.5

Determine these: $\bar{\phi}$, $\bar{c}$, A_f.

3-7. Solve the following problem, using the soil properties given. Determine the bearing capacity—both end-of-construction and long-term—for a 4-ft-diameter footing that is embedded 2 ft.

$$\gamma_t = 110 \text{ lb/ft}^3 \qquad \bar{\Phi} = 20° \qquad \bar{c} = 50 \text{ lb/ft}^2 \qquad A_f = 1.25$$

3-8. A circular area on the surface of an elastic mass of great extent carries a uniformly distributed load of 2,500 lb/ft^2. The radius is 20 ft. What is the intensity of vertical and horizontal stress at a point 15 ft beneath the center of the circle? At a point at the same depth beneath the edge of the circle?

3-9. A cylindrical habitat is placed on a bay bottom. Assume the base is flexible and the load is 100 lb/ft². Estimate the final settlement at the center and at the edge of the structure.

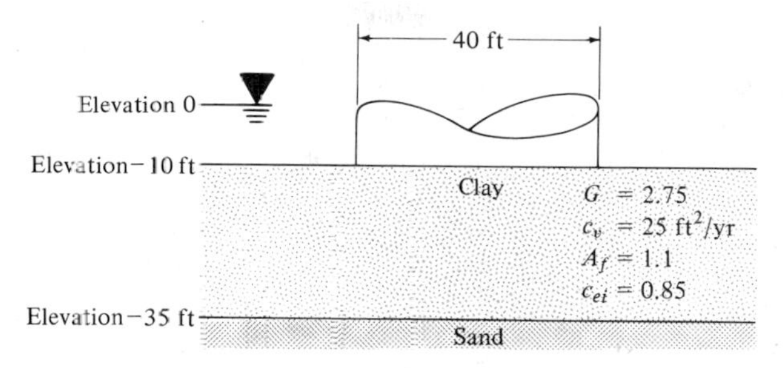

Fig. P3-9

3-10. Estimate the consolidation settlement for a 15-ft-diameter habitat resting on a 60-ft clay stratum. Assume a negative buoyancy of 100 lb/ft².

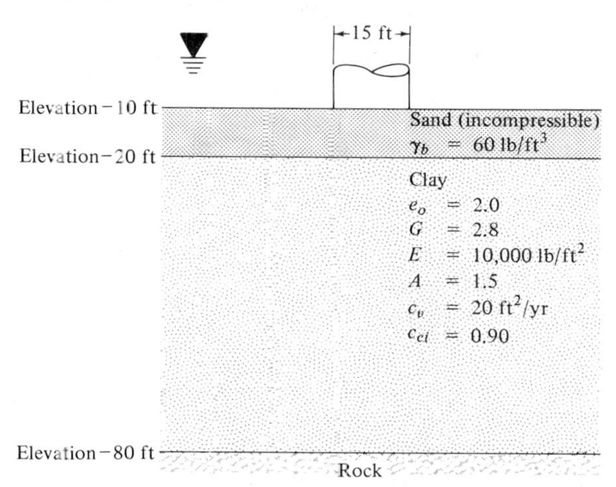

Fig. P3-10

3-11. The University of Rhode Island Portalab habitat stands on four 2-ft-diameter circular plates and under normal conditions has a negative buoyancy of 1,200 lb. After 4 months the average settlement is 0.1 ft and we assume that this occurs entirely in the upper 10 ft of soil. If the habitat should fill with water, the negative buoyancy would increase to about 8 tons. Predict the settlement after 4 months at the heavy condition. Assume $\gamma_b = 40$ lb/ft³.

3-12. A saturated sample of soil on the sea bottom is carefully scooped into an 18-in³ box by a diver. The undrained sample weighs 1.1 lb. It is oven-dried and the residue weighs 0.8 lb.

(*a*) Find γ_t, γ_b, G, n, and e.

(*b*) What is the stress on an object that is under 5 ft of this soil with the bottom at 10-ft depth?

3-13. A soil sample is placed in a triaxial apparatus such that drainage is prevented. The lateral total stresses are held at 10 psi and the vertical total stress is increased until failure of 20 psi occurs at σ. Previous tests on this soil gave a $\bar{c}$ of zero and a friction angle ($\bar{\phi}$) of 30°.

(a) What was the pore pressure at failure? Show a Mohr circle sketch of the test.

(b) What is A_f for this soil?

REFERENCES

Aas, G. (1965): *Proc. 6th Intern. Conf. on Soil Mechanics and Foundation Engineering*, Montreal, p. 141.

Arrhenius, G. (1963): "Pelagic Sediments," *The Sea*, vol. 3, Interscience, New York.

Baird, J. A. (1971): "The Shear Strength Properties of an Undisturbed Clay in Extension," M.Sc. thesis, University of Rhode Island, Kingston, R.I.

Bishop, A. W., and A. K. Eldin (1950): *Geotechnique*, vol. 2, pp. 13-32.

Bjerrum, L. (1967): Engineering Geology of Norwegian Normally Consolidated Clays as Related to the Settlement of Buildings, *Geotechnique*, vol. 18, pp. 83-118.

——, and I. Rosenquist (1957): Some Experiments with Artificially Sedimented Clays, Norwegian Geotechnical Institute, vol. 25, pp. 1-12.

Bryant, W. R. (1967): Shear Strength and Consolidation Characteristics of Marine Sediments from the Western Gulf of Mexico, *Proc. Intern. Res. Conf. on Marine Geotechnique*, University of Illinois Press, Urbana.

Crawford, C. B. (1965): *Can. Geotechnical J.*, vol. 2, pp. 90-115.

Del Flache, A. D., W. R. Bryant, and P. J. Cernock (1971): Determination of Compressibility of Marine Sediments from Compressional-Wave Velocity Measurements, *Third Offshore Technol. Conf.*, American Institute of Mining, Metallurgical, and Petroleum Engineers, Houston, Tex.

Emery, K. C. (1968): Relict Sediments on Continental Shelves of the World, *Am. Assoc. Petrol. Geol.*, vol. 52, pp. 445-464.

Ewing, M. (1964): Sediment Distribution in the Oceans: The Mid-Atlantic Ridge, *Geol. Soc. Am. Bull.*, vol. 75, pp. 17-36.

Guilcher, A. (1963): Continental Shelf and Slope, *The Sea*, vol. 3, Interscience, New York.

Heezen, B. C. (1966): Shaping of the Continental Rise by Deep Geostrophic Contour Currents, *Science*, vol. 152, pp. 502-508.

——, and C. B. Ericson (1954): *Deep-Sea Res.*, vol. 1, pp. 193-202.

——, and A. S. Laughton (1963): "Abyssal Plains, *The Sea*, vol. 3, Interscience, New York.

——, and H. W. Menard (1963): "Topography of the Deep Sea Floor," *The Sea*, vol. 3, Interscience, New York.

Heirtzler, J. R. (1968): Sea Floor Spreading, *Sci. Am.*, vol. 219, pp. 60-70.

Henkel, D. J. (1960): The Shear Strength of Saturated, Remolded Clays, *Proc. Am. Soc. Civil Engrs. Res. Conf. on Shear Strength of Cohesive Soils*, New York, pp. 533-554.

Hvorslev, M. J. (1948): Subsurface Exploitation and Sampling of Soils for Civil Engineering Purposes, *Report of Committee on Sampling and Testing*, Soil Mechanics Division, American Society of Civil Engineers, New York.

Keller, G. (1965): Deep Sea Nuclear Sediment Density Probe, *Deep-Sea Res.*, vol. 12, pp. 373-376.

Lambe, T., and R. Whitman (1969): "Soil Mechanics," Wiley, New York.

Leonards, G. A. (1964): *J. Soil Mech. Found. Eng., Trans. Am. Soc. Civil Engrs.*, vol. 90, pp. 133-155.

Morgenstern, M. R. (1967): *Marine Geotechnique*, vol. 18, pp. 83-118.

Nacci, V. A., and M. T. Huston (1969): Structure of Deep Sea Clays, *Proc. Conf. on Civil Eng. in the Oceans II*, American Society of Civil Engineers, Houston, Tex., pp. 599-619.

Noorany, T., and H. B. Seed (1967): *J. Soil Mech. Found. Eng., Trans. Am. Soc. Civil Engrs.*, vol. 90, pp. 49–80.

Parry, R. H. (1960): Triaxial Compression and Extension Tests on Remolded Saturated Clay, *Geotechnique*, vol. 4, pp. 166–188.

Preiss, K. (1966): Analysis and Improved Design of Gamma-Ray Backscattering Density Gages, *Highway Res. Board, H.R.B. Record No. 107*, pp. 1–12.

Rose, V. C., and J. R. Roney (1971): A Nuclear In-Place Measurement of Sediment Density, *Third Offshore Technol. Conf.*, American Institute of Mining, Metallurgical, and Petroleum Engineers, Houston, Tex.

Shepard, F. P. (1963): "Submarine Geology," 2d ed., Harper & Row, New York.

Skempton, A. W. (1954): *Geotechnique*, vol. 4, pp. 143–147.

Terzaghi, K. (1936): *Proc. 1st Intern. Conf. on Soil Mechanics*, Cambridge, Mass., pp. 161–165.

4
Hydrodynamics of Waterborne Bodies

T. Kowalski

4-1 RESISTANCE

4-1.1 Parameters of the drag problem

A body immersed in a fluid or floating in an interface between two fluids when set in motion experiences resistance. Power has to be expended to produce and maintain the motion. The motion of the body may consist of three phases:

1. Accelerating from zero or some datum speed
2. Constant speed
3. Decelerating from constant speed to zero or datum speed

Only constant speed phase will be considered here. It will be assumed that the nonsteady types of motion require less power than the constant speed phase. This means that there is a limit on the acceleration the body can attain with the power provided by the existing engines.

The resistance and propulsion of a body involve relative motions of the fluid. Thus, the relative motion between the body and the fluid produces the hydrodynamic forces and pressures which result in the drag. The relative motion produced by the propulsion devices exerts the thrust which propels the body through the fluid.

In studying fluid motions, the following fluid properties are important:

ρ = density	slugs/ft^3 or lb·s^2/ft^4
μ = dynamic viscosity	lb·s/ft^2
ν = kinematic viscosity	ft^2/s
p = pressure	lb/in^2 or psi
p_v = vapor pressure	psi
K = bulk modulus of elasticity	psi
U (or V) = velocity	ft/s or kn (1 kn = 1.69 ft/s)

There are two approaches to the analysis of the fluid motions:

1. Neglecting viscosity and compressibility of the fluid (perfect fluid) leads to theoretical or classical hydrodynamics treatment. A body deeply submerged in a perfect fluid will experience zero drag since there are no tangential forces and the normal pressure forces balance exactly over the surface of the body. Only when the body is brought up to the surface will the pressure distribution around the body cause deformation of the surface of the fluid, producing waves which will represent energy loss. Hence in a perfect fluid, wave making is the only drag component that can be calculated theoretically.
2. Neglecting compressibility of the fluid (real liquids at normal pressures) leads to applied hydrodynamics treatment. Owing to the viscosity of the fluid, there is a tangential shear stress on the surface of the body. The fluid is retarded layer by layer with the effect diminishing as the normal distance from the body increases. The extent of the retarded fluid defines the boundary layer within which all the viscous effects are confined. Outside the boundary layer the flow may be assumed to be that of a perfect fluid (potential flow). Moving aft along the surface of the body the boundary layer grows in thickness as more fluid is being retarded by the body. The diminishing effect of the retardation of the fluid produces a velocity profile across the boundary layer with a zero fluid velocity at the surface relative to the body and the velocity of the free stream U at the outer boundary of the layer. Since all the viscous effects are confined to the boundary layer, the viscous drag can be represented as the energy required to drag this extra volume of fluid.

The total drag of a body moving at a constant speed through a viscous, incompressible fluid is

$$D = D_f + D_w + D_e + D_a + D_{rw} + D_p + D_i \tag{4-1}$$

where D = total drag of the body, D_f = frictional (viscous) drag due to the tangential shear stresses existing in a viscous fluid caused by the relative motion between the body and the fluid, D_w = wave-making drag which arises only in the case of bodies moving at or near the interface of water and air. It is due to the variation of pressure distribution over the surface of the body resulting in the creation of waves on the surface of the water. These waves represent energy input that the body is supplying while in motion.

Considering the body shrunk to a point (a pressure point), its passage will produce two systems of waves (Havelock 1934):

1. A divergent system with the crests at an angle to the line of advance. There is one divergent wave crest on each side of the line of advance. These waves leave the body and travel along a line perpendicular to the crests.
2. A transverse system with the crests at right angles to the line of advance curving backwards to connect with the ends of the divergent wave crests.

As the speed increases, the wave pattern also moves at an increased speed, producing a fixed pattern of wave system with respect to the body.

A floating body will produce two major "pressure point" systems: one at the bow, which starts with a wave crest, and one at the stern, starting with a wave hollow. Other wave systems may begin at discontinuities on the body.

The system of divergent waves loses contact with the body and there is no subsequent interaction between the waves and the body. Hence all the energy expended in the creation of divergent waves is lost.

The system of transverse waves, however, moves along the same line of advance as the body and the transverse waves created by the bow will interact with the system produced by the stern. For deepwater waves, Chap. 2 showed that length was related to velocity squared. The waves created by the bow will thus vary in length with the speed squared and will react with the wave hollow of the stern transverse wave. Thus, if a bow wave crest coincides with the stern wave hollow, the resultant wave will theoretically be of zero height; hence the energy input required will be at a minimum. On the other hand, if the hollow of the bow wave arrives at the stern, the resultant will be an increased hollow representing an increased power requirement and in addition producing a substantial trim of the body. Figure 4-1 shows possible cases of the interaction between the two transverse wave systems superimposed on a typical graph of the wave-making drag D_w versus modified Froude number. The appearance of the humps and hollows in the drag curve indicates the interference effects between the bow and stern wave systems. It is clear that by a judicious choice of the speed and the length of the body during the design stages, a body can be made to operate in a "drag hollow" resulting in a lower overall drag.

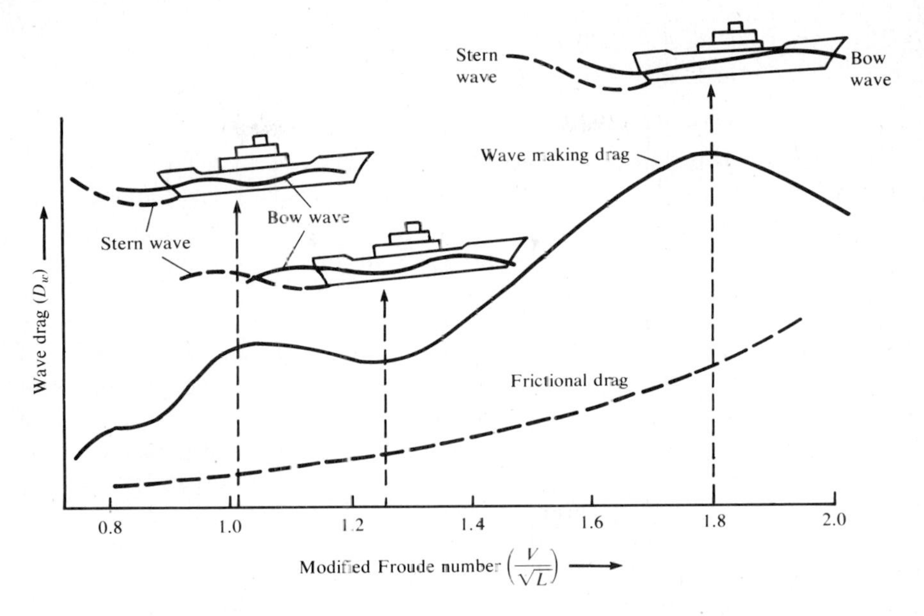

Fig. 4-1 Wave drag behavior as a function of bow and stern wave interaction.

There are two ways of influencing the wave-making drag:

1. At a given speed (usually the service speed) the length of the body can be designed so that a bow crest falls at the position of the stern hollow, thus producing a minimum wave drag for that speed. Once the body is designed for a particular speed, however, any departure from it will mean an increase of drag.
2. The lowest part of the bow can be enlarged into a bulb producing a speedup of the flow around it, hence a lower pressure. If the bulb is placed right under the bow wave crest, the height of the crest will be reduced; thus the body will use less energy in the creation of the bow wave. A given bulb design (shape and longitudinal position) will work best at a given speed. When the body is moving at a different speed, the reduction of the wave-making drag will not be as much and at some speeds it may even increase the drag. In addition the extra surface of the bulb will increase frictional drag.

D_e = eddy drag, which is caused by the appendages and any sharp projections on the surface of the body. Turbulent eddies are shed from the trailing edges of the appendages such as shaft brackets, bilge keels, etc. This represents energy losses.

D_a = air (or wind) drag, which represents the drag of the superstructure exposed to the relative air velocities.

D_p = pressure (or form) drag. When a body is moving through a viscous fluid, a boundary layer is formed on the surface of the body, growing in thickness in the downstream direction. Right behind the body a wake is formed equal in width to double the thickness of the boundary layer at the stern. Integrating the pressure distribution over the surface of the body, including the boundary layer and the wake, a net force will result acting in the opposite direction to the motion. This part of the drag is clearly dependent on the Reynolds number since in a nonviscous fluid there is no boundary layer and the streamlines close behind the body resulting in zero net pressure force, hence zero pressure drag. However, most methods of analysis combine pressure drag component with the wave-making component and treat them as dependent on Froude number. Since for a well-designed body the pressure component of the drag is quite small, this treatment does not introduce significant errors.

D_i = induced drag. This component exists only when the body or parts of it (hydrofoils, for example) develop lift forces.

D_{rw} = rough weather drag. In addition to being exposed to winds the body is acted upon by the waves in rough seas. The extra drag will consist not only of the direct wave action on the body, but also the drag due to excessive motions of the body caused by the rough seas.

The most important components of the total drag for a well-designed body are frictional, wave, and rough weather. Equation (4-1) assumes that the drag components are independent of each other and therefore can be added linearly to give the total drag. In fact there is cross coupling between the components, e.g., between frictional and wave-making drags. The present methods of drag calculation are not able to consider these cross-coupling effects except by means of correction factors.

The dependence of the drag components on the fluid parameters mentioned before will now be explored by means of dimensional analysis. Physical dependence, $D = f(\rho, \mu, U, L, p, g, K, p_v)$, can be expressed in terms of nondimensional groups obtained mathematically. For the description of this method see any text on fluid mechanics.

$$\frac{D}{\rho U^2 L^2} = \phi\left(\frac{UL\rho}{\mu}\,;\frac{U}{\sqrt{gL}}\,;\frac{p-p_v}{\rho U^2}\,;\frac{U}{\sqrt{K/\rho}}\right)$$

$$= \phi(N_R\,; N_F\,; \sigma; M) \tag{4-2}$$

where N_R = Reynolds number (inertia forces/viscous forces)

N_F = Froude number (inertia forces/gravity forces)

σ = cavitation number (resultant static pressure/dynamic pressure)

M = Mach number (inertia forces/compressibility forces)

Correlation between the components of the total drag and the nondimensional parameters is determined from the physical concepts of hydrodynamics. Thus, the frictional, air, eddy, and profile drag components will be some function of the Reynolds number. The wave-making component will depend on the Froude number. The rough weather component will most likely be a function of both Reynolds and Froude numbers. The complete theoretical calculation of the drag of a body, however, is still not possible. Experimental results have to be used to determine the actual functions suggested by theory, together with the necessary coefficients. The combination then of theory and experiment provides the required methods for the determination of the drag.

The experiments are, of necessity, performed on model scale (Todd 1951). To translate the results from the models to full size, scaling laws are needed. They are based on the similarity laws of model testing. For two systems to be dynamically similar they have to be:

1. *Geometrically similar.* All the ratios λ of their linear dimensions have to be the same:

$$\lambda = \frac{Ls}{Lm}$$

2. *Equal in force ratios.* This means that the model has to be tested at the same N_R, N_F, M, etc. This can be expressed as

$$\left(\frac{D}{\rho U^2 L^2}\right)_m = \phi(N_R, N_F, M, \sigma)_m \qquad \text{for model}$$

$$\left(\frac{D}{\rho U^2 L^2}\right)_s = \phi(N_R, N_F, M, \sigma)_s \qquad \text{full-size body}$$

If the model tests are conducted so that the right-hand sides are equal, then

$$\left(\frac{D}{\frac{1}{2}\rho U^2 L^2}\right)_m = \left(\frac{D}{\frac{1}{2}\rho U^2 L^2}\right)_s = C_T \tag{4-3}$$

so that

$$D_s = \frac{(\rho U^2 L^2)_s}{(\rho U^2 L^2)_m} \times D_m$$

This gives the full-size drag if the model drag is known. However, in practice it is impossible to satisfy all the force ratios at the same time.

Taking $\lambda = 60$, which represents a 10-ft model of a 600-ft ship, then N_R equality will require $(UL)_M = (UL)_s$ (assuming both model and ship operate in the same medium). $U_m = 60U_s$. N_F equality will require $U_m/\sqrt{gL_m} = U_s/\sqrt{gL_s}$; $U_m = U_s/\sqrt{60}$, but M equality will require $U_m = U_s$.

Judicial selection of the parameters that can be satisfied in a model test has to be made. The remaining parameters are accounted for by other means.

It appears from the above that Froude number equality is the easiest to satisfy, since then the model can be tested at lower speed than the full-size body. This is the basis of the Froude scaling laws.

4-1.2 Froude's methods of drag estimation from model tests

The basis of the scaling law is the assumption that the total drag can be split into separate components which can be added linearly with each component dependent on one of the nondimensional force ratios.

Froude's method combines the drag components in the following manner:

$$D = D_f + (D_w + D_e + D_p) + (D_a + D_{rw} + D_i) \tag{4-4}$$

$$D = D_f + D_r \tag{4-5}$$

where D_r = residuary drag = $D_w + D_e + D_p$, and the terms inside the second bracket are to be determined independently. The procedure is as follows.

1. Tow a geometrically similar model at a number of constant Froude numbers. These are called "corresponding speeds." Measure the total drag of the model at each speed.
2. Calculate the skin friction drag of the model from a friction line formulation for a flat plate. The one used extensively on the North American continent is Schoenherr's (1932) friction formula

$$\frac{0.242}{\sqrt{C_f}} = \log (C_f N_R) \tag{4-6}$$

   (See Table 4-1.)
3. Calculate the residuary portion of the model drag $D_r = D - D_f$.
4. Scale up the residuary part of the drag in accordance with Froude number equality. Considering the residuary drag independent of Reynolds number, the nondimensional analysis gives

$$\frac{R_r}{\frac{1}{2}\rho U^2 L^2} = f(N_F)$$

   For the same Froude number

$$\left(\frac{R_r}{\frac{1}{2}\rho U^2 L^2}\right)_m = \left(\frac{R_r}{\frac{1}{2}\rho U^2 L^2}\right)_s$$

Table 4-1 Skin friction drag coefficients derived from Schoenherr's formula

$$\frac{0.242}{\sqrt{C_f}} = \log (C_f N_R)$$

N_R \ n	$C_f \times 10^3$				
	5	6	7	8	9
1×10^n	7.179	4.410	2.934	2.072	1.531
2	6.138	3.878	2.628	1.884	1.408
3	5.624	3.600	2.470	1.784	1.343
4	5.294	3.423	2.365	1.718	1.299
5	5.058	3.294	2.289	1.670	1.266
6	4.874	3.193	2.229	1.632	1.240
7	4.727	3.112	2.179	1.601	1.219
8	4.605	3.043	2.138	1.574	1.201
9	4.500	2.968	2.103	1.551	1.186

Hence

$$(D_r)_s = (D_r)_m \frac{\rho_s}{\rho_M} \frac{U_s^2 L_s^2}{U_m^2 L_m^2}$$

and since U^2 varies as L,

$$(D_r)_s = (D_r)_m \left(\frac{\rho_s}{\rho_M}\right) \lambda^3 \tag{4-7}$$

5. Calculate the skin friction drag of the full-scale body from the same friction line formulation as for the model and obtain $(D_f)_s$.
6. The total drag in smooth water will then be

$$D = (D_f)_s + (Dr)_s$$

or in terms of drag coefficients using Eq. (4-3)

$$C_T = C_f + C_r \tag{4-8}$$

The method can be summed up as consisting of formula calculation of the frictional part of the total drag and experimental measurement of the residuary part of the drag.

The total drag obtained in this way is generally too low. The reason for this lies in the shortcomings of the scaling laws. Generally a correction is applied to the full-scale total drag by addition of the so-called *roughness allowance* ΔC_f. This correction has some physical justification as it is based on the fact that the model is tested at a substantially lower Reynolds number than the full-scale body. At a lower Reynolds number the boundary layer on the model surface is relatively thicker and it submerges larger roughness sizes. The full-scale body operating at a much higher Reynolds number (relatively thinner boundary layer) appears to the flow as having a much rougher surface. Hence the frictional part of the drag must be increased. A standard roughness allowance, $\Delta C_f = 0.0004$, is used with the Schoenherr friction line formulation in the American towing tanks.

The above method served reasonably well until the advent of supertankers, when additional correction had to be made owing, it seems, to the inability of the Schoenherr formula to extrapolate the friction line to the required very high Reynolds numbers of the order of $N_R = 10^9$. It is quite obvious that the problem of drag estimation has not been solved yet. Attempts have been made from time to time to improve the method. Examples are the Telfer (1927) and Hughes (1952, 1954) approaches.

The following example will show the use of the Froude method.

Example 4-1 (Part 1) A 10-ft model of a 160-ft barge is tested in a towing tank at a range of corresponding speeds up to 10 kn full scale. The wetted surface of the model is 22 ft² and the beam is 1.5 ft. The model drag is measured and a curve of drag versus speed is obtained. Given that the model drag equals 11.1 lb at a speed of 3 kn, find the drag of the barge at the corresponding speed. The model was tested in freshwater at 70°F, and the barge will be operating in seawater at a standard temperature of 59°F.

Solution Since the scale ratio $\lambda = 16$, the corresponding speed of the barge, at the same Froude number, will be $3\sqrt{16} = 12$ kn.

$$R_t = R_f + R_r = 11.1 \text{ lb}$$

and in coefficient form, using Eq. (4-3),

$$C_t = C_f + C_r = \frac{11.1 \text{ (lb)}}{\frac{1}{2} \times 1.936 \text{ (lb·s}^2/\text{ft}^4) \times (3 \times 1.69)^2 \text{(ft/s)}^2 \times 22 \text{ (ft}^2)}$$

$$C_t = 0.02028$$

Next calculating C_f from Eq. (4-6) at the respective Reynolds number,

$$N_R = \frac{3 \times 1.69 \text{ (ft/s)} \times 10 \text{ (ft)}}{1.055 \times 10^{-5} \text{(ft}^2/\text{s)}} = 4.80 \times 10^6$$

and

$$C_f = 0.00332$$

Hence

$$C_r = (C_t - C_f)_{\text{model}} = 0.02028 - 0.00332 = 0.01696$$

The residuary drag coefficient C_r will be the same for the full-scale barge since the model was tested at the same Froude number.

Therefore for the barge, $C_t = (C_f + C_r)$ and the remaining unknown is the skin friction coefficient for the full-size barge. This is found in exactly the same way as the C_f for the model.

Reynolds number for the barge:

$$N_R = \frac{12 \times 1.69 \text{ (ft/s)} \times 160 \text{ (ft)}}{1.2817 \times 10^{-5} \text{ (ft}^2/\text{s)}} = 2.53 \times 10^8$$

The kinematic viscosity for seawater at 59°F was obtained from tables of physical properties of water. Equation (4-6) gives the required friction coefficient C_f = 0.001825. Since the barge will, most probably, not be as smooth as the model, a roughness allowance of 0.0004 is added to the total drag coefficient. Thus for the barge $C_T = 0.01696 + 0.001825 + 0.0004 = 0.019185$ and $R_T = \frac{1}{2}\rho U^2 S C_T$, where S = wetted surface area of the barge; so S = wetted surface area of the model $\times \lambda^2$.

$$R_T = \tfrac{1}{2} \times 1.9905\ (\text{lb}\cdot\text{s}^2/\text{ft}^4) \times (12 \times 1.68)^2 (\text{ft/s})^2 \times 22\ (\text{ft}^2) \times 16^2 \times 0.019185$$

$$R_T = 44{,}227.5$$

To calculate the effective horsepower necessary to propel the barge at 12 kn,

$$\text{ehp} = \frac{12 \times 1.69\ (\text{ft/s}) \times 44{,}227.5\ (\text{lb})}{550\ (\text{ft}\cdot\text{lb/s})/(\text{hp})} = 1{,}631\ \text{hp}$$

This is the horsepower required to overcome the resistance of the barge in smooth water without appendages (so-called *naked hull drag*) and with zero wind. Wind drag will be calculated in Sec. 4-1.3.

There is no satisfactory way of treating the drag of appendages like bilge keels, shaft brackets or bossings, rudders, etc., developed at the present time. To minimize the drag due to appendages, lines of flow are determined on the surface of the body from model tests. The appendages are located so as not to disturb the streamlines. Methods to account for the appendage drag are empirical; use factors from previous tests, and a combination of tests plus empirical correction (Comstock 1967).

4-1.3 Air drag

The estimation of this component of the total drag is based on the assumption that air drag varies with relative wind velocity squared. $D_a \propto U_r^2$. Two formulas in general use for the estimation of the wind drag are:

1. Taylor's (1943) formula for wind direction from dead ahead:

$$D_a = 0.004(\tfrac{1}{2}B^2)U_r^2 \tag{4-9}$$

where B = beam of the ship (largest transverse dimension of the body).

2. Hughes's (1930) formula for wind direction at angle θ to the direction of motion:

$$D_a = K\rho U_r^2 \left[\frac{A_L \sin^2\theta + A_T \cos^2\theta}{\cos(\alpha - \theta)}\right] \tag{4-10}$$

where K = coefficient

ρ = mass density of air

A_L = longitudinal projection of the above-water part of the body (profile area exposed to wind)

$A_T = 0.3\, A_h + A_s$

A_h and A_s = transverse projected areas of the above-water part of the main hull and of the superstructure respectively

α = angle of action of resultant wind force measured from the direction of motion (see Fig. 4-2)

If the body is heading directly into the wind, $\theta = \alpha = 0°$, $K = 0.6$, and the Hughes formula becomes $D_a = 0.002\ U_r^2 A_T$. This is the same as Taylor's formula except that it is based on the transversely projected area A_T instead of the beam squared B^2.

Example 4-1 (Part 2) For maximum wind drag the barge is steaming into the wind which is blowing at 25 kn. Calculate the added horsepower necessary to maintain the 12-kn speed.

Solution Using Eq. (4-9):

$$D = 0.002 \times 1.5^2 \times 16^2 (\text{ft}^2) \times (12 + 25)^2 \times 1.69^2 (\text{ft/s})^2 = 4.504 \text{ lb}$$

Therefore additional horsepower required to overcome the wind drag is

$$\text{ehp}_a = \frac{12 \times 1.69 \text{ (ft/s)} \times 4505 \text{ (lb)}}{550 (\text{ft} \cdot \text{lb/s})/(\text{h})} = 167 \text{ hp}$$

which represents a 13 percent increase in power.

4-1.4 Drag estimation for specific body shapes

Although model testing is the only reasonably accurate method of drag estimation of waterborne bodies at the present time, a number of useful empirical or semiempirical design methods are also available. However, it is quite obvious that the design of most waterborne craft and their drag calculation is still part art and part science. In spite of the existing substantial accumulation of helpful data, many designers rely heavily on their previous experience and to quite an extent on their "intuition."

The following references describe some of the useful design methods.

Fig. 4-2 Nomenclature of the wind drag equation (4-10).

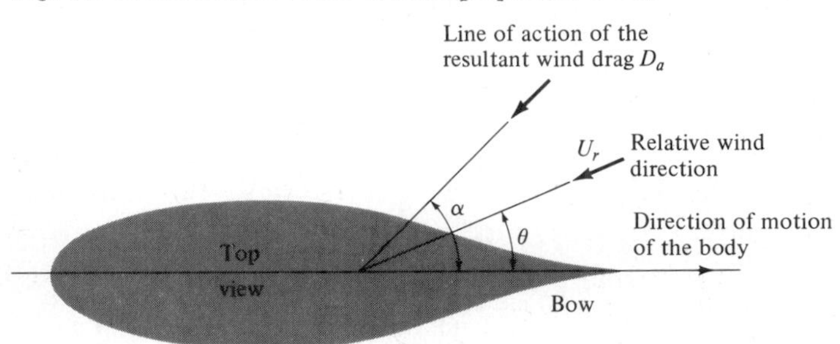

A. Merchant-ship-type bodies Results of resistance tests of many methodical variations on a basic form are given in graph form so that an optimum shape can be chosen for a specific range of operational requirements. These tests were performed by D. W. Taylor and are known as the Taylor Standard Series (Gertler 1954).

B. Small surface craft A number of useful references are available. Round bottom craft design methods are given in graph and data sheet forms by Clement (1963, 1964). Planing-type hull design is discussed by Clement et al. (1963) and by Savitsky (1964). Planing catamarans are dealt with by Clement (1962).

C. Submerged bodies The drag consists only of frictional and form components as long as the body moves deeply submerged (depth more than three diameters of the body). The frictional part of the drag can be calculated from skin friction formulas like Eq. (4-6). Form drag is very often not important if the body is well streamlined. A very useful reference book is one by Hoerner (1965), in which a great many design parameters are given.

D. Separate shapes Tests have been conducted for many years measuring the drag of a number of structural shapes like cylinders, spheres, angles, I beams, etc. The calculated drag coefficients have been published in numerous handbooks and textbooks, Hoerner (1965) probably giving the best such collection. The drag coefficients have to be used with care, for most of them were obtained from tests of individual two-dimensional sections at a specific Reynolds number range. Corrections have to be applied for three-dimensional structures, oblique fluid directions, and screening effects between the structural members. The above reference provides the necessary information for these corrections.

Aguirre and Boyce (1973) present a more accurate method of calculating forces on built-up structures than a simple summing up of individual forces. Using a concept of solidity ratio (shadow area of the structural members to the area of the structure's boundaries), the drag coefficient is estimated directly for the whole structure.

4-2 PROPULSION

4-2.1 Propulsion parameters

To propel a body through a fluid, a propulsion device has to provide a thrust which is derived from the change of momentum of the fluid. The thrust must overcome the drag of the body at a given speed plus the transmission, power conversion, and the interaction losses. These losses are the outcome of imperfect conversions of energy through the propulsion system.

In taking the thermal energy of the fuel burned in the engine and tracing the different stages of energy conversion until the power is converted into thrust, the following steps can be identified:

1. *Conversion of chemical to mechanical power* This is done in a number of ways depending on the type of machinery. Two systems generally in use are:
   (*a*) Conversion of chemical or nuclear energy into thermal energy of steam by means of a boiler.
   (*b*) Conversion of chemical energy into thermal, then directly into mechanical energy inside an engine. This mechanical energy may be produced by a reciprocating engine or by a turbine.

   Another step in the transmission of energy may be introduced where the thermal engine drives a generator that drives an electric motor, which finally provides the mechanical power. The chemical or nuclear to electric conversion can be bypassed in the propulsion system on board a moving body if batteries are used to drive the electric motors. The batteries will have to be recharged, however, and the first part of the conversion cycle will still have to be used. These steps in the energy conversion cycle are the least efficient.
2. *Transmission of mechanical energy* The losses in this part are due to mechanical losses in the bearings, reduction gearing, thrust blocks, and watertight glands. This represents the smallest part of energy loss in the propulsion system.
3. *Conversion of the mechanical power delivered to the propelling device into thrust* The thrust-producing device suffers losses in the conversion process and also because of the presence nearby of the body and its appendages. This is the second largest component of energy losses.

Figure 4-3 shows the losses in the propulsion system together with the associated efficiencies. From the figure the elements of propulsive efficiency can be identified. The starting point is usually taken as the power supplied by the engines P_s (the engine efficiencies are normally considered in the power production analysis) and the end point as the power used in overcoming the drag of the body P_E, the so-called effective horsepower. The propulsive efficiency can then be expressed in terms of the components of separate efficiencies:

$$\eta = \frac{P_D}{P_S} \frac{P_{TO}}{P_D} \frac{P_{TB}}{P_{TO}} \frac{P_E}{P_{TB}} = \eta_{tr}\eta_o\eta_r\eta_h \tag{4-11}$$

The additional terms are defined as:

P_{TO} = power converted into propulsive thrust "in the open" without any boundaries near the propulsive device

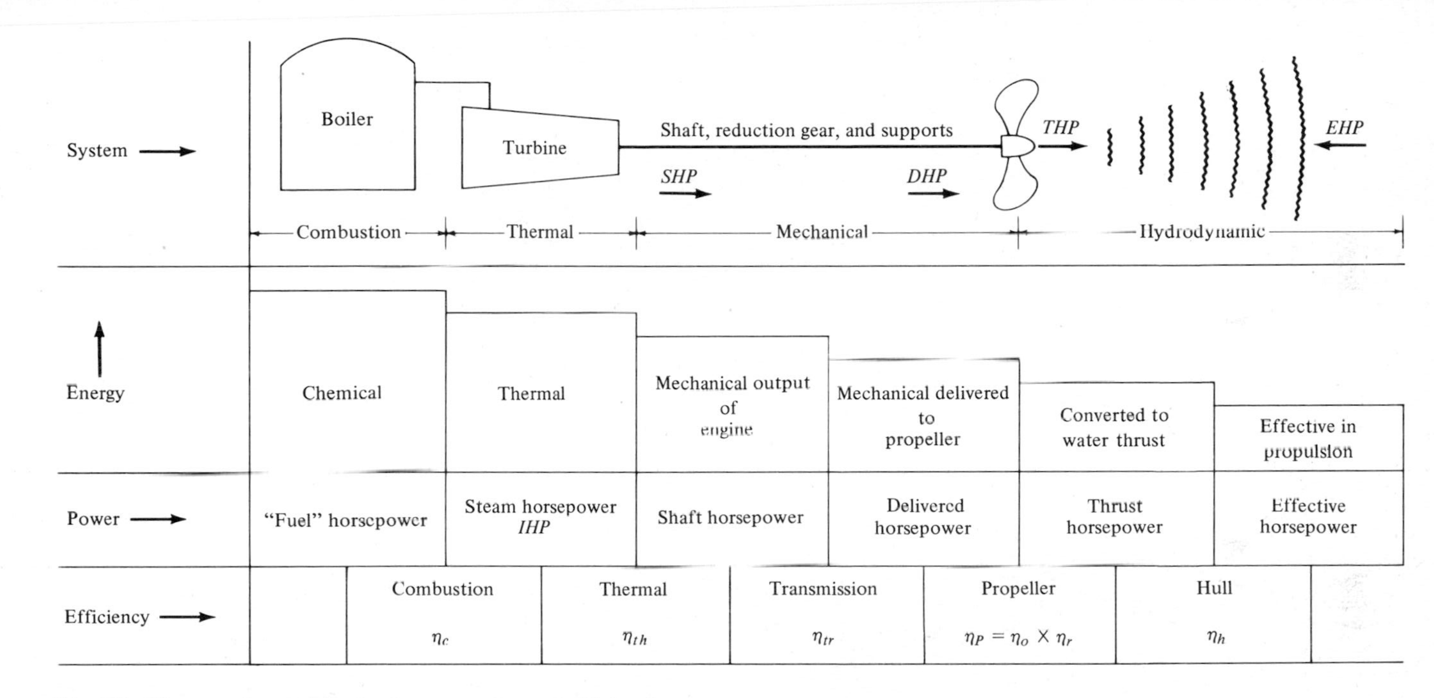

Fig. 4-3 Diagram of propulsive systems, energies, and efficiencies.

P_{TB} = power converted into thrust when the propulsive device is located in its working position with respect to the body; this is known as the "behind" condition

P_D = power delivered to propeller

η_{tr} = transmission efficiency; represents mechanical losses between engines and propeller

η_o = propeller efficiency "in the open"; represents performance of the propeller working in an infinite extent of fluid

η_r = relative rotative efficiency which shows the difference in the performance of the propulsion device between open water operations and behind the body operation

η_h = hull efficiency which gives the interference effects between the propulsion device and the body being propelled. There are two effects to be considered (taking a propeller working behind a ship)

First is the effect of the body on the propeller. A boundary layer formed on the surface of the body will produce a wake which will alter the water inflow velocities into the propeller. The wake may be uniform if produced by a body of revolution, or nonuniform if the body is not symmetrical about the axis of the propeller. The variation of the inflow velocities will result in the variation of the lift developed along the radius of the blade and also along the circumferential path described by a given section of the blade as it rotates through one revolution. The design of the propeller blade can take into account the symmetrical wake distribution by varying the angle of attack of the blade with the radius. However, since the circumferential variation can only be averaged, the performance of the propeller will be below optimum.

Second is the effect of the propeller on the body. The propeller blade rotating through the water at an angle of attack develops lift. Associated with the lift is a pressure field which exists on both sides of the blade. This pressure field interacts with the surfaces of the body around it and produces a resultant suction force which augments the drag of the body. More common consideration of this effect is to regard the interference as a "loss" of thrust of the propeller. Thus, if T is the thrust developed by the propeller and D the thrust available to overcome the body's drag, then $(T - D)/T = t$; t is defined as the thrust reduction fraction.

Now combining the two above effects, the hull efficiency may be expressed as

$$\eta_h = \frac{D \times V}{T \times V_A} = \frac{1 - t}{1 - w} \tag{4-12}$$

where

$$w = \text{Taylor wake fraction} = \frac{V - V_A}{V} \tag{4-13}$$

and where V = speed of body

V_A = resultant speed of water through propeller disk (including the wake effect)

Since the transmission efficiency represents purely mechanical losses, it is usually treated separately from the other components which are based on hydrodynamical considerations. Thus:

Propulsive efficiency = transmission efficiency × quasi-propulsive efficiency

In propeller design the quasi-propulsive efficiency (QPC or η_D) is used.

A number of propulsive devices are used for waterborne craft. Some of the more common are propeller (water), airscrew, water jet, paddle wheels, and vertical axis propeller.

The most efficient and therefore most used is the propeller. A propeller consists of the hub and the blades. The blades are the working part of a propeller. They rotate and advance through the water and develop lift and drag similarly to the wings of a plane. The hub serves as a mounting for the blades and as an attachment to the driving shaft.

4-2.2 Geometry of a propeller

Diameter of a propeller (D) is equal to the diameter of a circle described by the tip of any blade during one revolution without motion ahead.

Helix is the path of the tip, or any point on the blade, as the propeller rotates at constant angular velocity and moves forward at constant speed.

Pitch (P) is the distance the propeller would move in one revolution if it were operating in a solid medium (like a bolt in a threaded hole).

Speed of advance of the propeller (V_A) is the actual forward distance moved per unit time.

Slip of the propeller (S) is the difference between the pitch and the actual advance per one revolution.

Face is the side of the propeller facing aft.

Back is the other surface, which develops the major part of the thrust. The pressure on the back of the propeller is lower than the ambient, and is subject to cavitation. The design of the propeller blade concerns itself mainly with the shape of the back surface.

Blade area is used as an indication of thrust developed by the blades. It is also used in the consideration of cavitation problems and the strength calculations. There are two ways of defining the blade area:

1. Projected blade area: the area calculated from the projection of the blade outlined on a transverse plane A_p
2. Expanded blade area: the true area of the blade A_E

The area of the propeller is the blade area times the number of blades. This area excludes the area of the hub. Dividing by the area of the propeller disk $(\pi/4)D^2$ gives nondimensional blade area ratios.

The shape outlines of propeller blades range from the narrow width of an air propeller to the very wide overlapping widths of a heavily loaded naval propeller.

There are two basic shapes of the sections of the blade along the circumferential cut of the blade. One is an airfoil shape with a nearly flat face section which is efficient from the lift development point of view. It suffers, however, from susceptibility to cavitation. The second is a circular back and flat face section which, although less efficient as a lift-producing section, is less apt to cavitate. It is also easier to manufacture and is stronger. Many very successful propeller designs consist of airfoil sections up to about 80 percent radius and circular back section near the tip.

Some propellers are designed with *variable pitch*; this means that the pitch of the sections varies along the radius. This is equivalent to a change in angle of attack of the sections and is designed to take into account the variation of the inflow conditions. *Variable pitch* should not be confused with *controllable pitch* propellers in which the whole blade is rotated, changing the angle of attack of all the sections the same amount.

The design of a propeller consists of two parts:

1. Determination of the geometrical shape of the blades so that they can be manufactured
2. Determination of the performance of the propeller so that the thrust, torque, and efficiency can be calculated

A number of propeller theories have been developed to achieve the above objectives.

4-2.3 Propeller theories

Design of propellers has evolved from a very simple momentum-type theory to the present-day circulation theory, which permits a complete design to be attempted. However, the resulting propellers are still not completely successful, owing to some simplifying assumptions used. Laboratory tests performed with models are still the mainstay of the propeller design.

A. Momentum theory of propellers This theory treats the propeller as a pressure-increasing device which accelerates the fluid ahead of the propeller and imparts momentum to the fluid in the race. There is no rotation of the fluid and the fluid is considered to be ideal (incompressible with zero viscosity). Half the increase of velocity of the fluid in the race can be proved to have occurred ahead of the propeller owing to the suction exerted by the propeller. The velocity increase is expressed by the axial inflow factor a, which is the velocity increase divided by twice the forward velocity of the propeller V.

The theory gives the maximum possible or ideal efficiency η_{ideal}, and the thrust T,

$$\eta_{\text{ideal}} = \frac{1}{1 + a} \qquad (4\text{-}14)$$

$$T = 2\rho A V^2 a(1 + a) \qquad (4\text{-}15)$$

It can be seen that for a given speed the efficiency will increase if the inflow factor a is decreased, but this will reduce the available thrust unless the area of the propeller disk A is increased. Hence for maximum efficiency the propeller diameter should be as large as physical conditions allow.

The theory does not consider the shape of the propeller and therefore it will not help in the design of the sections, blade outlines, etc. (Rankine 1865).

B. Blade element theory This is an advance on the momentum theory since it examines the sections of the propeller blade, calculates the lift and drag components developed by each section, and then integrates the components to produce the values of lift and drag of the whole blade and hence of the propeller.

The blade section is considered a part of an infinite wing rotating and advancing through the fluid at an angle of attack. The lift and drag developed by the section are resolved into the thrust and torque components.

The blade element theory gives poor estimates of propeller performance since it oversimplifies the physical conditions. Many modifications have been made to the basic blade element method, and the theory has been improved quite considerably.

C. Vortex or circulation theory of propellers The following main improvements have been made by taking into account:

1. Effects of induced velocities produced by the vortex system of a finite blade instead of infinite wing. The induced velocities modify the relative velocity of the inflow and the angle of attack of the blade section.
2. Effects of viscosity which modify angle of attack, lift, and drag.
3. Camber correction, which takes into account the width and the curvature of the blade section. The above-mentioned induced velocities are based on the lifting-line representation of the vortex system of a narrow blade. Camber correction allows for the lifting surface effects which curve the flow over the wide blade sections and effectively reduce the camber of the airfoil, thus reducing the lift.

4. Effects of heavily loaded propellers on the induced velocities. The radial induced component produces a contracted race behind the propeller, altering its performance characteristics.
5. Modifications to inflow velocities by the wake of the body.

Even with all the above corrections practical experience and model experiments are necessary to adjust or introduce correction factors to obtain accurate thrust and torque values. Goldstein (1929) and Lerbs (1952) give the background to the above theory.

4-2.4 Propeller design

Although theoretical calculation methods have been substantially improved, most of the propeller design is based on the results of model tests. These have been conducted over the past years so that now there is a substantial accumulation of valuable performance data.

To be useful in the design of different sizes of propellers operating under different conditions, the data have to be presented in nondimensional form. The coefficients derived by dimensional analysis methods are:

Thrust coefficient: $$K_T = \frac{T}{\rho N^2 D^4} \tag{4-16}$$

Torque coefficient: $$K_Q = \frac{Q}{\rho N^2 D^5} \tag{4-17}$$

Advance coefficient: $$J = \frac{V}{ND} \tag{4-18}$$

Cavitation number: $$\sigma = \frac{P - Pv}{\frac{1}{2}\rho V^2} \tag{4-19}$$

Efficiency: $$\eta_o = \frac{K_T}{K_Q} \times \frac{J}{2\pi} \tag{4-20}$$

In addition to the above system there is a very useful set of coefficients which expresses the propeller parameters in such a way that the propeller diameter is not required to be known *a priori*. This set is known as the Taylor coefficients.

$$B_P = \frac{N(P_D)^{0.5}}{(V_A)^{2.5}} \tag{4-21}$$

and

$$\sigma = \frac{ND}{V_A} \tag{4-22}$$

where N = rpm, P_D = power delivered to the propeller, in horsepower, and V_A = speed in advance, in knots, where $V_A = V(1 - w)$.

The model tests were conducted in a number of research establishments in different countries, but the results are always presented in one of the above systems of coefficients. The tests usually cover a systematic variation of number of blades, blade area ratio, and advance coefficient.

The most extensive and readily available results are the Wageningen B-Screw Series (van Lammeren et al. 1969). They apply to the design of most of the propellers used in merchant-type ships, motorboats, and other commercial applications.

Another set of important published data is based on the model test results of British Admiralty Experiment Works (Gawn 1953).

There are many other sources of data for the design of specific propellers, but they are usually considered proprietary information or are not readily available in the open literature.

Some useful and moderately easy design methods based on the circulation theory of propellers or experimental results have been developed and are listed here:

Large blade area propellers are treated by de Groot and Hoffman (1949).

Design method based on circulation theory is given by Eckhardt and Morgan (1955) and Kerwin and Leopold (1964).

Design of a propeller adapted for the operation behind the body is described by van Manen and van Lammeren (1955).

4-2.5 Propeller cavitation

Whenever the resultant ambient pressure at a point in the liquid equals or falls below the vapor pressure, the liquid will begin to "boil" and cavitation bubbles will appear. The bubbles move to areas of higher pressure and collapse. If they collapse while in contact with the propeller, the surface will become eroded. The accumulation of bubbles on the propeller surface will form a vapor cavity. Since the water is no longer in contact with the blade, this part of the blade will not produce any thrust and the propeller performance will be reduced.

The effects of cavitation on the propeller are thus reduced thrust and efficiency, erosion of propeller surface, and noise generated by the collapse of cavitation bubbles.

A typical sequence in the development of cavitation on a propeller begins with a tip vortex; spreads down the leading edge on the back of the propeller to the root of the blade; progresses along the back to the trailing edge until the whole back surface is enveloped by a cavity. A cone vortex is also formed at the end of the hub. At this time the thrust is wholly generated by the positive pressure acting on the face of the propeller.

Operation under cavitating conditions should be avoided, and in the design of a propeller this has to be checked. Most of the information on the cavitation characteristics of propellers is obtained from model tests. These are conducted in cavitation tunnels where the ambient pressure can be reduced to scale properly the cavitation number.

Information on cavitating propellers is given by Burrill and Emerson (1962) and Newton and Rader (1960). Supercavitating propellers, in which a reduction in efficiency is accepted for greater thrust, are discussed by Posdunine (1944), Tulin (1964), and Tachmindji and Morgan (1958).

4-2.6 Special types of propulsion devices

Although the screw propeller is the most common propulsion device for waterborne craft, other devices are also being used. The decision to use other types of propulsion is based not on the maximum efficiency consideration, since the screw propeller is at present the most efficient propulsion device, but on operational requirements. Thus when the craft has to operate in very shallow waters, a water-jet propulsion system, a paddle wheel, or an air propeller may be considered. When maneuverability is of utmost importance, a vertical axis propeller system could be used.

A. Water-jet propulsion system Propulsion is obtained by drawing the water from ahead, imparting extra velocity to it, and expelling the water jet at higher velocity at the stern. The thrust is then realized from the change of momentum of the water jet. Thrust = $\rho Q \Delta V$, where Q is the rate of flow of water through the jet pump and ΔV is the difference of the velocity of the water jet and the forward speed of the craft.

The efficiency of the system is (for ideal fluid and zero entry losses)

$$\eta = \frac{\text{output}}{\text{output} + \text{losses}} = \frac{1}{1 + \Delta V/2V} \tag{4-23}$$

which is the same as for the ideal propeller. When the losses due to friction in the water passages and the entry losses are included, the jet propulsion efficiency becomes much lower. The above analysis is based on the water entering the jet pump at the speed of the craft. If the pumping system is required to accelerate the water in the jet from zero velocity (as when the water is drawn from the side of the craft), then the overall efficiency is further reduced.

The water-jet propulsion will be of some benefit, however, in special applications like the propulsion system of a hydrofoil craft. At the speeds of the hydrofoil boat a supercavitating propeller has to be used and the power has to be transmitted through two 90° turns resulting in a low overall efficiency. A water-jet propulsion will in this case be competitive. Schuster et al. (1960) discussed in detail the problems of water-jet propulsion.

B. Vertical axis propellers This type of propulsive device consists of a flat horizontal disk with vertical blades which are attached to the disk near its outer edge. As the disk is rotated by the engine, the blades adjust their angle of attack with respect to the direction of advance, and thrust is produced. The action of the resultant thrust can be set in any desired direction. This then provides a propulsion and steering system. When two such propellers are fitted, one forward and one aft, the ship can be rotated about its own axis and a very precise station-keeping capability results. Mueller (1955) describes this system of propulsion.

C. Shrouded propellers When a propeller is working inside a nonrotating nozzlelike ring, certain advantages ensue. The propeller blade can have a wide tip, and by reducing the clearance between the tip and the nozzle, the tip vortex is eliminated. This increases the thrust developed by the outer portion of the blade. If the nozzle is properly designed, it will develop lift with a component in the direction of the thrust. The shroud also provides protection for the propeller. The disadvantages are extra weight and cost and extra frictional drag. Van Manen and Superina (1959) describe the design of such propellers.

4-3 STABILITY

4-3.1 Introduction

The stability of a system is defined as its ability to return to the original position after it has been slightly disturbed from that position. We distinguish between statical stability (when the system is acted upon by a steady force or moment applied gradually for a usually short duration of time) and dynamic stability (when the system is subjected to a disturbance applied suddenly or repeatedly).

There are three conditions of stability which are defined by the behavior of the system after the disturbing forces or moments are removed:

1. If the system returns to its original position of equilibrium, the system is said to be in *stable* equilibrium.
2. If it seeks a different position of equilibrium, it is said to be *unstable*.
3. If it remains in the disturbed position, it possesses *neutral* stability.

Thus a pendulum suspended at a point A, with the weight vertically below A, will always return to this position after being disturbed; hence this configuration corresponds to a position of stable equilibrium. When the weight is raised above A, the system is unstable. Finally if the pendulum is resting on a horizontal table, and can rotate about point A, any position of the weight will be a position of equilibrium. The pendulum in this case will be in neutral equilibrium.

4-3.2 Floating body in equilibrium

A body is said to be in equilibrium if

$$\Sigma \text{ forces} = 0 \tag{4-24}$$

$$\Sigma \text{ moments} = 0 \tag{4-25}$$

The body in Fig. 4-4 is in equilibrium under the action of forces shown:

$$W + B + F_{\text{reaction}} + F_{\text{cable}} = 0 \tag{4-26}$$

If the body is floating freely, then $F_{\text{reaction}} = F_{\text{cable}} = 0$ and $W = -B$. That is, the weight of the body is balanced by the buoyancy force acting in the opposite direction and $\Sigma F = 0$.

If the lines of action of the forces acting on the body intersect at one point or the body is acted upon by two equal and opposite couples, then Σ moments $= 0$.

A freely floating body will always seek a position of equilibrium and will satisfy Eqs. (4-24) and (4-25). This is the principle on which the stability of floating bodies is based.

Reserve buoyancy is defined as the volume of watertight body above the load waterline multiplied by the specific weight of the water. This provides an indication of the extra weight that can be put on board without sinking the body. The reserve buoyancy of a conventional type of submarine is approximately 10 percent and of a typical displacement-type ship is in excess of 100 percent.

A freely floating body can undergo six motions: three linear and three rotational. The motions are defined with respect to a system of three orthogonal axes:

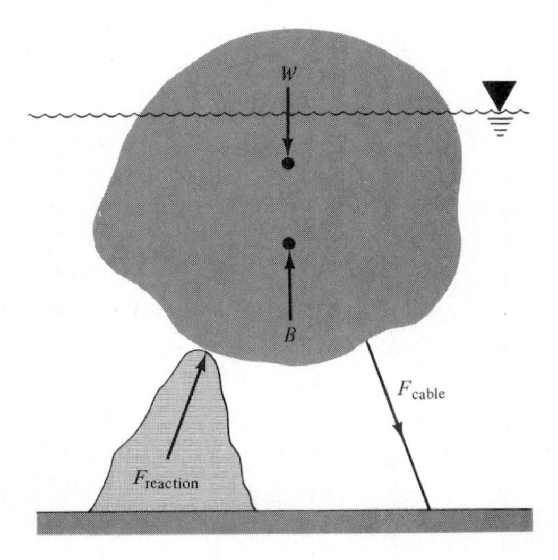

Fig. 4-4 Various forces acting on a partly floating body.

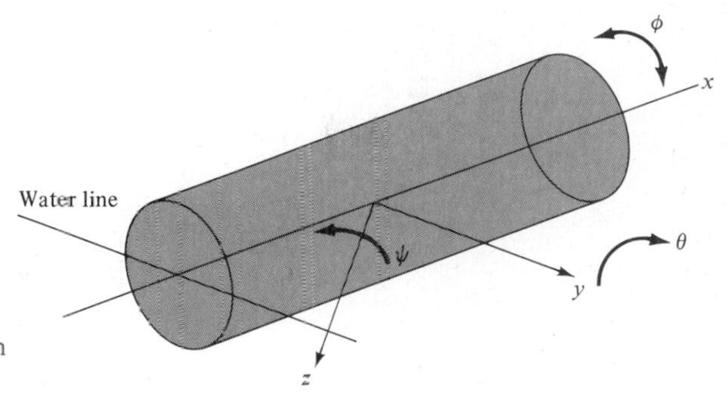

Fig. 4-5 Coordinate system of a floating body.

x axis, along the longest dimension of the body; motion is called *surge*.
y axis, in the horizontal plane at right angles to x axis; motion is called *sway*.
z axis, in the vertical plane at right angles to xy plane; motion is called *heave*.

Rotation about x axis is called *roll* and is measured by an angle ϕ.
Rotation about y axis is called *pitch* and is measured by an angle θ.
Rotation about z axis is called *yaw* and is measured by an angle ψ.

The origin of the coordinate system is usually located in the plane of the design water plane and in the middle of the length of the body. The directions of the motions are indicated in Fig. 4-5.

A floating or submerged body can possess all three types of stability depending on the direction of the disturbance. Thus a floating body will:

Have neutral stability in surge, sway, and yaw
Be stable in heave
Be stable or may be unstable in roll and pitch

A submerged body, if neutrally buoyant, will have neutral stability in the vertical motion.

The stability of interest, therefore, will be in rolling and pitching motions.

To discuss these two types of motion, consider a freely floating body at an interface between water and air. In this case the weight of the body is exactly balanced by the buoyant force, and the two act along the same vertical line. Since these are the only external forces acting on the body, the body is in equilibrium. If an upsetting moment inclines the body through an angle, a new waterline W_1L_1 will be established; see Fig. 4-6. The lines of action of the forces will now separate, and a restoring moment will appear. Thus $W = B$; and the righting moment $= W \times GZ = W \times GM \times \sin \phi$. Considering GM positive when M lies above G, we can define the three types of stability employing GM as a criterion of stability:

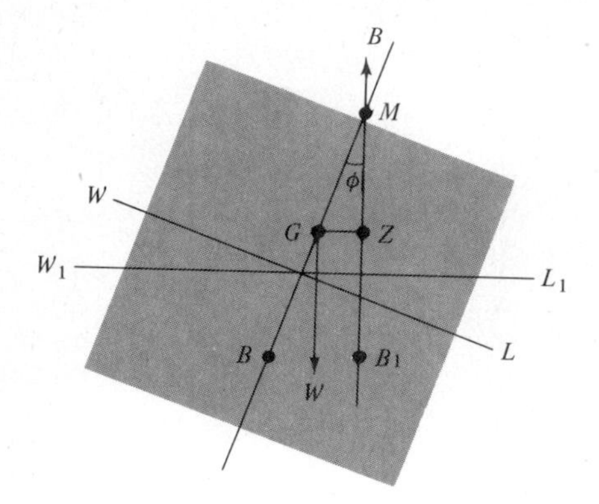

Fig. 4-6 Locating the metacenter of a floating body.

If $GM > 0$, the body is in stable equilibrium.
If $GM = 0$, the body is in neutral equilibrium.
If $GM < 0$, the body is unstable.

The point M, which is called *metacenter*, is defined as an intersection of two lines of action of the force of buoyancy at two inclinations of the body a small angle apart. The metacenter will lie on the centerline of the body for small angles of inclination from the upright position. The intersection point moves away from the centerline at larger angles of inclination.

In addition to being a very important parameter in stability calculations, the metacenter has also a physical significance; it can be thought of as a point of suspension of an equivalent simple pendulum of weight W and length GM whose period of oscillations is equal to that of the body in unresisted free rolling motion.

4-3.3 Statical stability

The determination of stability requires the calculation of either GM or GZ. To obtain GM, we need to find the position of the center of gravity G and the metacenter M. The positions of these two points are governed by different considerations.

G is determined from the distribution of weights contained within the body; it is the center of gravity of the total body (both below and above the waterline), including the weights carried on board.

B, on the other hand, depends only on the shape of the wetted part of the body; it is the centroid of the displaced volume of the water.

Denoting the lowest point of the body, usually on the centerline, by K, we find KG by taking moments of all the structural and other weights about K. KM

is calculated by noting that $KM = KB + BM$. KB locates the centroid of the immersed part of the body, and BM is obtained from the formula

$$BM = \frac{I}{V} \tag{4-27}$$

where I = area moment of inertia of the water plane (second moment of area) about an axis of rotation, ft^4; V = volume of displacement, ft^3. Thus

$$GM = KM - KG = KB + BM - KG \qquad \text{or} \qquad GM = KB + \frac{I}{V} - KG \tag{4-28}$$

The derivation of Eq. (4-27) is based on the consideration of movement of the center of buoyancy as the body is inclined through a small angle.

Figure 4-7 shows a transverse section of a body floating on the surface of water and being inclined to an angle ϕ by an external moment. In the inclined condition the body is floating at a new waterline W_1L_1 but at the same displacement or weight W. Center of gravity remains at G, but center of buoyancy B moves to B_1 owing to the transfer of emerged volume of wedge WOW_1 to the immersed volume of wedge LOL_1. Since the buoyancy remains the same, the volumes of the two wedges must be equal, v say. The horizontal transfer of the wedge volume is gg_1; hence the transfer moment $= vgg_1$. This has to be equal to the transfer moment of the whole body's buoyancy from B to B_1.

$$BB_1 V = gg_1 v; \qquad BB_1 = gg_1 \frac{v}{V}$$

From Fig. 4-7, $BB_1 = BM \tan\phi$; and

$$gg_1 v = \int_0^L \tfrac{1}{2} y (y \tan\phi) 2(\tfrac{2}{3}) y \, dx$$

$$= 2\int_0^L \tfrac{1}{3} y^3 \, dx \tan\phi$$

$$= I_{xx} \tan\phi$$

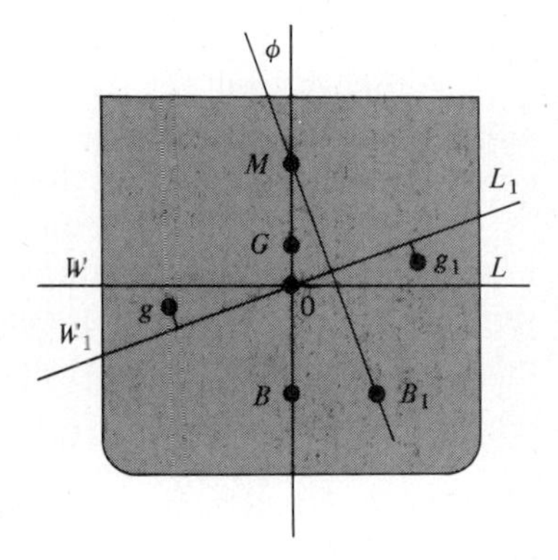

Fig. 4-7 Nomenclature used in the derivation of Eq. (4-27).

Thus

$$BM \tan\phi = \frac{I_{xx} \tan\phi}{V} \qquad BM = \frac{I_{xx}}{V}$$

The metacentric height GM is used as the criterion of stability for small inclinations of the body from its equilibrium position, usually up to approximately 10°. Up to this angle $\sin\phi \doteq \tan\phi \doteq \phi$ rad and the metacenter is assumed to remain fixed on the centerline of the body. This is reasonably correct for normal ship-shaped types of bodies. The stability at small angles or the initial stability is then given as righting moment = $W \times GM \times \phi$.

This assumption also linearizes the equation of motion, giving

$$I_v\ddot{\phi} + C_t\dot{\phi} + W \times GM\phi = 0 \tag{4-29}$$

where C_t is the damping coefficient and I_v is the virtual mass moment of inertia. The virtual mass moment of inertia includes an added mass term which exists whenever the body undergoes accelerated motion. Pitching, rolling, and heaving motions undergo cyclic accelerations; hence the mass in motion is augmented by an added mass part giving virtual mass: $m_v = m(1 + a)$, where the coefficient a depends on the shape of the body and the type of the motion. The corresponding virtual mass moment of inertia can be expressed as

$$I_v = I(1 + A)$$

where A is the added mass moment of inertia coefficient in rotational motion.

$$\omega_n^2 = \frac{WGM}{I_v}$$

Taking $I_v = mk_g^2$,

$$\omega_n = \sqrt{\frac{gGM}{k_g^2}} \tag{4-30}$$

Thus a body having a larger metacentric height will oscillate with higher circular frequency ω_n; that is, its motion will be less comfortable and will result in higher centrifugal accelerations.

Although from the stability point of view, the larger the value of GM, the safer is the body, taking the comfort of the crew and the structural considerations into account, there is an upper limit imposed on the value of GM.

As was mentioned before, a floating body is inherently stable in heaving motion. Thus under the influence of wave action the body will oscillate in a vertical direction about the equilibrium waterline, in addition to the rolling motion described above. The equation of heaving motion is

$$(m_v)\ddot{z} + c\dot{z} + kz = F(t) \tag{4-31}$$

where m_v = virtual mass of the body = $m(1 + a)$

a = added mass coefficient in heaving motion (equal to 0.5 for a sphere and 1.0 for a cylinder)

m = mass of the oscillating body
c = viscous damping coefficient
$k = \gamma A_w$ = spring constant
$F(t)$ = wave forcing function
$\ddot{z}, \dot{z}, z$ = linear acceleration, velocity, and displacement in heaving motion
γ = specific weight of water in which the body is floating
A_w = area of the water plane of the body

The undamped natural frequency ω_n is given by

$$\omega_n = \sqrt{\frac{k}{m_v}} \text{ rad/s} = \sqrt{\frac{\gamma A_w}{m_v}} \tag{4-32}$$

The natural frequency can be decreased by reducing the water plane area of the body as is done in the so-called flip-ship to reduce the possibility of synchronism with the waves at sea. A circular cylinder floating vertically has the area of its circular cross section for water plane. This will give a minimum ω_n and a large tuning factor (ratio of forcing frequency to natural frequency), resulting in a small magnification factor for the resulting motion of the cylinder in heave.

Example 4-2 A buoy of square cross section, side a ft and length 1 ft, where 1 is greater than a, is floating in water of specific gravity 1.0. The buoy is built of homogeneous material of specific gravity 0.5. Determine the way the buoy will float.

Solution Assume the buoy floats with two sides horizontal and two vertical as shown in Fig. 4-8. To be stable in this position, the buoy must have its GM positive.

From $GM = KB + I/V - KG$ [Eq. (4-28)], where $KB = \frac{1}{4}a$; $KG = \frac{1}{2}a$; $I = \frac{1}{12}la^3$; and $V = \frac{1}{2}aal$.

$$GM = \frac{1}{4}a + \frac{\frac{1}{12}la^3}{\frac{1}{2}la^2} - \frac{1}{2}a = \frac{1}{6}a - \frac{1}{4}a = -\frac{1}{12}a$$

Since GM is negative, the buoy is unstable and will not float in this position. To be stable, the buoy must float with the water plane along one of the diagonals of the square cross section.

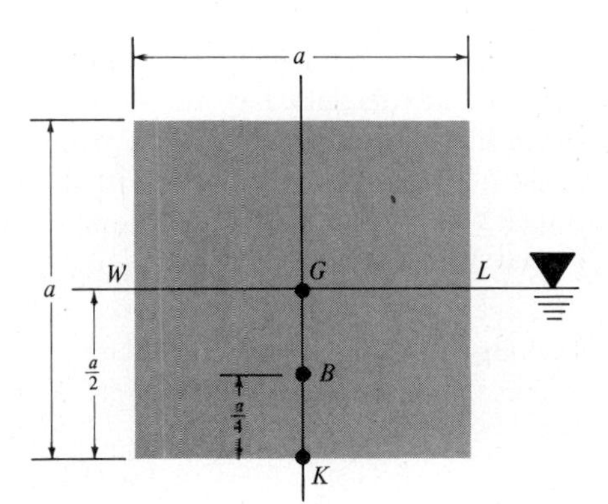

Fig. 4-8 Assumed stable position for homogeneous body in Example 4-2.

To show that this is a stable condition, calculate GM again. The diagonal of the square cross section has a length of $a\sqrt{2}$ ft. Thus $KB = \frac{1}{3}a\sqrt{2}$; $KG = \frac{1}{2}a\sqrt{2}$; $I = \frac{1}{12}l(a\sqrt{2})^3$; and $V = \frac{1}{2}la^2$;

$$GM = \frac{1}{3}a\sqrt{2} + \left[\frac{\frac{1}{12}la^3(\sqrt{2})^3}{\frac{1}{2}la^2}\right] - \frac{1}{2}a\sqrt{2} = a\sqrt{2}\left[\frac{2}{3} - \frac{1}{2}\right] = a\frac{\sqrt{2}}{6}$$

GM is positive; therefore the buoy is in stable equilibrium.

The buoy could be made stable in the first assumed position if it was ballasted so that the center of gravity moved below the metacenter, that is, if KG was smaller than KM. This would require $KG < \frac{5}{12}a$.

The above calculations apply to inclinations about the x axis. As for the stability about the y axis, the only change appears in the calculation of the area moment of inertia, which is now $I = \frac{1}{12}al^3$. Since l dimension is larger than a, the value of the inertia component will be substantially greater and this will have a predominant effect on the value of GM.

In the first position of the buoy, the metacentric height will be

$$GM = \frac{1}{4}a + \frac{\frac{1}{12}al^3}{\frac{1}{2}la^2} - \frac{1}{2}a = \frac{1}{6}\frac{l^2}{a} - \frac{1}{4}a$$

Assuming $l = 2a$, $GM = [\frac{1}{6}(l/a)^2 - \frac{1}{4}]a = \frac{5}{12}a$, and the buoy is stable about the y axis inclinations.

In general, for bodies elongated in the x direction, longitudinal stability (rotation about y axis) is much greater than transverse stability (rotation about x axis) owing to the much larger area moment of inertia which varies like yx^3 in the former case and like xy^3 in the latter case.

Although GM is the criterion of stability, physically it is the *righting moment* that returns the body to its original position.

$$\text{Righting moment} = WGZ = WGM \sin\phi \tag{4-33}$$

Thus GM is an indication of stability as long as M stays in the same place, that is, at small angles of inclination. GM is therefore a criterion of *initial stability*. For large angles of inclination GZ is the stability parameter.

4-3.4 Stability at large angles

As a stable body is inclined, at a constant displacement, by an external moment, a righting moment will appear as shown in Fig. 4-7. The righting moment will increase at first, but beyond a certain angle of inclination it will start decreasing until it becomes zero, at which angle the body will become unstable. A typical righting moment curve drawn to the angle of inclination is shown in Fig. 4-9. Since the weight of the body remains constant throughout the inclination, the curve of GZ will be similar except for the vertical scale having different values.

From the GZ curve (curve of statical stability) the following important stability parameters can be obtained.

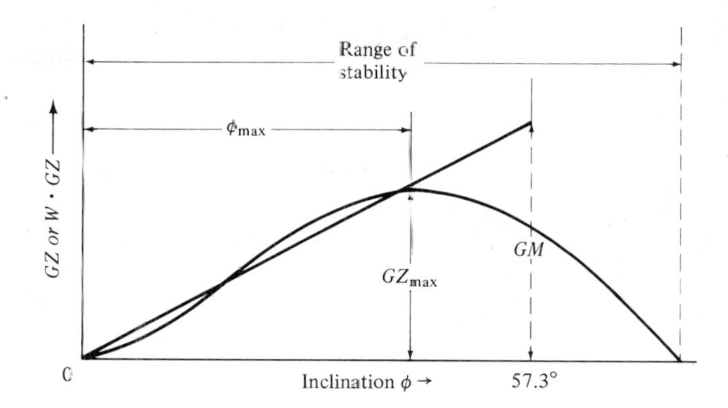

Fig. 4-9 Righting moment curve as a function of inclination.

Range of stability is defined by the largest angle of inclination from which the body will return to its original position, after the inclining moment is removed. This happens when GZ becomes zero. The value of this parameter is used to compare different designs with regard to their stability.

Maximum righting lever is the value of GZ at the maximum point of the curve. Associated with the lever is the angle of maximum righting lever. These two parameters describe more accurately the stability of the body. If a steady moment acting on the body heels it past the angle of maximum righting lever, the diminishing GZ will not be able to prevent the body from capsizing. However, as long as the heeling moment is removed before the body reaches the angle defining the range of stability, the body will return to its upright position.

Metacentric height GM can be found by drawing the tangent at the origin to the GZ curve. It intercepts the ordinate at 57.3° (1 rad) at a height which is equal to GM.

The shape of the GZ curve at moderate angles relates to roll behavior. A hollow curve indicates small restoring moments resulting in larger rolling angles. A convex shape indicates that the body will better resist the inclining moments.

The curve of statical stability therefore gives all the important information regarding the safety of the body. The construction of the GZ curve will be discussed next.

For a given body at a constant displacement, the curve can be defined as

$$GZ = f(\phi) = A + B\phi + C\phi^2 + D\phi^3 + \cdots$$

$$GZ = 0 \qquad \text{when } \phi = 0$$

Therefore $A = 0$. If the body is symmetrical about the axis of inclination, all even powers of ϕ must vanish:

$$GZ = B\phi + D\phi^3 + E\phi^5 + \cdots$$

and if ϕ is small,

$$GZ = B\phi + D\phi^3 = GM\phi + D\phi^3$$

The first term gives the tangent to the GZ curve at the origin. The second term gives the deviation from the tangent.

For example, taking a wall-sided body,

$$GZ = (GM + \tfrac{1}{2}BM \tan^2\phi) \sin\phi \tag{4-34}$$

When ϕ is small, $\sin\phi = \tan\phi = \phi$; and $GZ = GM\phi + \frac{1}{2}BM\phi^3$.

Taking a circular cross-section body, $GZ = GM \sin\phi$. When ϕ is small,

$$GZ = GM\phi \tag{4-35}$$

Values of GZ depend on the position of the center of gravity of the body (influenced by the distribution of all the weights within the body) and the location of Z (dependent on the external shape of the immersed portion of the body).

Two approaches on which most of the calculations for GZ curves are based are these:

1. Considering the transfer of buoyancy due to the immersed and the emerged wedges, see Fig. 4-10,

$$GZ = BR - BG \sin\phi$$

where $BRV = h_1 h_2 v$

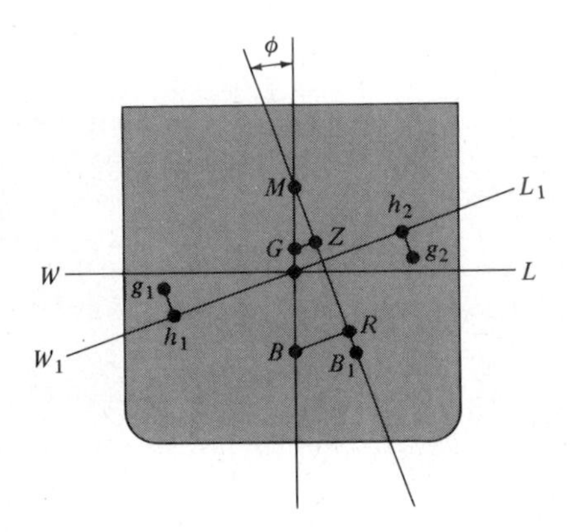

Fig. 4-10 Nomenclature used in Eq. (4-36), Attwood's formula.

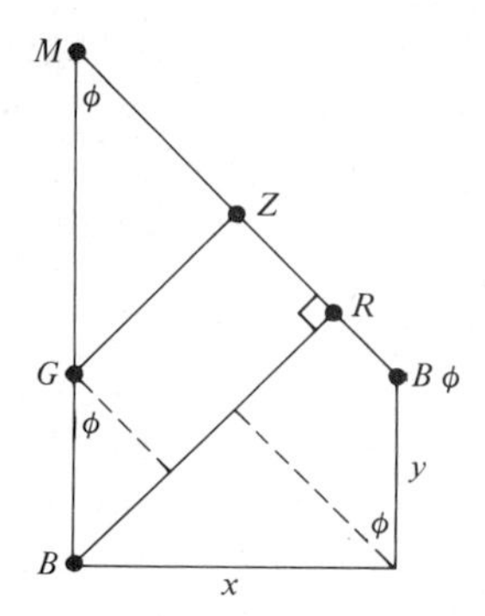

Fig. 4-11 Nomenclature used in Eq. (4-37), Reech's formula.

$$GZ = h_1 h_2 \frac{v}{V} - BG \sin\phi \qquad (4\text{-}36)$$

This equation is known as Attwood's formula. B, G, and V are known from the upright position of the body and the volume of the wedges (which are equal) and their centroids have to be calculated at each angle of inclination.

2. Considering the volume of the immersed body at different angles of inclination, the position of the center of buoyancy is calculated and GZ obtained from the following formula (see Fig. 4-11):

$$GZ = x \cos\phi + y \sin\phi - BG \sin\phi \qquad (4\text{-}37)$$

This is known as Reech's formula. Calculation of GZ consists of finding the coordinates x and y of the center of buoyancy at each angle of inclination.

The above two formulas require the determination of the exact waterline at each angle of inclination that gives the constant displacement. Very often this is not easily obtained and a trial-and-error process has to be used, as follows:

1. At a given angle of inclination draw a number of practical waterlines which should bracket (some above, some below) the required displacement.
2. For each waterline determine the displacement and the center of buoyancy.
3. Draw a line through the center of buoyancy perpendicular to the waterline.
4. Denoting the intersection of the mid-waterline and the original middle line by S, we calculate the righting arm from (see Fig. 4-12):

$$GZ = SZ - SG \sin\phi \qquad (4\text{-}38)$$

which gives the GZ at one displacement.

Repeating the above steps for other waterlines at the same angle of inclination will result in a set of GZ-vs.-displacement values. When plotted, cross-curves of stability will be obtained. Drawing a cross-curve at the required displacement gives a stability curve, GZ versus ϕ, as shown in Fig. 4-12.

The basis for determination of GZ is the calculation of an auxiliary lever SZ. There are a number of methods which can be used, depending on the shape of the body, the availability of instruments (planimeters or integrators), or computers. They are fully described by Comstock (1967) and Gillmer (1970).

In the above calculations of the righting lever GZ it was assumed that the position of the center of gravity G either was known or could be calculated from the distribution of weights in the body. However, when the body is more complicated than a regular mathematical figure, or the distribution of the weights within the body is difficult to determine, the position of the center of gravity has to be obtained by experiment. This is called an inclining experiment.

4-3.5 Inclining experiment

The experiment consists of the measurements of the angle of inclination of the body under the influence of a known inclining moment. The inclining moment is produced by shifting a weight through a horizontal distance within the body. The angle is measured with a long pendulum or an inclinometer. In addition the displacement of the body is calculated from the immersion or the draft of the body.

The body is inclined through a small angle so that the metacenter M remains on the centerline. Figure 4-13 shows the positions of the relevant parameters. Owing to the shift of the inclining weight w through a distance t, the center of gravity G will move to G_1, which will lie on the vertical line passing through the new position of the center of buoyancy B_1.

The moment shift produces an angle of inclination ϕ, and $wt = WGG_1 = WGM \tan \phi$, where weight W includes the inclining weight w.

Fig. 4-12 Finding a stability curve by trials.

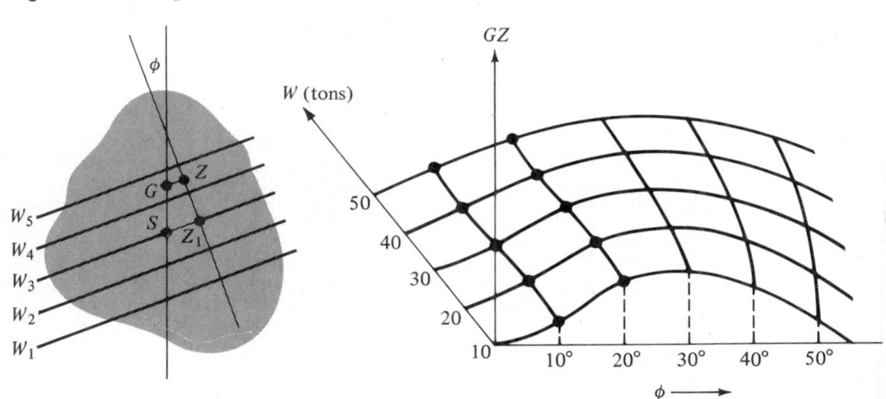

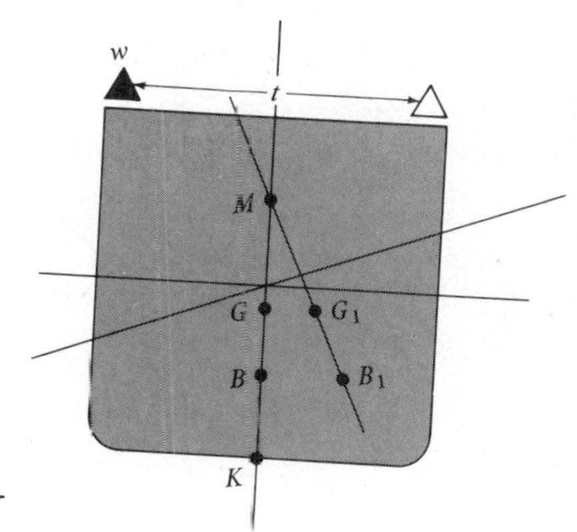

Fig. 4-13 Diagram of an inclining experiment.

$$GM = \frac{wt}{W \tan \phi} \tag{4-39}$$

Hence $KG = KM - GM$.

Since the inclining weights are usually not part of the weights carried on board, a correction to the height KG is made by taking moments about K.

Thus the center of gravity G_1 of the body alone will be

$$KG_1 = \frac{WKG - wKg}{W - w} \tag{4-40}$$

Once the body is designed and built, the position of the metacenter M is fixed and therefore the stability can be influenced only by the variation of the position of the center of gravity. The shift of G within the body can best be analyzed by considering the total movement as a sum of the shifts parallel to the x, y, and z axes. The effect on the stability of the shift of the center of gravity is considered next.

4-3.6 Effect on stability of shifts of weights

Whenever the position of a weight is changed within the body, the center of gravity of the body shifts along a parallel straight line to the line of movement of the shifted weight. The distance the center of gravity is moved can be calculated by taking moments about a chosen datum line.

A. Vertical shift of weight Owing to the shift of a weight w through a vertical distance z, the center of gravity of the body will move to G_1. From Fig. 4-14 $WGG_1 = wz$:

$$GG_1 = \frac{wz}{W} \quad \text{and} \quad G_1Z_1 = GZ - GG_1 \sin \phi \tag{4-41}$$

Therefore the correction to the stability curve is a sine correction as shown. If the weight shift was in a downward direction, the sine correction would be

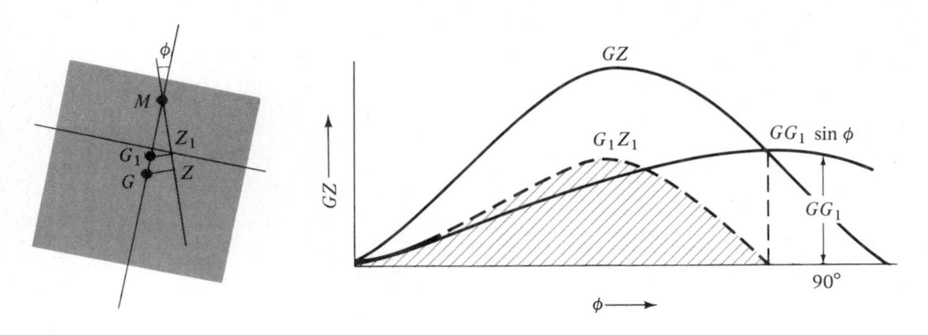

Fig. 4-14 The reduction of stability due to an upward shift of weight.

additive, improving the stability. Assuming upward shift of the weight the stability is affected by the reduction of the maximum righting lever, the range of stability, and the initial metacentric height GM. The vertical movement of the center of gravity can also be caused by icing of the rigging and the superstructure. In such a case the weight of the ice is an addition to the total weight of the body.

The corrective action, to restore original stability, is to shift another weight vertically downward.

B. Horizontal shift of weight When the weight w is now moved horizontally through a distance y, the body's center of gravity will move to G_2. From Fig. 4-15, $WG_1G_2 = wy$. The body will list at an angle ϕ and G_2 will lie vertically above the new center of buoyancy B_2.

$$G_2Z_2 = G_1Z_1 - G_1G_2 \cos\phi = GZ - GG_1 \sin\phi - G_1G_2 \cos\phi \quad (4\text{-}42)$$

A cosine correction results and reduces the stability irrespective of the direction of shift. A body assumes a new stable position about an angle of list given by the intersection of the original GZ curve and the cosine correction curve. However,

Fig. 4-15 The reduction of stability due to a horizontal shift of weight.

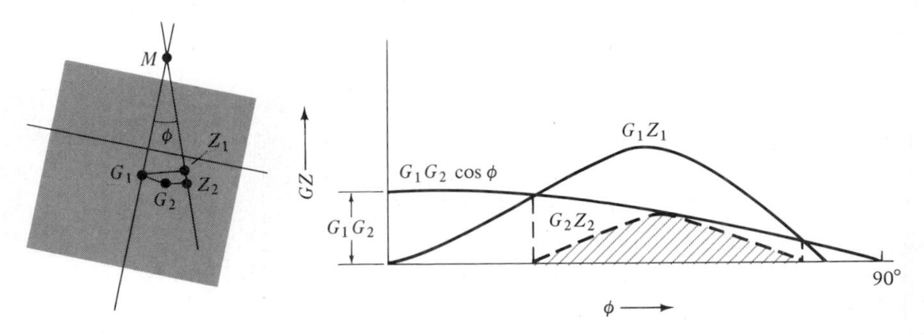

the stability on the opposite side to the shift of weight will be increased because it will take a higher upsetting moment to incline the body to that side.

The corrective action in this case is to move another weight to the opposite side.

C. Free surface effect The presence of liquids with free surfaces, that is, liquids whose surfaces incline with the motions of the body (remaining parallel to the water surface), reduces the stability of the body. While the body inclines, the weight of the liquid moves to the lower side with the line of action of the gravity force intersecting the centerline at the liquid's own metacenter m. The liquid acts as if suspended from point m. This results in a virtual rise of the center of gravity of the liquid from its actual position to the virtual point m. Thus the stability curve will have a sine correction as indicated below.

From Fig. 4-16,

$$GG_v V\gamma_1 = bmv\gamma_2$$

where V = volume of displacement of the body
γ_1 = specific weight of the water in which the body is floating
$bm = i/v$
i = area moment of inertia of free liquid surface
v = volume of the liquid
γ_2 = specific weight of the liquid

$$GG_v = \frac{i\gamma_2}{V\gamma_1} \qquad \text{and} \qquad G_vZ_v = GZ - \frac{i\gamma_2}{V\gamma_1}\sin\phi \tag{4-43}$$

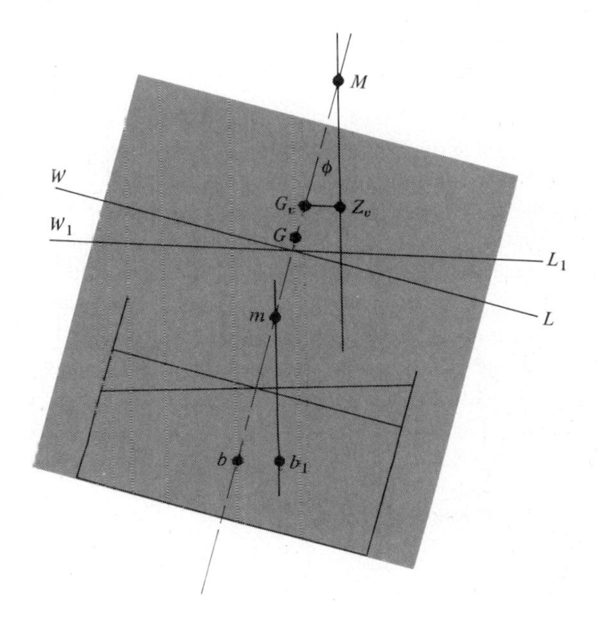

Fig. 4-16 Nomenclature of a floating body with interior liquid tank having a free surface.

also

$$G_vM = GM - GG_v \tag{4-44}$$

The reduction of stability is independent of the volume of the liquid, but depends very strongly on the dimension of the free surface. Since the area moment of inertia i varies as the third power of the transverse dimension of the surface, subdividing the compartment is very beneficial. The following example illustrates this effect.

Example 4-3 A rectangular barge 100 ft long, 30 ft wide, and floating at 5-ft draft has a metacentric height $GM = 2$ ft. Find the GM when the barge is flooded with water to a depth of 5 ft so that it will float at a deeper draft.

Solution Neglecting the thickness of the hull materials, the new draft equals 10 ft. From Eq. (4-44):

$$G_vM = 2 - GG_v$$

$$GG_v = \frac{(\frac{1}{12})100 \times (30)^3}{10 \times 30 \times 100} = 7.5 \text{ ft}$$

$$G_vM = 2 - 7.5 = -5.5 \text{ ft}$$

Therefore the barge will be unstable in this position. The remedy is to subdivide the barge in the direction parallel to the axis of rotation. With two equally spaced bulkheads, the free surface will consist of three 10- by 100-ft independently inclining water surfaces.

Now

$$GG_v = \frac{3 \times (\frac{1}{12})100 \times 10^3}{10 \times 30 \times 100} = 0.83$$

$$G_vM = 2 - 0.83 = 1.17 \text{ ft}$$

which is positive and therefore the barge will be stable.

D. Swinging weights effect Similar to the effect of a free surface on the stability of a floating body, the presence of weights on board that can swing about their point of suspension produces a virtual rise of the center of gravity. When a weight is free to swing, its line of action will pass through the point of suspension and will incline with the inclination of the body. The effect will be as if the swinging weight's center of gravity was located at the suspension point. This will raise the overall center of gravity to its virtual position G_v, where a new position is calculated by taking moments of all the weights about a datum line.

By lashing up the weights so that they cannot swing, the reduction in metacentric height from GM to G_vM will be prevented.

4-3.7 Dynamical stability

Until now the statical stability was being considered, that is, inclinations due to forces or moments applied very slowly. When a body is inclined by a suddenly applied force or moment, like a gust of wind, the investigation of stability has to consider the energy aspect of the dynamic situation.

Considering the inclination of the body at constant displacement and with zero frictional losses, the dynamical stability is defined as the work necessary to incline the body to a given angle ϕ:

$$\text{Dynamical stability} = \int_0^{\phi} \text{moment } d\phi \int_0^{\phi} WGZ\, d\phi = W \int_0^{\phi} GZ\, d\phi \qquad (4\text{-}45)$$

The last expression represents the area under the GZ curve multiplied by the weight of the body.

4-3.8 Stability while submerging

The transverse stability of a body that is submerging can best be considered in three stages:

1. On the surface just before submergence: As the body submerges, the draft increases and so does the volume of displacement. The moment of inertia of the water plane decreases and since $BM = I/V$, BM tends to zero. This means that the criterion of statical stability for floating bodies GM becomes GB and the same requirement for stability pertains. As long as GB is positive, that is, as long as B lies above G, the body is stable.
2. At the point of submergence: The criterion of stability GB positive applies. If this is not the case, the body should traverse this stage very quickly and rely on the momentum or control surfaces to carry out the submergence or emergence maneuver safely.
3. Fully submerged: Again the stability requires the center of buoyancy B to be above the center of gravity G. The stability, depending only on the position of B and G, is the same regardless of the axis of inclination.

Figure 4-17 shows the variation of the stability parameters during a submerging maneuver.

The vertical stability of a submerging body is best considered in two stages.

1. On the surface: While floating on the surface, the body is inherently stable in a vertical direction (in heaving motion).
2. Submerged: If weight of the body is nearly equal to the buoyancy, the body without motion ahead (that is, without help of control planes) will either rise or sink. The body has no inherent stability.

If weight is greater than buoyancy, the body will start sinking unless it settles on a layer of more dense water.

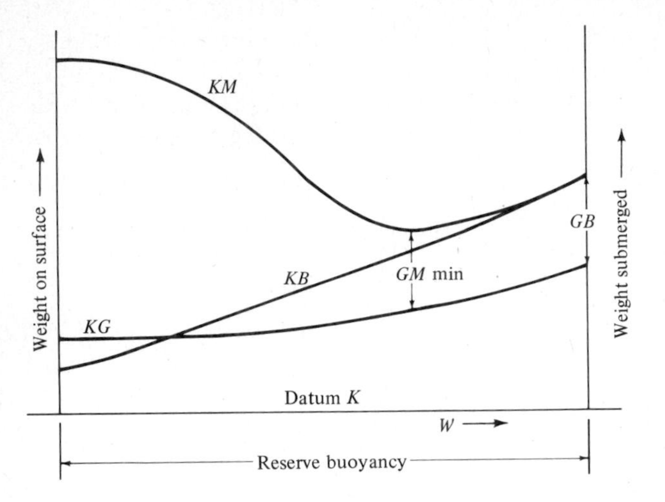

Fig. 4-17 Variation of stability parameters during submergence.

If the body is constructed out of "stiff" materials or has enough stiffeners that it will retain its volume under increasing pressure, the higher density of water at greater depths will result in an inherently stable situation. Stable in this connotation means that the body will rise to the surface when stopped.

If the greater depth pressure contracts the body more than the water, the body will be unstable in vertical motion and unless it encounters a higher density layer, it will sink.

PROBLEMS

4-1. A rectangular pontoon is used as a working platform for divers in a channel where the current is flowing at 1.5 kn. Given the following data, determine the pull in the anchor cable. Pontoon: length, 81 ft; beam, 20 ft; draft, 5 ft. Assume that $C_r = 0.0262$ and the angle of anchor chain with horizontal = 30°.

4-2. A model of a hydrofoil boat was tested in a towing tank and the following results were obtained:

V, kn	0.84	1.30	0.52	2.24	2.47	2.63	2.93	3.48	3.74	4.21	4.27
D, lb	0.009	0.48	0.070	0.222	0.299	0.355	0.481	0.605	0.756	0.840	0.943

Model	*Full-scale body*
Length = 3.44 ft	Length = 110 ft
Wetted area = 2.49 ft²	
Tested in freshwater at 66°F	Operating in salt water at 59°F

Draw an ehp curve for the full-scale body.

(*Note*: The above data points represent actual experimental results and not smoothed, corrected data. Caution has to be exercised when using these results, especially at the low-speed range.)

4-3. If a head wind is blowing at 25 kn in the same direction as the current in Prob. 4-1, calculate the extra tension in the anchor cable.

4-4. (*a*) Explain the origin of form and wave drags.

(*b*) Define slow, moderate, and fast speeds for a surface body.

(*c*) A surface-piercing 60° *V*-type hydrofoil is used to provide the lift for a body moving at 45 kn through calm water. A symmetrical foil is used with a chord of 2 ft and the immersed length of 10 ft. Draw vectors to show all the forces acting on the foil and describe their origin.

4-5. A model habitat 3 ft long and 1 ft wide is subjected to towing tests in a saltwater tank. The full-scale device has dimensions 30 by 10 ft and we wish to tow it at 8 kn in salt water.

(*a*) What is the corresponding speed in the tank, assuming water temperatures are the same?

(*b*) The model drag is measured at 3.1 lb and its wetted surface is 6.4 ft². If the estimated skin friction coefficient is 0.004, what is the residual drag coefficient of the model and prototype?

(*c*) If the skin friction coefficient for the full-scale unit is established at 0.002, what towing horsepower is estimated for the full-size habitat at the 8-kn speed?

4-6. Given:

Model propeller		*Full-scale propeller*	
Diameter	18 in	Diameter	24 ft
Revolutions	960 rpm	Revolutions	240 rpm
Speed of advance	(V_A) 20 fps		
Thrust	35.1 lb		
Torque	120 lb·in		
Tested at 65°F in freshwater		Operating at 59°F in seawater	

Calculate the efficiency and the thrust horsepower of the full-scale propeller at the corresponding speed.

4-7. For operation of a pontoon in shallow water a water-jet propulsion has to be used. Two fire hoses with 3-in-diameter nozzles are rigged to propel the pontoon at 5 kn. The effective horsepower ehp = 30 hp and the pontoon is operating in salt water of density 1.991 lb·s²/ft⁴.

Calculate the ideal efficiency of the system assuming no pumping losses and with the inlet for the pumps facing forward.

4-8. Draw a righting moment curve for a spherical buoy made of homogeneous material of specific gravity = 0.75. How will the curve be modified if the specific gravity of the sphere is reduced to 0.5?

4-9. A raft is constructed of two triangular prisms connected by a horizontal platform. Determine the stability of the raft when the prisms are filled with fuel of specific gravity 0.87, up to the waterline. The fuel has free surface. The dimensions of the raft are

Length	60 ft
Distance between centerlines of the prisms	5 ft
Draft	4 ft
Width of prism at 4-ft draft	2 ft

Assume the raft is floating in freshwater and the platform is above the water.

4-10. A spherical buoy, 2 ft in diameter, is floating half submerged with $KG = 0.5$ ft. A radio antenna is to be added at a height of 10 ft above the top of the sphere. The weight of the antenna is 10 lb. Check whether the buoy will stay upright.

4-11. A submersible is constructed out of three watertight spheres streamlined by aluminum enclosures. The spaces between the enclosure and the spheres are self-flooding. The spheres are touching each other and are arranged in line.

To control the trim of the submersible, mercury is pumped between the end spheres. The mercury is contained in rectangular tanks 3 ft wide by 6 ft long. When the submersible is at a level trim, the depth of the mercury liquid in the forward tank is 1 ft and in the aft tank 0.5 ft.

Given the following particulars, determine the stability of the submersible (metacentric height with the free surface):

Diameter of spheres	8 ft
KG of submersible	2 ft
Specific gravity of mercury	13.6
Specific gravity of seawater	1.025

Assume: (*a*) Water can flood in and out of the spaces between the spheres and the enclosure without time lag.

(*b*) Mercury tanks have a free surface.

(*c*) The mercury tanks are centered at the bottom of the spheres.

4-12. Given the floating submarine hull as shown, submerged in seawater:

(*a*) Show the location of the metacenter and the center of buoyancy. Is the unit stable?

(*b*) An instrument pod of mass 50 lb/ft is to be located 10 ft above the water surface. Where is the new center of buoyancy and is the unit now stable?

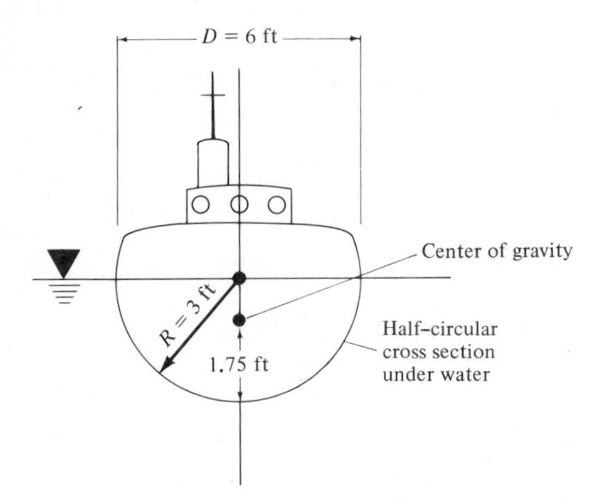

Fig. P4-12

4-13. A hollow, circular spar buoy with outer diameter of 0.5 ft and submerged length of 12 ft is weighted at the bottom to stand upright in the water. Assume the added mass in the heave mode is almost zero. What is the frequency of oscillation in the heave mode? If the device is located in 15 ft of water, what will be the velocity and length of waves (from Chap. 2 considerations) to excite it in the heave direction?

4-14. A long, solid homogeneous pontoon float is made of a syntactic material and has a cross section of an equilateral triangle 5 ft on a side. Can it stably float with one point down, if in this position the water plane is 4 ft above the submerged vertex? What is the righting moment in this position?

4-15. The University of Rhode Island habitat is made of thin sheet steel (assume weightless) and ballasted by inside concrete bunks, as shown in the cross-section sketch. The concrete weighs 150 lb/ft³. Where will the water plane be? Estimate the righting moment by breaking the concrete sections into four pieces and assuming their masses are at the center of the respective sections. If a diver weighing 200 lb clambers up on one side, what will be the angle of heel?

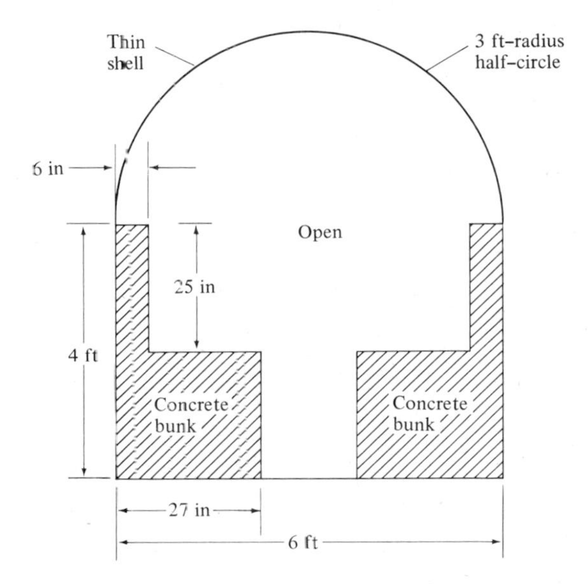

Fig. P4-15

4-16. For the habitat in Prob. 4-15, estimate the natural bob frequency (in heave) and the natural roll frequency. Which mode is more likely to be excited in shallow, inshore waters?

REFERENCES

Abbott, I. H. (1969): "Theory of Wing Sections," Dover, New York.

Aguirre, J. E., and R. R. Boyce (1973): Estimation of Wind Forces on Offshore Drilling Platforms, *Trans. Roy. Inst. Naval Architect*, vol. 115.

Baker, G. S. (1937–1938): Development of Hull Form of Merchant Vessels, *Trans. Northeast Coast Inst. Engr. and Shipbuilders*, vol. 54.

Burrill, I. C., and A. Emerson (1962–1963): Propeller Cavitation: Further Tests on 16 Inch Propeller Models in the King's College Cavitation Tunnel, *Trans. Northeast Coast Inst. Eng. Shipbuilders*, vol. 79.

Clement, E. P. (1962): Graphs for Predicting the Ideal High-Speed Resistance of Planing Catamarans, *Internal Shipbuilding Progr.*, vol. 9, no. 99.

—— (1963): How to Use the Transactions of the Society of Naval Architects and Marine Engineers Small Craft Data Sheets for Design and for Resistance Prediction, *Soc. Naval Architect. Marine Eng., Tech. Res. Bull.*, no. 1-23.

—— (1964): Gtaphs for Predicting the Resistance of Round-Bottom Boats, *Internal Shipbuilding Progr.*, vol. 11, no. 114.

—— and D. L. Blount (1963): Resistance Tests of a Systematic Series of Planing Hull Forms, *Trans. Soc. Naval Architect. Marine Eng.*, vol. 71.

—— and J. D. Pope (1961): "Stepless and Stepped Planing Hulls, Graphs for Performance Prediction and Design," *Naval Ship Research and Development Center Report 1490*, Carderock, Md.

Comstock, J. P. (ed.) (1967): "Principles of Naval Architecture," Society of Naval Architects and Marine Engineers, New York.

de Groot, D., and F. J. Hoffmann (1949): "Designing of Screws of Large Blade Area by Means of the Gawn Series in $B_P - \delta$ Form," Publication no. 85, Netherlands Ship Model Basin, Wageningen.

Eckhardt, M. K., and W. B. Morgan (1955): A Propeller Design Method, *Trans. Soc. Naval Architect. Marine Engr.*, vol. 97.

Gawn, R. W. L. (1953): Effect of Pitch and Blade Width on Propeller Performance, *Trans. Inst. Naval Architect.*, vol. 95.

Gertler, M. (1954): "A Re-analysis of the Original Test Data for the Taylor Standard Series," *Naval Ship Research and Development Center Report 806*, Carderock, Md.

Gillmer, R. C. (1970): "Modern Ship Design," U.S. Naval Institute, Annapolis, Md.

Goldstein, S. (1929): On the Vortex Theory of Screw Propellers, *Proc. Roy. Soc. London*, Ser. A, vol. 63.

Havelock, R. H. (1934): Wave Patterns and Wave Resistance, *Trans. Inst. Naval Architect.*, vol. 76.

Hoerner, S. F. (1965): "Fluid-Dynamic Drag," published by the author, Midland Park, N.J.

Hughes, G. (1930): Model Experiments on the Wind Resistance of Ships, *Trans. Inst. Naval Architect.*, vol. 72.

—— (1952): Frictional Resistance of Smooth Plane Surfaces in Turbulent Flow, *Trans. Inst. Naval Architect.*, vol. 94.

—— (1954): Friction and Form Resistance in Turbulent Flow and a Proposed Formulation for Use in Model and Ship Correlation, *Trans. Inst. Naval Architect.*, vol. 96.

Kerwin, J. E., and R. Leopold (1964): A Design Theory for Subcavitating Propellers, *Trans. Soc. Naval Architect. Marine Eng.*, vol. 72.

Lerbs, H. (1952): Moderately-Loaded Propellers with a Finite Number of Blades and an Arbitrary Distribution of Circulation, *Trans. Soc. Naval Architect. Marine Eng.*, vol. 60.

Mueller, H. (1955): Recent Developments in the Design and Application of the Vertical Axis Propeller, *Trans. Soc. Naval Architect. Marine Eng.*, vol. 63.

Newton, R. N., and H. P. Rader (1960): Performance Data of Propellers for High-Speed Craft, *Trans. Roy. Inst. Naval Architect.*, vol. 102.

Posdunine, V. L. (1944): On the Working of Supercavitating Propellers, *Trans. Inst. Naval Architect.*, vol. 86.

Rankine, W. J. M. (1865): On the Mechanical Principles of the Action of Propellers, *Trans. Inst. Naval Architect.*, vol. 7.

Sarchin, T. H., and L. L. Goldberg (1962): Stability and Buoyancy Criteria for U.S. Naval Surface Vessels, *Trans. Soc. Naval Architect. Marine Eng.*, vol. 70.

Savitsky, D. (1964): "Hydrodynamic Design of Planing Hulls," *Marine Technology, 1*, Society of Naval Architects and Marine Engineers, New York.

Schlichting, O. (1934): Ship Resistance in Water of Limited Depth. Resistance of Sea-Going Vessels in Shallow Water, *Jahrb. STG*, vol. 35, pp. 127–143.

Schoenherr, K. E. (1932): Resistance of Flat Surfaces Moving through a Fluid, *Trans Soc. Naval Architect. Marine Eng.*, vol. 40.

Schuster, S. (1960): "On Certain Problems of Water Jet Propulsion," Jahrbuch Schiffbau Technische Gesellschaft, Berlin, Germany, 1960. Also Naval Ship Research and Development Center, Translation no. 306, August 1962, Carderock, Md.

Tachmindji, A. J., and W. B. Morgan (1958): "The Design and Estimated Performance of a Series of Supercavitating Propellers," Second Symposium on Naval Hydrodynamics, Office of Naval Research, U.S. Navy, Washington.

Taylor, D. W. (1943): "The Speed and Power of Ships," U.S. Government Printing Office, Washington.

Telfer, E. V. (1927): Ship Resistance Similarity, *Trans. Inst. Naval Architect.*, vol. 69.

Todd, F. H. (1951): The Fundamentals of Ship Model Testing, *Trans. Soc. Naval Architect. Marine Eng.*, vol. 59.

Tulin, M. P. (1964): Supercavitating Propellers: History, Operating Characteristics and Mechanisms of Operation, *Hydronautics, Inc., Technical Report 127-6*.

van Manen, J. D., and A. Superina (1959): The Design of Screw Propellers in Nozzles, *Internal Shipbuilding Progr.*, vol. 6, no. 55.

van Lammeren, W. P. A., J. D. van Manen, and M. W. C. Oosterveld (1969): The Wageningen B-Screw Series, *Trans. Soc. Naval Architect. Marine Eng.*, vol. 111.

—— and W. P. A. van Lammeren (1955): The Design of Wake-Adapted Screws and Their Behavior Behind the Ship, *Inst. Eng. Shipbuilders Scotland*, vol. 99.

5

Ocean Corrosion

C. Petersen and G. Soltz

5-1 INTRODUCTION

In this chapter we will cover the most important aspects of marine corrosion, and practical examples will be used to amplify some problems. Corrosion is a very costly problem and as such it deserves the attention of the modern ocean engineer. Conservative estimates give the yearly loss in the United States due to corrosion as 10 billion dollars (Fontana and Greene, 1967) and, surprisingly, it is rising.

The term "corrosion" is used to denote the destruction process affecting a specific family of engineering materials—metals. A broader term would be material destruction or deterioration, and this would encompass the wastage of any type of material. *Corrosion is deterioration due to an electrochemical reaction.* This definition will be expanded later in this chapter.

It is the task of every engineer to minimize this problem. Progress in modern engineering is controlled largely by the development of stronger, more durable materials and by improvements in methods of corrosion control. Modern engineers are also able to integrate material specifications more accurately into their designs. In the early days, however, this was not true and they usually wisely left a considerable margin for safety in their calculations. Those conservative safety margins not only took care of any initial design miscalculations, but later they also absorbed the time-dependent losses in material strength caused by corrosion. In the modern aerospace industry, where weight considerations have always been of paramount importance, the structural safety factors are usually as low as 1.5 times the minimum required strength. Compared to this, marine hull designs, even relatively modern ones, usually call for safety factors up to 5 times the minimum required strength. In some of the present-day work, and in most future marine designs, however, this conservative attitude may well disappear.

Reduction in safety factors usually occurs because of industry's desire to reduce material or design costs. However, in the case of submersible vehicles, where material costs usually are not the main factor, engineers are forced to reduce the safety factor in order to maintain some degree of buoyancy. As the trend toward a reduction in engineering safety factors continues, it becomes much more important that the engineers who are designing high-performance structures can calculate their effective strength accurately. They must realize that, in time, corrosion of the structure can take place and if it does, it can negate much of their earlier strength calculations. This corrosion factor is particularly true in the case of some of the new higher-strength alloys that today's designers are looking at. New alloy development often tempts the engineer into improving his designs around their stated mechanical properties, but this can be dangerous if mechanical testing has been carried out in air only. Tests in air only can be very misleading, as they may indicate completely different material properties from those in tests conducted in the ocean environment, where corrosion plays an important role. Figure 5-1 illustrates such a change in property by comparing a fatigue test conducted both in and out of a corrosive medium. Note the drastic loss of strength. Any performance differences that exist in the material must be fully understood by a designer before he adopts a new material. Therefore, unless the corrosion resistance of an alloy is satisfactory along with its superior mechanical strength, it may be of little practical engineering use.

It is therefore imperative that the ocean engineer understand the factors involved in corrosion. It is only through a better understanding of the basic corrosion principles that we can avoid poor (from the standpoint of corrosion) marine designs in the future. Although some corrosion is inevitable, it usually can be predicted and designed around at the drawing board. There is a vast difference, however, between designing around corrosion and economically eliminating corrosion. The latter task should usually be undertaken by a materials specialist before final decisions are made. If possible, this type of person should also be involved in the initial design.

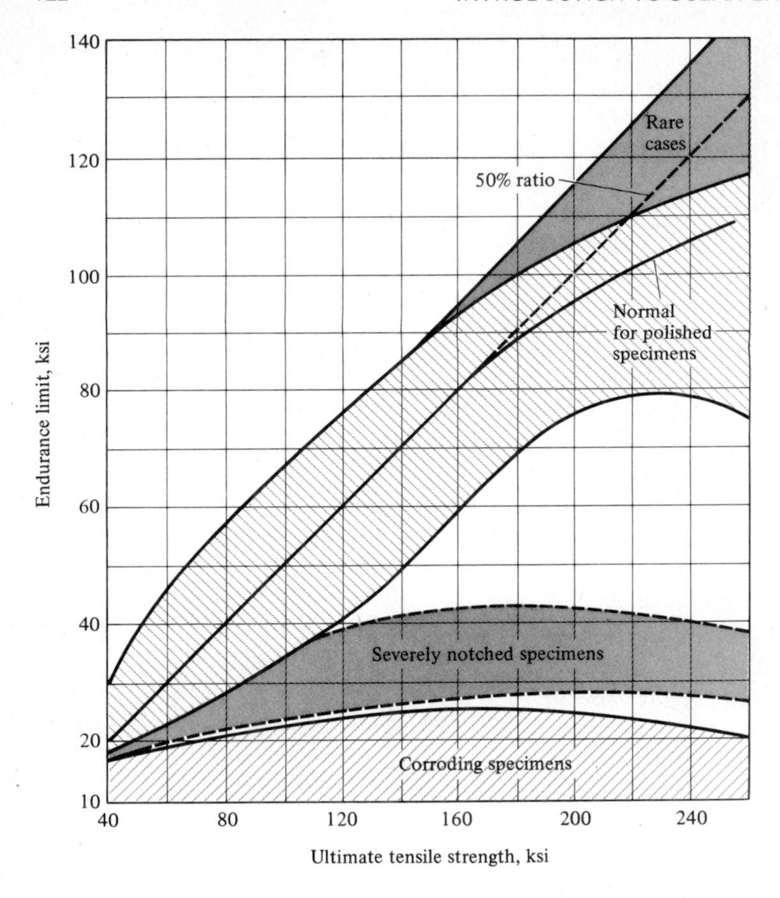

Fig. 5-1 Relation between the endurance limit and ultimate tensile strength of various steels showing the effect of corrosion on strength. (Munse 1964.)

5-2 CORROSION PRINCIPLES

5-2.1 Electrochemistry

For an electrochemical reaction to occur, four factors must be present at the same time. These are:

1. An *anode* area, that is, a *metal* area that corrodes by oxidation (loss of electrons).
2. A *cathode* area, that is, a metal or semiconducting surface where a reduction reaction occurs (uses electrons from the anode).
3. An *electrolyte* in contact with the anode and cathode (ions are transferred via this path).

4. An *electron* flow path between the anode and cathode (this must be a metal; electrons flow from anode to cathode).

If all the above factors are not present simultaneously, then you will not have corrosion. This is the key to designing out corrosion. *If you eliminate one factor by your design, you eliminate corrosion.*

We can categorize all corrosion reactions by six electrochemical reactions. At the anode there is oxidation and loss of metal. This reaction is shown by the following basic formula:

$$M \rightarrow M^{Z^+} + Ze \qquad (5\text{-}1)$$

where M is only metal; Ze is the number of electrons lost; and Z is the metal's valence. Actual examples are:

Magnesium	$Mg \rightarrow Mg^{2+} + 2e$
Iron	$Fe \rightarrow Fe^{3+} + 3e$
Aluminum	$Al \rightarrow Al^{3+} + 3e$

The *electrons which are generated must be used up simultaneously since there cannot be a charge buildup*. This electron uptake occurs at the cathode, where a reduction reaction occurs. There are five probable cathodic reactions that are related to the corrosion process (one or more of these may occur at the same time):

1. Acid	$2H^+ + 2e \rightarrow H_2$ (gas)	hydrogen reduction	(5-2)
2. Oxygenated acid	$4H^+ + O_2 + 4e \rightarrow 2H_2O$	oxygen reduction	(5-3)
3. Neutral or near-neutral solution	$O_2 + 2H_2O + 4e \rightarrow 4OH^-$	oxygen reduction	(5-4)

Note that this is the most common and important reaction in marine corrosion.

4. Metal reduction	$M^{Z^+} + e \rightarrow M^{(Z-1)^+}$	(5-5)
5. Metal deposition	$M^{Z+} + Ze \rightarrow M^0$	(5-6)

Reaction (5-4) is the one that controls most marine corrosion, drawing the needed oxygen from the surrounding atmosphere. This is one reaction the ocean engineer should be especially familiar with.

It should be noted that no metal damage occurs in any of these cathodic reactions, and as a result the cathode area is usually ignored by designers. This tendency to ignore the cathode area's true significance is one of the most common and important mistakes made by engineers. Remember, *if you can stop the cathodic reaction, you can stop corrosion.*

There are many electrochemical reactions, but under a given set of conditions only a few of them will be favored. In order to know which ones will be favored, the science of thermodynamics must be called upon.

5-2.2 Thermodynamics

Thermodynamics was evolved to handle the flux of energy and heat in a physical or chemical reaction. One of the parameters developed was *free energy*. Free energy indicates in which direction a reaction will go. This is the first thermodynamic relation which the engineer will need to apply. The equation for Gibbs' free energy is of the form:

$$\Delta G = -nFE \tag{5-7}$$

where G is Gibbs' free energy; n is the number of electrons transferred; F is Faraday's constant (96,500 A/s); and E is the potential of the reaction in question. If the value obtained for ΔG is positive, the reaction will not occur. If the value of ΔG is negative, then the reaction is possible.

All the values needed to calculate ΔG are known except E, the reaction potentials. These potentials are due to the exchange of electrons through a certain resistance. This resistance is the sum total of the metal's resistance, the resistance of the electrolyte, and the resistance on the metal-electrolyte interface. The potential is called the electrode potential and occurs whenever a metal is placed in an electrolyte. To determine the value of E, we will first look at the simplest case—one metal in a standard solution containing 1 g-ion/l of the same metal's ions (Fig. 5-2). As soon as the metal is placed in solution, a minute amount of it is immediately ionized. This reaction causes the metal to have an excess of electrons which must be used up by a simultaneous reduction reaction. The reduction reaction would consist of the metal ions in solution taking on electrons and plating out. Equilibrium conditions would soon occur so that the net loss of material would be zero, and the net current would be zero. Such a set of reactions, however, would develop a potential owing to the current flow through the metal. This current is called the exchange current, and the potential developed is called the rest potential, or half-cell potential. This potential, however, can by measured only in relation to another electrode situated in the same electrolyte. The potential difference measured, therefore, is not an absolute potential, and an international reference cell has been formulated so that there is some standardization (Fig. 5-2*b*).

The hydrogen electrode has been internationally recognized as the zero reference standard. This has the form:

$$\tfrac{1}{2}H_2 \rightarrow H^+ + e \qquad E = 0.00 \text{ V}$$

and is obtained by bubbling hydrogen gas through a platinum tube immersed in

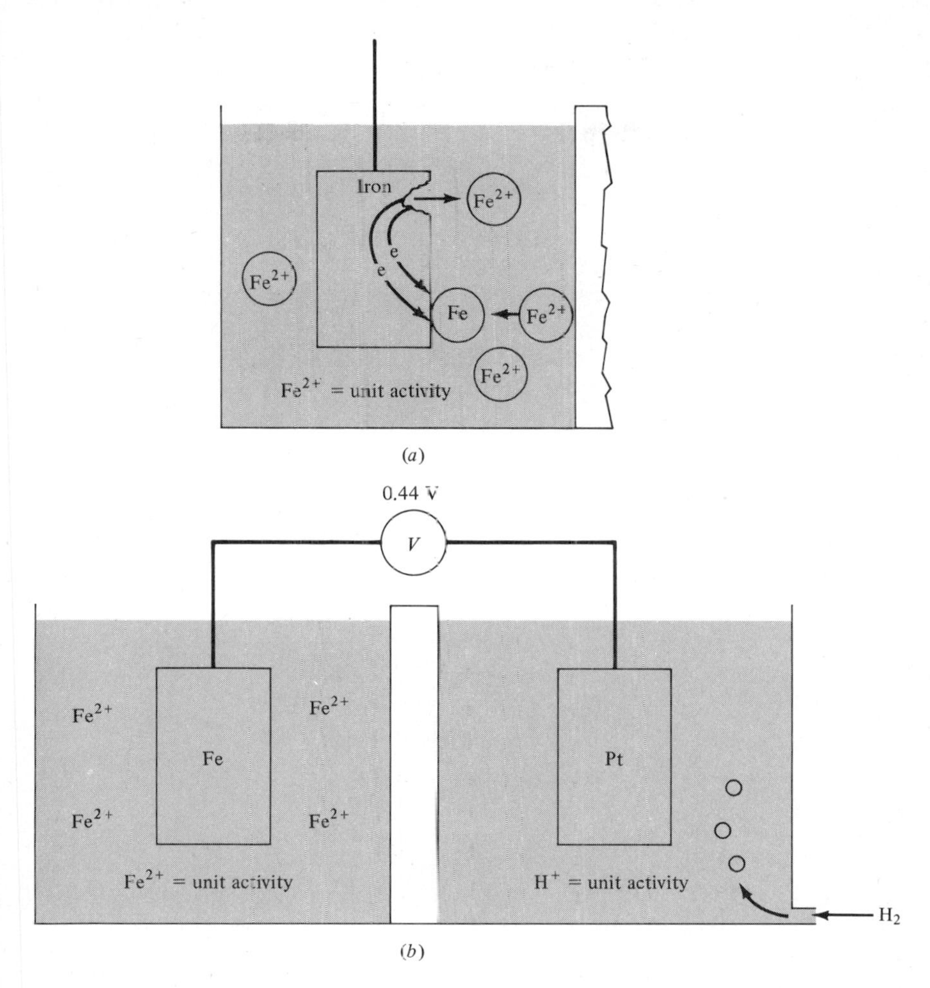

Fig. 5-2 (*a*) Single electrode in solution of its own ions; (*b*) iron electrode and hydrogen reference electrode used to measure emf.

pure water. The platinum acts solely as the electron path and does not enter into the reaction.

Making use of this reference cell, a series of standard oxidation-reduction potentials was made up. This standard emf series is shown in Table 5-1.

These half-cell potentials can now be used to determine which reaction is more likely to occur. The cell potential E that is to be used in Eq. (5-7) is the potential difference between the cathode and anode in an electrochemical cell made of two half-cells. The electrochemical cell composed of two half-cells will have a potential equal to the difference between these half-cells. The more active

Table 5-1 The electromotive series
E_0 = electrode potential at 25°C, in volts, relative to a standard hydrogen electrode: ions at unit activity

Electrode	*Reaction when electrode is cathode*	E_0, *volts*
K, K^{++}	$K^+ + e = K$	−2.922
Ca; Ca^{++}	$\frac{1}{2}Ca^{++} + e = \frac{1}{2}Ca$	−2.87
Na; Na^+	$Na^+ + e = Na$	−2.712
Mg; Mg^{++}	$\frac{1}{2}Mg^{++} + e = \frac{1}{2}Mg$	−2.34
Be; Be^{++}	$\frac{1}{2}Be^{++} + e = \frac{1}{2}Be$	−1.70
Al; Al^{3+}	$\frac{1}{3}Al^{3+} + e = \frac{1}{3}Al$	−1.67
Mn; Mn^{++}	$\frac{1}{2}Mn^{++} + e = \frac{1}{2}Mn$	−1.05
Zn; Zn^{++}	$\frac{1}{2}Zn^{++} + e = \frac{1}{2}Zn$	−0.762
Cr; Cr^{3+}	$\frac{1}{3}Cr^{3+} + e = \frac{1}{3}Cr$	−0.71
Ga; Ga^{3+}	$\frac{1}{3}Ga^{3+} + e = \frac{1}{3}Ga$	−0.52
Fe; Fe^{++}	$\frac{1}{2}Fe^{++} + e = \frac{1}{2}Fe$	−0.440
Cd; Cd^{3+}	$\frac{1}{2}Cd^{++} + e = \frac{1}{2}Cd$	−0.402
Ni; Ni^{++}	$\frac{1}{2}Ni^{++} + e = \frac{1}{2}Ni$	−0.250
Sn; Sn^{++}	$\frac{1}{2}Sn^{++} + e = \frac{1}{2}Sn$	−0.136
Pb; Pb^{++}	$\frac{1}{2}Pb^{++} + e = \frac{1}{2}Pb$	−0.126
Pt; H_2 (*g*); H^+	$H^+ + e = \frac{1}{2}H_2$	0.000
Normal calomel electrode	$\frac{1}{2}Hg_2Cl_2 + e = Hg + Cl$	0.2802
Cu; Cu^{++}	$\frac{1}{2}Cu^{++} + e = \frac{1}{2}Cu$	0.345
Pt; Fe^{++}, Fe^{3+}	$Fe^{3+} + e = Fe^{++}$	0.771
Ag; Ag^+	$Ag^+ + e = Ag$	0.800
Hg; Hg^{++}	$\frac{1}{2}Hg^{++} + e = \frac{1}{2}Hg$	0.854
Pt; Cl_2 (*g*); Cl^-	$\frac{1}{2}Cl_2(g) + e = Cl^-$	1.358
Pt; Pt^{++}	$\frac{1}{2}Pt^{++} + e = \frac{1}{2}Pt$	1.2 (approx)
Oxygen reduction in acid	$\frac{1}{2}O + H^+ + e = \frac{1}{2}H_2O$	1.229
Au; Au^{3+}	$\frac{1}{3}Au^{3+} + e = \frac{1}{3}Au$	1.42
Au; Au^+	$Au^+ + e = Au$	1.68

Sources: (1) O. A. Hougen, K. M. Watson, and R. A. Ragatz, "Chemical Process Principles," part II, p. 1034, Wiley, New York, 1959. (2) M. Henthorne, Electrochemical Corrosion, *Chem. Eng.,* June 14, 1971, p. 103.

metal will act as the anode, and the more noble metal the cathode. The potential difference between the two half-cells is the driving force that moves electrons from the anode to the cathode area, and it is this ΔE that we can place in the free energy equation to determine the reaction's direction.

Example 5-1 Determine the potentials and free energies for the following reactions:

1. $Fe + H_2SO_4 \rightarrow Fe^{++} + SO_4^{--} + H_2$

2. $2Ag + H_2SO_4 \rightarrow 2Ag^+ + SO_4^{--} + H_2$
3. $2Ag + O_2 + H_2SO_4 \rightarrow 2Ag^+ + SO_4^{--} + H_2O$

Solution Make use of the values given in Table 5-1 and Eq. (5-7).

1. Equation 1 above is merely the sum of two half-cell reactions:

$Fe = Fe^{++} + 2E$ and $2H^{++} + 2e = H_2$

The SO_4 does not really enter into the reaction. Therefore the potential needed in Eq. (5-7) is $E = E(H/H^+) - E(Fe/Fe^{++}) = 0.44$ V.

$$\Delta G = -(1)(96{,}500)(0.44) = -42{,}460$$

Therefore the reaction is spontaneous.

2. The half-cell reactions involved here are similar:

$Ag = Ag^+ + e$ and $H^+ + e = H$

$$E = E\left(\frac{H}{H^+}\right) - E\left(\frac{Ag}{Ag^+}\right) = 0.00 - (0.80) = -0.800 \text{ V}$$

$$\Delta G = -(1)(96{,}500)(-0.800) = +77{,}104$$

Since ΔG is positive, the reaction is not possible.

3. This reaction involves the same metal dissolution half-cell as above in equation 2, but the other half-cell is different:

$H^+ + \frac{1}{2}O + e = \frac{1}{2}H_2O$

$$E = E\left(\frac{O}{O^{--}}\right) - E\left(\frac{Ag}{Ag^+}\right) = 1.229 - 0.800 = 0.430 \text{ V}$$

$$\Delta G = -(1)(96{,}500)(0.430) = -41{,}495$$

This time ΔG is negative and the reaction is spontaneous. This points out the effect of adding an oxidizer to the electrolyte.

In general the cell voltage = E (cathode) − E (anode). Which half-cell is the anode and which is the cathode is determined by the total reaction. For example, the metals in all the above reactions were being oxidized; therefore they were the anodes.

The emf series in Table 5-1 has its value, but in the real world we very seldom work with solutions at unit activity. In order to determine the actual potential of a system in which the reactants are not at unit activity, the Nernst equation is applied:

$$E_{metal} = E^0 + 2.3\frac{RT}{ZF}\log\frac{M^{Z^+}}{M} \tag{5-8}$$

where E_{metal} = exact half-cell potential of the pure metal under the actual environment conditions

E° = emf potential at 25°C

R = 8.3143 J/°mol

T = absolute temperature

Z = electrons transferred in the reaction

M^{Z^+} = activity of metal in solution

M = activity of metal (considered to be unity)

Table 5-2 Galvanic series of metal and alloys in quiet seawater

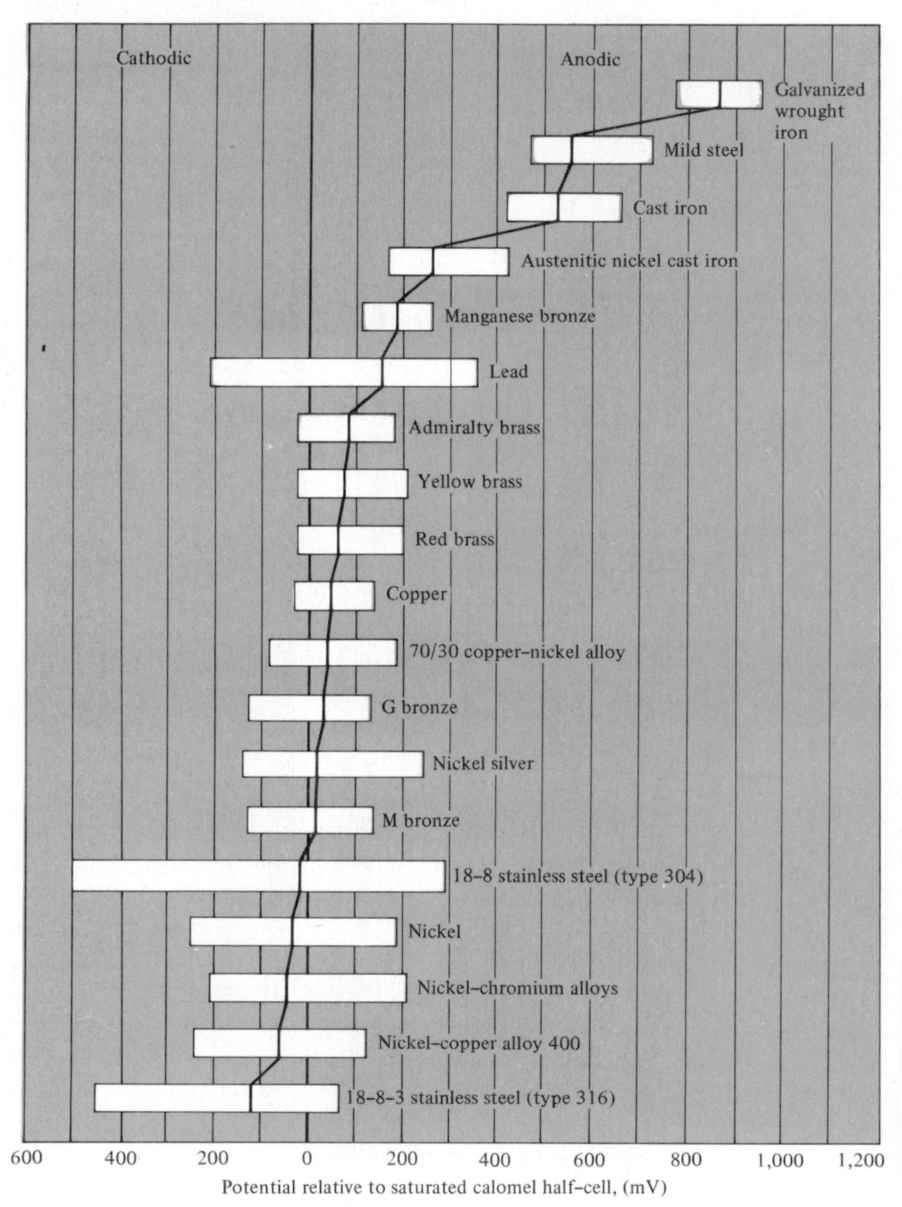

Example 5-2 Determine the potential of a platinum electrode in a 10^{-5}-g-ion/l solution of platinum ions at 25°C.

Solution Using the Nernst equation (5-8) we can calculate the new potential.

$$E = E_0 + 2.3 \frac{RT}{ZF} \log \frac{M^{Z^+}}{M} \qquad E_0 = +1.2 \text{ V (from Table 5-1)}$$

$$M^{Z^+} = 10^{-5} \qquad M = 1 \qquad Z = +2$$

$$E = -1.2 \text{ V} + 2.3 \frac{(8.3143)(298)}{(2)(96{,}500)} \log (10^{-5}) = +1.2 - 0.1475 = +1.0525 \text{ V}$$

The standard emf series and the Nernst equation are not very useful to the working ocean engineer, however, since they do not give the information he usually needs, namely, the potential in seawater. Thus a much more useful metal activity series is the galvanic series of metals and alloys in seawater (Tables 5-2 and 5-3). A

Table 5-3 Galvanic potentials in flowing seawater
Velocity = 13 ft/s except where noted.

Metal or alloy	*Temperature, °C*			*Volt* vs. saturated calomel*		
Zinc	26			−1.03		
Mild steel	24			0.61		
Gray cast iron	24			0.61		
Austenitic cast iron†		14			0.47	
Copper	24			0.36		
Admiralty brass	24.6			0.36		
Gunmetal	24			0.31		
Aluminum brass	24.6			0.29		
Admiralty brass		11.9			0.30‡	
Lead-tin solder (50-50)		17			0.28	
90/10 copper-nickel alloy (1.4 Fe)			6			0.24
90/10 copper-nickel alloy (1.4 Fe)		17			0.29	
90/10 copper-nickel alloy (1.5 Fe)	24			0.22		
70/30 copper-nickel alloy (0.51 Fe)			6			0.22
70/30 copper-nickel alloy (0.51 Fe)		17			0.24	
70/30 copper-nickel alloy (0.51 Fe)	26.7			0.20		
Nickel-copper alloy 400	22			0.11		
Nickel	25			0.10‡		
Titanium	27			−0.10		
Graphite	24			+0.25		
Platinum		18			+0.26‡	

*All values negative vs. saturated calomel reference electrode except those for graphite and platinum.

†Austenitic nickel ductile cast iron Type D-2 (3.0 C, 1.5–3 Si, 0.7–1.25 Mn, 18–22 Ni, 1.75–2.75 Cr).

‡Seawater velocity = 7.8 ft/s.

large potential difference between two metals would indicate a larger free energy and hence a greater tendency to corrode, and a small difference would indicate less tendency, but it will tell you nothing about how fast a reaction will occur. It is the speed at which this reaction occurs which determines how much metal wastage takes place. Therefore, in order to determine how fast the reaction occurs, the engineer must turn to kinetics and determine the corrosion current.

5-2.3 Electrode kinetics

To determine the feasibility of a certain corrosion reaction, we have considered the behavior of independently operating anodes and cathodes. Owing to the use of a high-impedance voltmeter in measuring potentials, there is virtually no current passing between the anodes and cathodes. In the real world, however, the anode and cathode are connected and electrons are free to pass between them, giving rise to current flow.

When electrons are allowed to pass from anode to cathode and current flows, the potential of both the anode and cathode changes. This is simply a result of Ohm's law $E = IR$, where R is fixed by the metal and electrolyte. This change in potential is called *polarization* and causes a reduction in the potential difference between anode and cathode.

Once a metal has come to equilibrium in the corroding medium, it will have a corrosion potential (E_{corr}) and a corrosion current density (I_{corr}) between anode and cathode areas. The corrosion potential may easily be determined by the setup shown in Fig. 5-3. Note, however, that this potential is different from the earlier thermodynamic half-cell potential since we are not working with isolated anodes and cathodes.

The corrosion current density, as stated earlier, gives rise to the determination of metal loss. By using Faraday's law, the current density can be converted into a corrosion rate.

$$R_{\text{MPY}} = 1.3 \times 10^5 \, \frac{(I)(V)(MW)}{(Z)(\rho)(A)} \tag{5-9}$$

where R_{MPY} = corrosion rate, mils/year
I = current density, A/cm^3

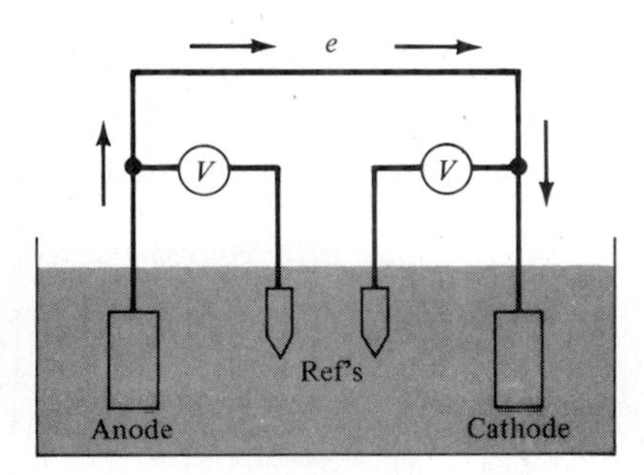

Fig. 5-3 Finding electrode potentials when a current flows.

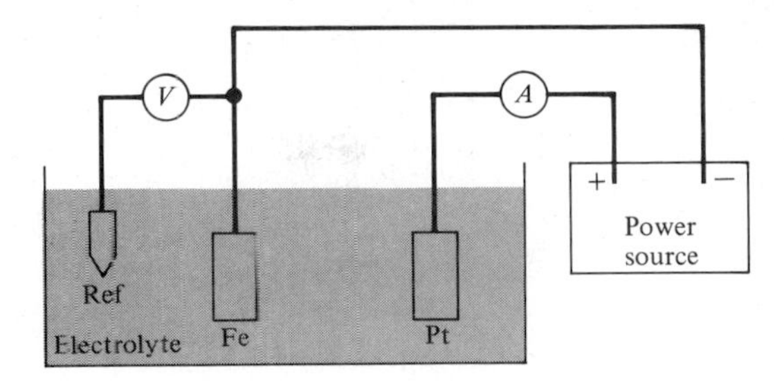

Fig. 5-4 Electron supply to a corroding specimen.

V = volume of the specimen, cm^3
MW = molecular weight, g
Z = metal's valence
ρ = specific gravity, cm^3
A = surface area, cm^2

Measuring this corrosion current directly, however, is not usually feasible since it is often the current passing between minute, almost inseparable, anodes and cathodes on the same metal's surface. Therefore, the current is measured indirectly.

Looking at a typical pair of anode and cathode reactions, we can briefly explain the principle involved.

Anode: $Fe \rightarrow Fe^{++} + 2e$

Cathode: $Fe^{++} + 2e \rightarrow Fe^0$

Two electrons are generated by the corrosion of the iron and two are used up by the formation of the stable iron metal. Suppose we could supply the electrons needed for the cathodic reaction by some means other than the corrosion of iron at the anode. Since the anode's electrons are no longer needed, the corrosion would cease and the piece of iron would behave entirely like a cathode. This is the principle involved: make the specimen behave predominantly like an anode or cathode by using an external electron source or sink. The equipment needed to do this is shown in Fig. 5-4. Since the anode and cathode are now separated, it is possible to measure the current flow. Suppose we wished to study the behavior of iron in seawater, and a freely corroding specimen had a potential of –0.4 V. If we take this specimen and, using the apparatus shown in Fig. 5-4, make it have a potential (since current affects potential) of –0.8 V, a current will be flowing between the iron and the inert electrode. We can measure this current, and this is the current which would be passing between the anode and cathode if the iron had a corrosion potential of –0.8 V.

This process is repeated over a wide range of potentials until a curve such as that shown in Fig. 5-5 is obtained. As one approaches the corrosion potential, it is important to note that the test specimen has both anodes and cathodes on

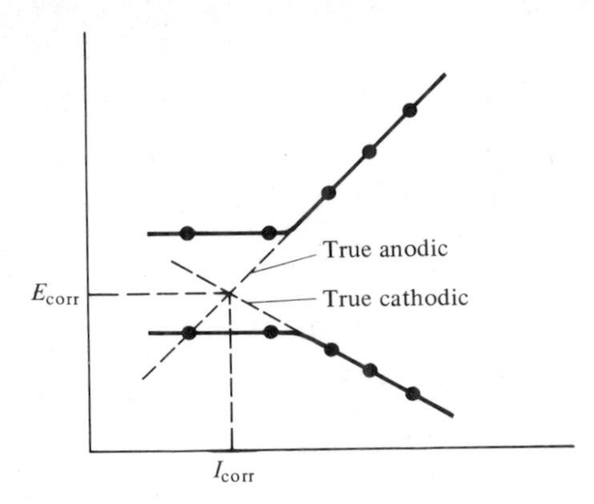

Fig. 5-5 Finding corrosion current and voltage graphically.

its surface and as a result the measured curve is a combination of the true anodic and true cathodic curve.

We have now obtained the information needed in order to calculate rates. In actual practice the process of developing the curve takes less than an hour. Once we know the corrosion potential of the iron in seawater, we can go to our curve, find the current density, and calculate weight losses.

Example 5-3 An iron buoy has a potential of −0.4 V. A curve of potential-vs.-log current density has been constructed for iron in seawater and is shown here. Find the corrosion rate for the iron buoy. It has 12 in OD and 11 in ID.

Solution Find −0.4 V on the potential axis and proceed across until you intersect the curve. Then drop down to the current axis and find the current density. As seen from the curve, $I = 10^{-4}$ A/cm³. Calculate the actual metal volume and exposed surface area.

$$V = 0.5236(\text{OD}^3 - \text{ID}^3) = 207.87 \text{ in}^3$$

$$A = 3.14(\text{OD}^2) = 452.39 \text{ in}^2$$

The density of iron is 7.86 g/cm³ and its molecular weight is 55.847 g. Therefore, using Eq. (5-9), $R_{MPY} = 53.8988$ mils/year.

Note: The volume and area must be converted to centimeters.

Fig. Example 5-3

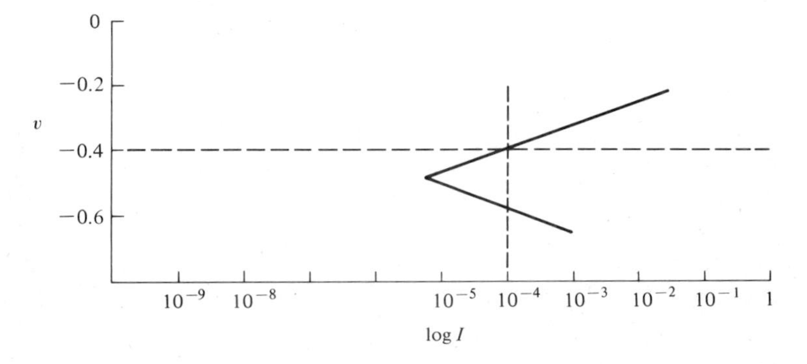

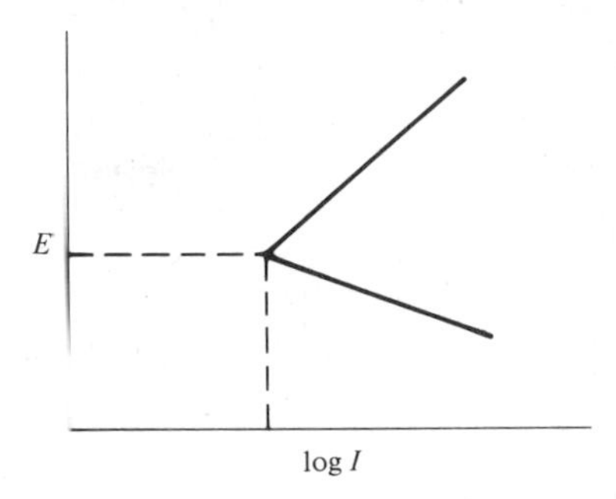

Fig. 5-6 Graph of activation polarization.

Remember that this process must be repeated for each type of corrodent, and the curves *cannot* be carried over from one electrolyte to the next.

In constructing these curves, the engineer should be familiar with a few basic types. Basically two types of polarization curves are found:

1. *Activation polarization* The curve is essentially linear on the semilog plot of potential vs. log I. This means that the rate at which the corrosion takes place is governed solely by the slowest step in the electrochemical reactions involved (Fig. 5-6).
2. *Concentration polarization* The curve increases or decreases linearly until a maximum current density is reached, at which point the current density remains constant over a wide range of voltage changes. This means that either the anodic or cathodic reactions are consuming or evolving more products than the electrolyte surrounding the electrode can transport. The rate is controlled by diffusion rates through the electrolyte. This knee in the curve may be eliminated by stirring the electrolyte (Fig. 5-7).

A third type of electrode polarization that can occur on some but not all metals, and in some but not all environments, is *passivation*. This feature appears only on the anodic curve, and is obtained only after some corrosion has taken place. This corrosion provides for the formation of a metastable film which

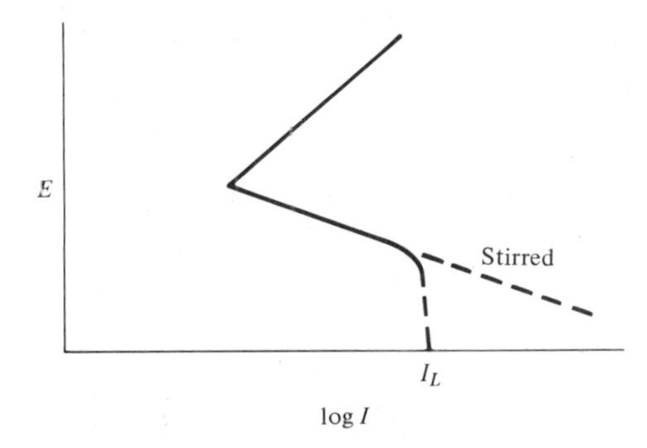

Fig. 5-7 Graph of concentration polarization.

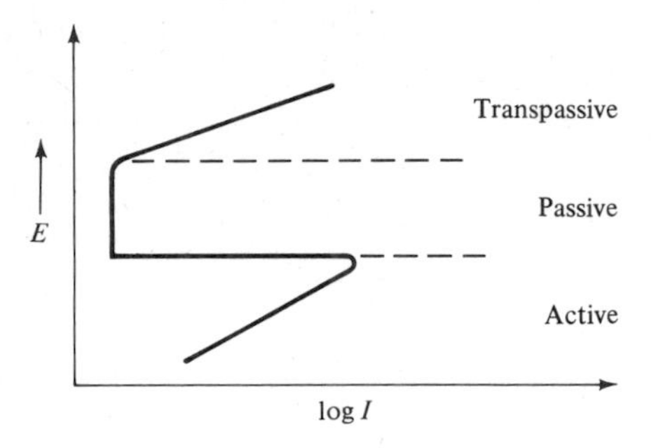

Fig. 5-8 Graph showing passivation.

limits further corrosion. Figure 5-8 shows such a curve. Note the decrease in corrosion current. It is the principle of passivation which makes stainless steel so useful in many environments. The problem with passive films in the ocean environment is that if they are destroyed, they usually do not re-form, and a more severe kind of corrosion occurs. Stainless steels are ideal in air but can be very poor in ocean work.

The practical importance of electrode polarization is best illustrated by the Evans polarization diagrams (Fig. 5-9). These diagrams are similar to the diagrams seen earlier except that two electrodes are involved instead of one. Also, only one side of each electrode's reaction is shown (note the dotted lines for reactions not usually considered). The upper solid line is the cell's cathodic half-cell polarization line. The lower solid line is the anodic half-cell polarization line. Note that at some point the anodic and cathodic lines intersect. This point of intersection indicates the corrosion potential and maximum electron flow of the system. Note that *the total electron flow shown for the anode and cathode reactions must always be equal.*

We have now seen how to determine the thermodynamic direction of a reaction and the rate at which a metal corrodes. For many materials, and for

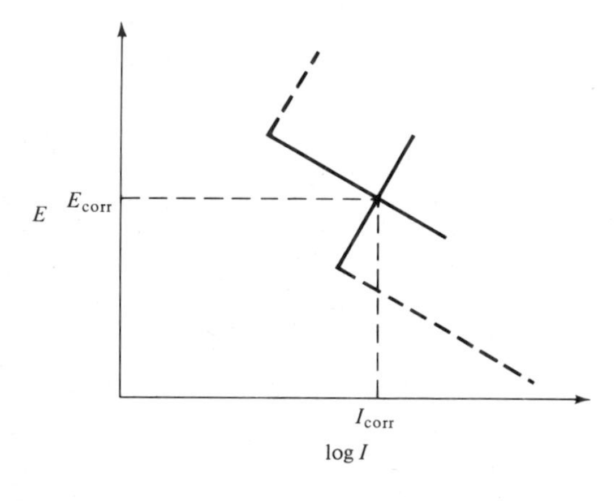

Fig. 5-9 The basic Evans diagram.

several marine environments, some of this work has already been carried out and we need only to consult the proper tables to obtain the desired values. Once these values have been obtained, the engineer can incorporate them into his design. The most valuable sources of information on this marine work are "Corrosion Handbook" (Uhlig, 1948) and literature put out by various industries. Many of the tables from the International Nickel Company's "Guidelines for Selection of Marine Materials" (1966) are presented here for your benefit and use. It should be pointed out that a number of useful materials are missing, that the values given for corrosion rate are very sensitive to water velocities, and also that seawater from area to area may vary sufficiently to warrant your own test work.

A typical example of how to use these data is given in the following example.

Example 5-4 Determine the thickness of steel needed to ensure safe operation of a hydrofoil for 10 years, given the following information: From strength of materials considerations we found that a ¼-in steel was needed to withstand the stresses encountered. A factor of safety of 1.25 is to be added to the final thickness. The vehicle travels at 130 ft/s.

Fig. 5-10 General wasting (immersed in quiet seawater) less than 2 ft/s.

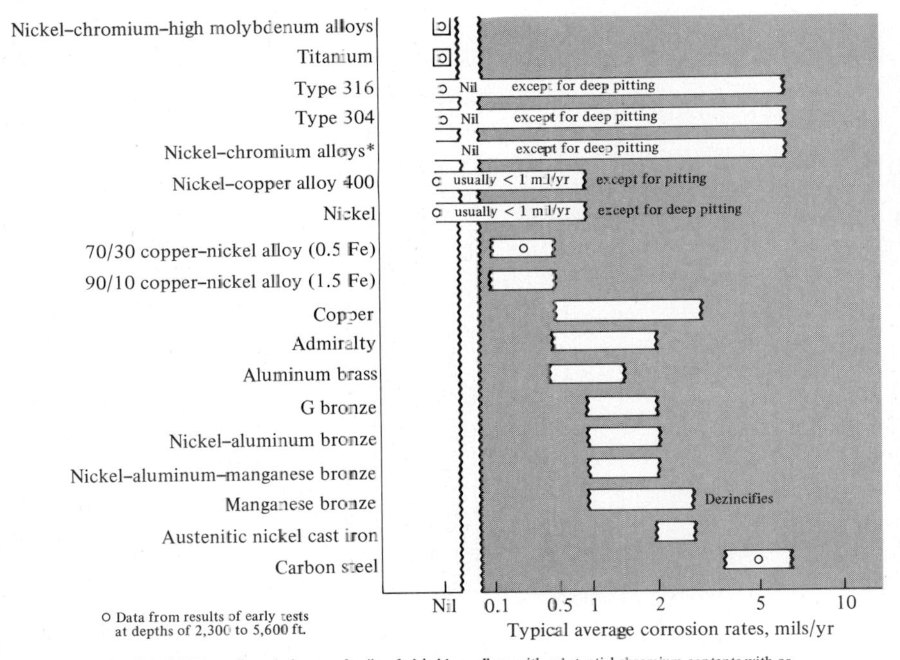

*Nickel–chromium alloys designate a family of nickel base alloys with substantial chromium contents with or without other alloying elements all of which, except those with high molybdenum contents, have related seawater corrosion characteristics.

Solution The ¼-in determined for stress resistance is not nearly enough wall thickness when we begin to consider corrosion. Using Fig. 5-12 we find that carbon steel corrodes at a rate of around 300 mils/year at 130 ft/s. Thus the amount of steel that will be lost in 10 years is 3,000 mils or 3 in. The wall thickness of steel now needed to counteract both stress and corrosion is up to 3.25 in. Adding the safety factor gives a new wall thickness of $1.25\ (0.25 + 3.0) = 4.0625$ in.

Oftentimes, however, the work that the engineer gets involved in is an after-the-fact type of job. Once corrosion has occurred, the engineer is called in to solve the problem. He is no longer interested in polarization curves or electrode potentials, but in why and how it occurred and how it can be stopped. Thus the engineer must be able to diagnose the problem by observing the external manifestations of the corrosion, or by means of some micrographic work. To assist him in this, all types of corrosion have been broken down into several forms. A description of each form follows.

5-3 TYPES OF CORROSION COMMONLY FOUND IN THE MARINE ENVIRONMENT

A. General corrosion manifests itself by a continuous layer of corrosion over the whole specimen surface. In order for this to occur, the anodes and cathodes

Fig. 5-11 Seawater velocity (pipe and tube ranges) and wasting produced.

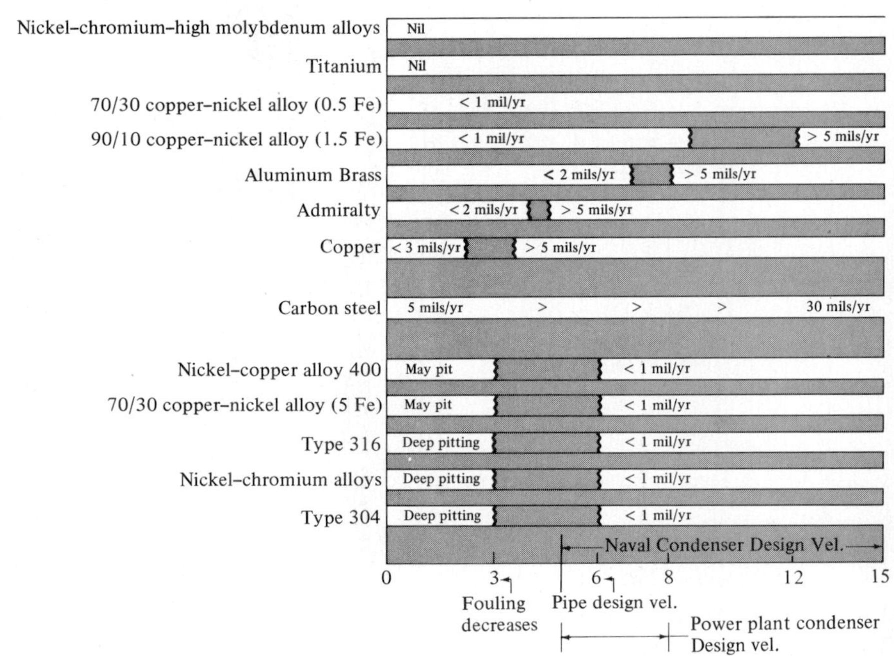

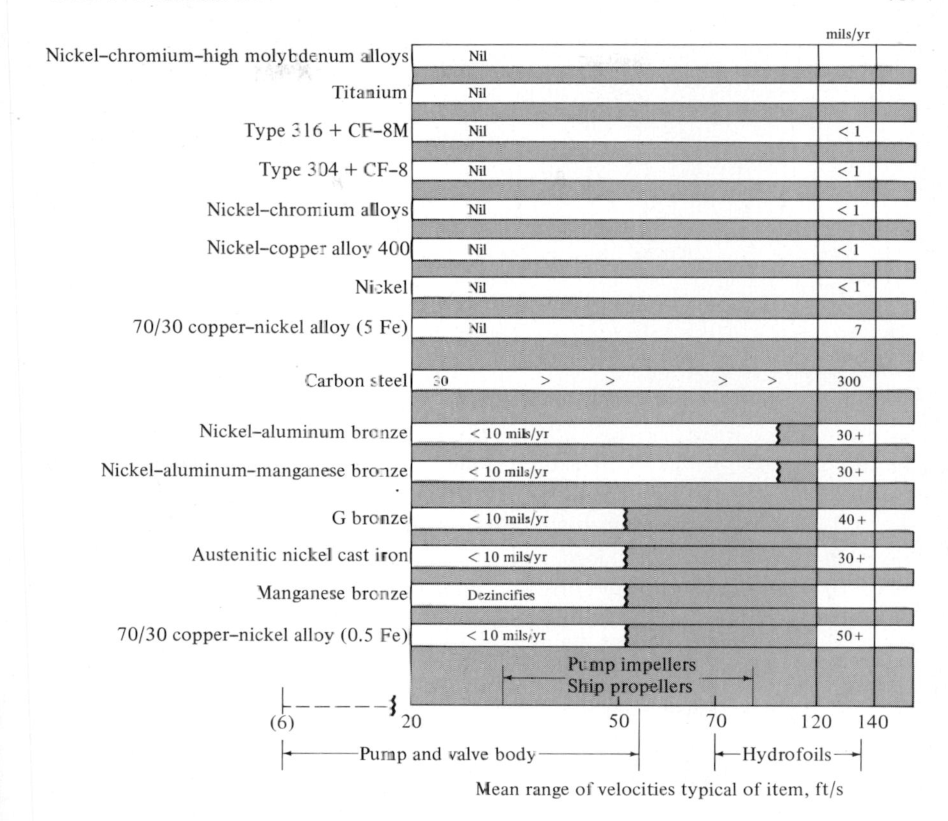

Fig. 5-12 Seawater velocity (pump to hydrofoil range) and wasting produced.

cannot stabilize but must be continuously shifting. This type of corrosion is found most often in the atmosphere rather than in a submerged environment. Since the attack is uniform, this is the easiest type of corrosion to design for, and most corrosion rates listed in tables are for this type (Figs. 5-10, 5-11, 5-12).

B. Galvanic corrosion occurs when two dissimilar metals are connected either directly or by a metallic path, and are immersed in the same corrodent. A potential difference will cause an electron flow into the more noble metal, a shift in potential along the polarization curve, and hence a change in I_{corr}. The external manifestation of this is that one metal in a couple will corrode faster than normal, and the other more noble metal will corrode slower or even cease corroding altogether. The previously discussed galvanic series can be used to estimate driving forces and potentials. It is this type of corrosion which allows a ship made of steel to be protected by zinc anodes bolted to the hull, but it is also the reason an aluminum superstructure may deteriorate rapidly on a steel-hulled ship. *In general try and use as few different metals as possible, and*

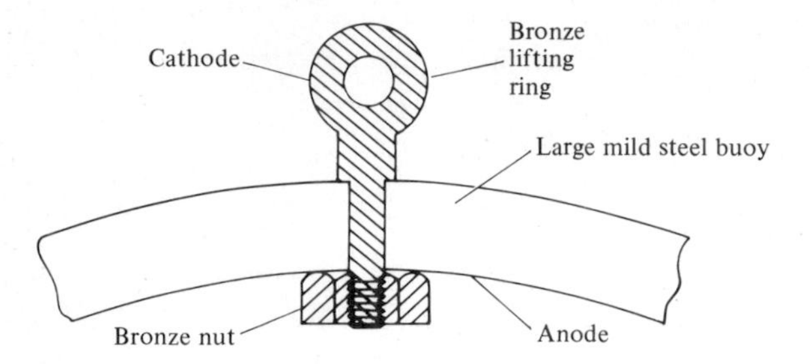

Fig. 5-13 Small and important attachments should be cathodic with reference to the main item, as here.

isolate the ones you do use. Also, if two dissimilar metals must be connected (e.g., at weld), make the anode the larger of the two (Fig. 5-13).

C. Intergranular corrosion is a more microscopic form of corrosion. It is the result of potential differences between the grain boundaries of the metal, and the grain bodies. If the grain boundaries are anodic to the grain body, an attack along these boundaries will occur (Fig. 5-14). If the grain body is anodic, you will have essentially uniform corrosion since the bodies are much larger than the boundary areas. Perhaps the most prevalent form of this type of corrosion is "weld zone" decay where, owing to the welding process, a change in structure occurs and excessive attack results near the weld.

D. Crevice corrosion is a form of localized corrosion that occurs at areas of limited oxygen availability. This usually occurs at slightly open joints (crevices) and under deposits which allow electrolyte penetration. It is usually found under such items as nuts, bolt heads, washers, or even barnacles. It is the localized nature of the attack which can speed up the material loss and eventually cause perforations and possible mechanical failure (Fig. 5-15).

Fig. 5-14 Depiction of intergranular corrosion. (Gall 1970.)

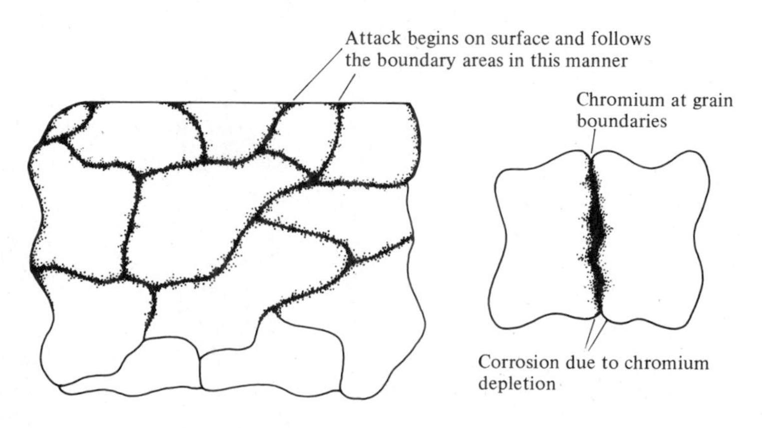

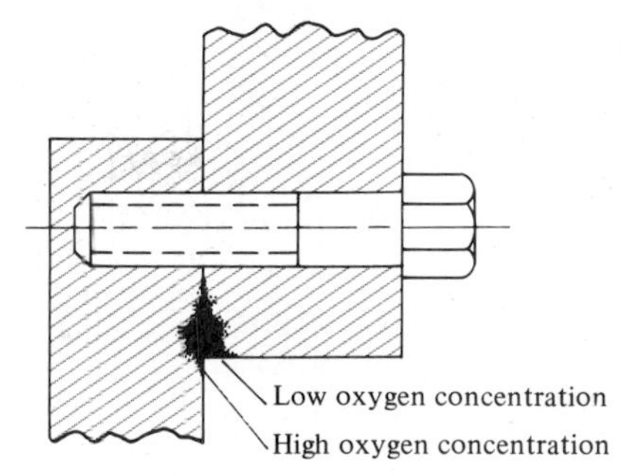

Fig. 5-15 Depiction of crevice corrosion. (Gall 1970.)

E. Pitting corrosion is another form of localized attack very similar to crevice corrosion except that it does not require an existing pit in order for attack to occur. As a result of this, attack can occur at any spot. This makes the pits a possible hazard, since one could penetrate a hull anywhere, or could mechanically weaken a structure without actual perforation. They are also difficult to detect because the surface attack is so slight. Pitting is prevalent on stainless steels in quiet water and becomes less prevalent as velocity increases (Fig. 5-16).

F. Dealloying corrosion A number of metal *alloys* are susceptible to corrosion which attacks only one of the alloy's elements. The more active element is usually removed from the alloy, leaving the more noble alloy element. The exact mechanism is not fully understood, but the process does not outwardly change the shape of the metal. It manifests itself by a very porous structure which is very dissimilar to the parent alloy. Because of this porosity, the strength of the metal is greatly reduced, as is the hardness. This attack, very common to cast iron in seawater, is called "graphitization" since only a graphite matrix is left. It is also prevalent in brasses containing more than 15 percent zinc and is called "dezincification" since only the zinc is removed.

G. Erosion corrosion is due to the environment's velocity; it is usually found at bends or elbows in piping. On some metals any increase in velocity of the electrolyte will increase the corrosion rates, owing to the removal of semiprotective corrosion products. Iron corrodes in this manner, and is therefore a poor choice if used in flowing seawater without some type of protection. In other metals a certain velocity must be exceeded before these products begin to be removed; this is called the "critical" velocity.

H. Cavitation corrosion is another form of corrosion caused by velocity, but the mechanism is different. A water vapor bubble is formed at the surface of the metal owing to a pressure drop caused by a high velocity. When the bubble collapses, water rushes in to fill the void and strikes the surface. Forces are large

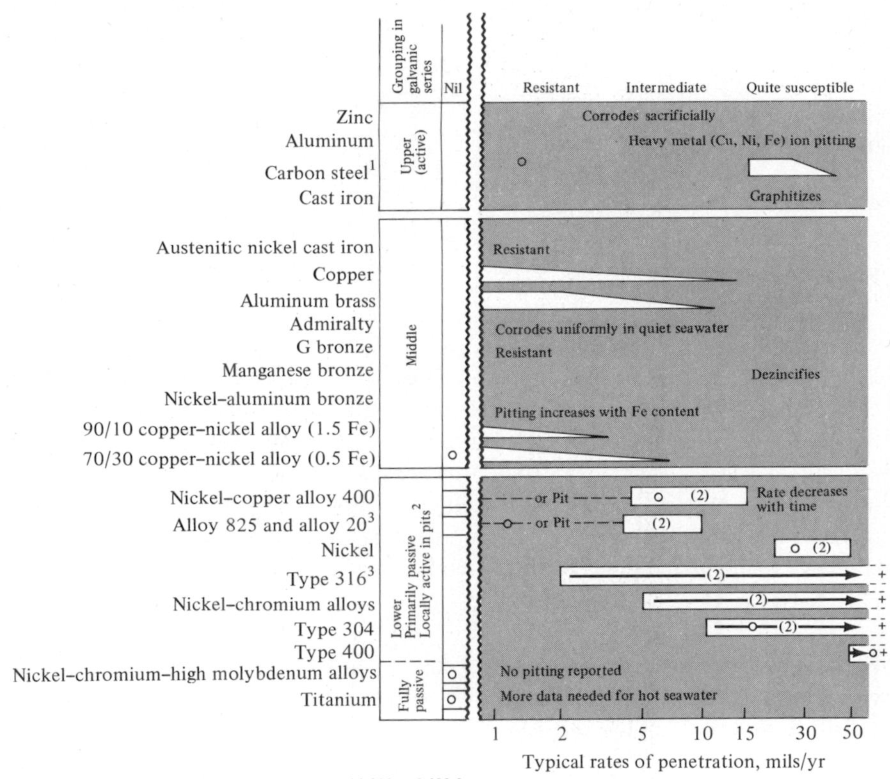

o Data from results of early tests at depths of 2,300 to 5,600 ft.

(1) Shallow round bottom pits.

(2) As velocity increases above 3 ft/s pitting decreases. When continuously exposed to 5 ft/s and higher velocities these metals, except type 400 series, tend to remain passive without any pitting over the full surface in the absence of crevices.

(3) These grades have an advantage over type 304 stainless steel and related grades in that there is a substantial reduction in the number of pits, i.e., probability of pitting even though the depth of such pits as do occur is not greatly reduced.

Fig. 5-16 Pitting (immersion in seawater less than 2 ft/s).

enough that the metal, or at least its protective film, can be physically damaged. This attack usually occurs repeatedly at the same area owing to geometric and hydrodynamic factors, and over a period of time considerable damage can occur. A smooth surface minimizes the problem, but once the surface is roughened by the process, it tends to accelerate itself. The most predominant places where this attack occurs are on hydrofoils and propellers.

I. Stress corrosion cracking is due to the joint action of mechanical stress and the environment. The exact mechanism is still not known, but owing to the danger associated with this type of corrosion, much research is being carried out to define the mechanism.

The danger with stress corrosion cracking is that it can usually go on undetected, and failure of the structure is usually the first clue that there is a problem. The marine environment induces cracking in more metals than most other common environments, and as a result the engineer must be very careful in his choice of alloys for marine work.

J. Hydrogen embrittlement is manifested by cracking similar to stress corrosion, or by a bursting of the metal. Therefore, its avoidance is just as important as stress corrosion cracking. The attack is due to atomic hydrogen penetrating the metal and combining either with the metal to form a brittle phase or with another hydrogen to form molecular hydrogen (H_2) which exerts a large pressure. One of the problems is that the hydrogen can be introduced into the metal by corrosion ($H^+ + e \rightarrow H$) or by a cathodic protection system. Therefore, care must be exercised when applying cathodic protection.

5-4 CORROSION PREVENTION

The earlier parts of this chapter were devoted to explaining the basic causes and factors involved in corrosion. We have now reached a point where we can discuss methods of preventing this corrosion. We will briefly cover the important methods of corrosion prevention available to the engineer. As stated earlier, if you can eliminate one of the corrosion reactions (cathodic or anodic), you can stop corrosion. This is the basis behind all forms of corrosion prevention. The most desirable time to consider corrosion is in the design stage, and so we will first discuss some of the methods most easily applied at that point.

5-4.1 Material selection

The most commonly applied method of corrosion control is material selection. If one were to approach the problem from a purely technical standpoint and look at it only as a corrosion problem, the choice would be quite simple: choose the most resistant material. But design is usually a marriage of many requirements, none of which can be overlooked. Typical requirements and some of the procedures involved in making a proper choice have been summarized in Table 5-4. The list is not all-inclusive but is meant to be a helpful guideline. Almost any one of the selection considerations can be the primary requirement, and the task of choosing a material will vary from job to job.

A typical choice of material is usually a compromise between many of these factors, with economics often being the deciding factor. This chapter is dealing solely with corrosion and corrosion resistance, but the designer should be familiar with all the other properties of materials as well. In order to evaluate the performance of a material from a corrosion standpoint, there are a few items he needs to know; these are listed in Table 5-5.

Table 5-4 Checklist for material selection

Requirements to be met:
- Properties (corrosion, mechanical, physical, appearance)
- Fabrication (ability to be formed, welded, machined, etc.)
- Compatibility with existing equipment
- Maintenance
- Specification coverage (e.g., military specs)
- Availability of design data

Selection considerations:
- Expected total life of plant or process
- Estimated service life of material
- Reliability (safety and economic consequences of failure)
- Availability and delivery time
- Need for further testing
- Material costs
- Maintenance and inspection costs
- Return on investment analysis
- Comparison with other corrosion control methods

Table 5-5 Information necessary for estimating corrosion performance

Corrodent variables:
- Main constituents (identity and quantity)
- Impurities (identity and amount)
- Temperature
- pH
- Degree of aeration
- Velocity of agitation
- Pressure
- Estimated range of each variable

Type of application:
- What is function of part or equipment?
- What effect will uniform corrosion have on serviceability?
- Is size change, or appearance, or corrosion a problem?
- What effect will localized corrosion have on usefulness?
- Will stresses be present? Is stress corrosion cracking possible?
- Is design compatible with the corrosion characteristics of the material?
- What is the desired service life?

Experience:
- Has material been used in identical situation? With what specific results? If equipment is still in operation, has it been inspected?
- Has material been used in similar situation? What was performance, and specifically what are the differences between the old and new situations?
- Any pilot plant experience?
- What literature is available?

To go into the actual performance of materials metal by metal and alloy by alloy would be beyond the scope of this chapter. Many of the performance characteristics of commonly used materials have been given in previous figures, and much more information is available in outside texts; but if you are not sure of a material's performance, call in someone who is. This is an area where a little knowledge can be dangerous. One guideline that should be followed, no matter which material you choose, is to try to *keep the number of different metals down to a minimum.*

5-4.2 Good design

A second means by which corrosion can be combated at the design stage is the elimination of geometric configurations that can stimulate corrosion, such as natural crevices, stagnant areas, or stress risers. The engineer should review preliminary plans and look for obvious problem areas. These areas are then altered whenever practical. The criteria used to decide whether a geometric alteration should be made are

1. What effect will the change have on other parts?
2. What is the cost of alteration?
3. What is the cost of corrosion of a time period, if not altered?
4. What is the cost of other corrosion control methods?

The remaining corrosion control methods should be considered during the design stages, but they are also usually suitable for later installation. These methods may be used alone or in combination with each other. It should be noted that combined corrosion control systems are usually the most effective. These combined systems may be more expensive to begin with, but over a long period they can give results not obtainable from a single method.

5-4.3 Paintings and coatings

The most common type of corrosion control is paint. Millions of gallons of paint are used annually. A large ship may require 2,000 gal or more per drydocking in addition to all the routine painting done throughout the year. In recent times the use of long-life paints (usually called coatings) has increased. Some coatings have a useful life of over 10 years in the marine environment. If internal coatings are used, they can be installed better during construction than after. Compared to paint, coatings are usually very expensive on a per gallon basis. The cost factor is complicated by the fact that a coating's life is very dependent on surface preparation. The preparation, in fact, is far more costly than the cost of the coating used (usually four to six times as much). Therefore, if good surface preparation is used, only the best coatings should be used. If poor surface preparation is used, then paint should be applied. Most paint or coating systems are multicoat and can range in total thickness from 0.003 in to 0.050 in. The inorganic zinc coatings are presently the best for marine applications. They may be used as a single coat, 0.003 in thick, giving 10-year life to the system, for above-water applications. However, if used underwater, they must be top-coated. The main drawback with the inorganic zincs is the high degree of surface preparation needed to ensure a successful coating application. Inorganic zinc coatings are not resistant to environments with a pH other than 6 to 9 or for long-term submerged service. Many other types of coatings are also used in the marine industry and are summarized in Table 5-6. The choice of coating depends on factors such as

1. Surface preparation possible
2. Environment
3. Material cost
4. Weather conditions at time of application
5. Surface abrasion expected during service life
6. Life desired from the system
7. Type of metal being coated
8. Types of workers available for applying the coating system

5-4.4 Cathodic protection

The next most commonly used form of corrosion protection is cathodic protection. This type of protection is very useful for submerged areas, but it cannot be used above water. Cathodic protection is best used in conjunction with coatings or paint. If the two systems are used, it is very important to use a paint not affected by alkaline conditions.

There are two types of cathodic protection: (1) impressed and (2) galvanic.

The impressed current system requires external power and is usually permanent. This system is much more complex than galvanic protection. Galvanic protection makes use of the fact that aluminum, magnesium, and zinc will protect steel if attached to it in seawater. The main principle of cathodic protection is that when a metal receives electrons, it becomes a cathode. When a metal becomes a cathode, it can no longer readily ionize (corrode). Cathodic protection can be used with most metals. When used in conjunction with a proper paint system, it protects at holes in the paint. The most common use of cathodic protection is the installation of anodes at the stern of ships. These anodes reduce or eliminate the galvanic effect of the propeller.

Table 5-6 Paints and coatings

1. Oil paints–based on natural oils from plants and fish:
   a. Easy to apply.
   b. Relatively inexpensive.
   c. May require longer drying times.
   d. Permeable and recommended only for mild atmosphere.
2. Alkyd paints–resins obtained from the reaction of glycerin and phthalic anhydride:
   a. Have to be baked to dry (unless combined with oil paints).
   b. More corrosion-resistant than oil paints, but still not suitable for chemical service.
3. Emulsion or water-base paints–resin in a water vehicle:
   a. Little odor.
   b. Easy to apply.
   c. Easy to clean up.
4. Urethane paints–reaction of isocyanates with polyols:
   a. Good toughness and abrasion resistance.
   b. Corrosion resistance may approach that of vinyls and epoxies.
5. Chlorinated rubber–natural rubber chlorinated:
   a. Does not wet well.
   b. Dries quickly.
   c. Resistant to water and many inorganics.
   d. Temperature maximum–150°F.
   e. May be painted for better protection.
6. Vinyl paints–polymerization of compounds containing vinyl groups:
   a. More corrosion-resistant than oil or alkyd-based paints.
   b. Resistant to a variety of aqueous acids and alkaline media.
   c. Temperature maximum–150°F.
   d. Adherence and wetting can be poor.
   e. Adherence for the first 24 h or so is suspect.
7. Epoxy paints–reaction of polyphenols with epichlorohydrin:
   a. Amine-hardened epoxy coatings (hardness and resin–most resistant to chemicals).
   b. Polyamide-hardened epoxy is less resistant to acids but is tougher and more moistureproof.
   c. Epoxy-ester less corrosion-resistant but easier to apply.
   d. Coal tar-epoxy has good resistance to water, soil, and inorganic acids.
8. Silicone paints–high-temperature service (with modification up to 1200°F):
   a. Not very good against chemicals.
   b. Are water-repellent.
9. Coal tars–applied hot; used especially in underground applications
10. Zinc paint–metallic zinc dust in an organic or inorganic vehicle:
    a. Used in galvanic protection by the zinc to prevent pitting at holes in the coating.
    b. Effective in neutral and slightly alkaline solutions.
    c. Organic zinc paint requires less surface preparation and is easier to topcoat than inorganic counterpart is.
    d. Inorganic is more heat-resistant, however.

5-4.5 Inhibitors

Another method of corrosion control is the use of chemical inhibitors. Inhibitors are usually used only in closed systems, such as diesel engines, boilers, or tanks. If a closed system is involved, the use of inhibitors can be a very economical approach. There are five classes of inhibitors:

1. Absorption (affects anodic and cathodic reactions)
2. Hydrogen evolution poisons (affect only the hydrogen evolution reaction)
3. Scavengers (usually removes O_2 needed for the cathodic reaction)
4. Oxidizers (work with certain types of metals such as iron. If not used in sufficient quantity, they can also cause metal pitting; therefore "a dangerous type" of inhibitor)
5. Vapor phase inhibitor

Except with the last type of inhibitor, the metal being protected is usually in a solution. Some protection will usually still exist on a metal after it is removed from an inhibited solution. The amount of residual protection will vary greatly. The vapor phase inhibitors work only in confined spaces, e.g., sealed boxes or vessels (not too large). In conjunction with scavenger-type inhibition, the removal of water vapor can prove very effective. This method is used extensively to preserve naval vessels, for extended periods of time at relatively low cost, i.e., mothballing. Corrosion of most clean metal surfaces will almost cease below 50 percent relative humidity. It is very important that the surface be clean, as contaminants will greatly reduce the effectiveness of this method.

5-4.6 Anodic passivation

A final method of corrosion control is anodic passivation. This type of protection occurs naturally on metals such as stainless steel, titanium, and aluminum. It can also be induced by an impressed anodic potential on iron and certain other metals. Anodic protection is usually avoided in the marine environment, as the chloride ion adversely affects it. Chloride breaks down the passive film needed for this type protection and can cause severe pitting of passive metals such as stainless steel.

5-5 CONCLUSION

The role of corrosion in ocean engineering is usually negative. It is therefore important that we eliminate it wherever possible. This can best be accomplished by having a sound basic understanding of the mechanisms of corrosion and by applying the proper methods of control. The type of corrosion control chosen usually depends on economic considerations, and in some cases corrosion elimination may be considered not worth the costs involved. Usually the only reason that corrosion control outweighs economic considerations is for safety purposes. There can be no question of whether or not to use good corrosion practice if stress corrosion cracking of a vital part is possible.

It is hoped that in the future ocean engineers will be able, by understanding this chapter, to avoid "foolish corrosion." Properly designed corrosion control systems, however, should be developed under the advice of a corrosion specialist during the initial design stages, rather than by the design engineer himself.

PROBLEMS

5-1. You have a galvanic couple composed of copper and steel. In the solution in question the copper electrode has a potential of -0.25 V, and the iron has a potential of -0.67 V. Calculate the free energy of the galvanic couple and predict which metal will corrode faster. Show the electrochemical formula for the reactions which will take place.

5-2. There has been a catastrophic failure caused by a leak in a forward cargo tank. You are the consulting engineer called in to analyze the fracture for signs of corrosion. Describe how you would go about your investigation, and according to your knowledge of corrosion types, what forms of corrosion you would expect.

5-3. Using the Nernst equation:

$$E = E_0 + \frac{0.059}{2} \log \frac{M^{Z+}}{M} \qquad E_0 = +0.34 \text{ V} \qquad Z = 2$$

what is a copper electrode's potential in a 10^{-7} g-ion/l copper sulfate solution? What is the potential in a 10^{-5} g-ion/l solution?

5-4. Using the information from Prob. 5-3, show on the polarization curve given what effect this concentration change would have on a corrosion cell between this copper electrode and a zinc-zinc sulfate half-cell: Assume the zinc cell conditions remain constant, and assume the copper's cathodic Tafel slope remains constant.

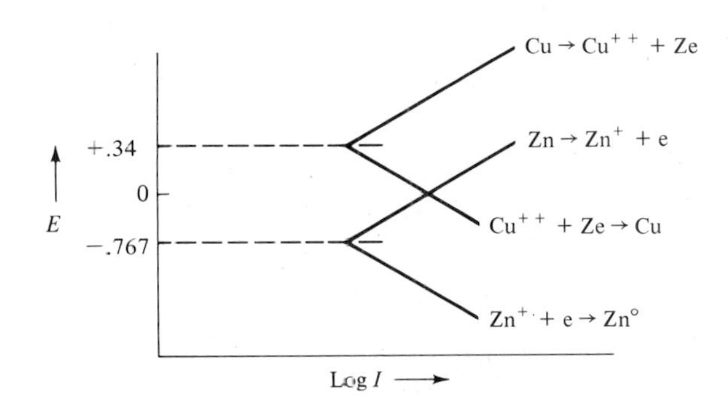

Fig. P5-4

5-5. Metal A and metal B are resting in an acid solution. Using a standard hydrogen electrode, the following data were obtained:

	Anodic curve		*Reduction curve*	
	Potential, mV	*Current, mA*	*Potential, mV*	*Current, mA*
Metal *A*	3	1.5	3	17.5
	4	2.0	4	7.3
	5	2.6	5	3.3
	6	3.4	6	1.5
Metal *B*	1	2.0	1	
	2	4.5	2	
	3	10.0	3	75.0
	4	22.5	4	34.0
	5	50.0	5	15.5
	6	100.0	6	7.0

Plot these values on semilog paper and determine the galvanic potential and current as well as the individual corrosion potentials and currents.

5-6. You are designing condenser tubes for a ship condenser. Using strength of materials theory, compute the wall thickness necessary to contain the steam at 15 psi differential pressure in 2-in-ID 70/30 Cu-Ni tubes and then calculate how thick your corrosion allowances would have to be if you wanted them to last 20 years with a flow of 7 ft/s.

5-7. What is the life expectancy of an iron pipe used in a reactor if its wall thickness is 0.50 in and 0.25 in is required solely for strength? The fluid velocity is 12 ft/s.

5-8. Explain what is meant by polarization of an electrode. Use diagrams to help explain if necessary.

5-9. If you had to rivet two pieces of metal together, would it be better to have the rivets anodic or cathodic to the plate? Why?

5-10. Paints are good for preventing corrosion, but if you had to choose between painting only the anode or only the cathode, which should you paint? *Hint*: What happens if the paint is porous or you chip it a little?

5-11. Six new standard scuba bottles (steel) were tested by the University of Rhode Island Scuba Safety Project. Conditions were as follows:

No. 1: 2,200 psi air, 500 ml salt water, bottle horizontal
No. 2: Replication of (same as) no. 1
No. 3: 2,200 psi air, 500 ml salt water, bottle vertical
No. 4: 100 psi air, 500 ml salt water, bottle horizontal
No. 5: 2,200 psi air, 500 ml freshwater, bottle vertical
No. 6: 100 psi air, 500 ml freshwater, bottle vertical

The results of a 100-day life test at 105°F were as follows:

No. 1: Drastic corrosion over one-half the wall thickness
No. 2: Very serious corrosion but not as bad as no. 1
No. 3: Substantial corrosion but bottle still usable
No. 4: Minor corrosion of a pitting nature
No. 5: Slight corrosion with a fragile corrosion ring of products
No. 6: No sign of corrosion

Do these results seem in keeping with the materials and discussions of this chapter? Discuss and make reference to appropriate sections. Note that 500 ml of water forms a pool in the bottom of a vertical cylinder but spreads out into a thin sheet when the bottle lies on its side.

REFERENCES

Fontana, M., and N. Greene (1967): "Corrosion Engineering," McGraw-Hill, New York.

Gall, H. J. (1970): "Corrosion Control for Manned Space Flight Network Facilities," NASA, Washington.

Henthorne, M. (1971): Electrochemical Corrosion, *Chem. Eng.*, June.

—— (1971): Polarization Data Yield Corrosion Rate, *Chem. Eng.*, July.

—— (1972): Materials Selection for Corrosion Control, *Chem. Eng.*, March.

International Nickel Co. (1966): "Guidelines for Selection of Marine Materials," New York.

Munse, W. H. (1964): "Fatigue of Welded Steel Structures," Welding Research Council, New York.

Uhlig, H. H. (1948): "The Corrosion Handbook," Wiley, New York.

6

Selection of Materials for Ocean Application

Herman Sheets

6-1 EFFECTS OF THE OCEAN

The selection of ocean engineering materials is based upon the effects of the ocean environment and the contemplated use of the materials. In the past, the major applications for materials in the ocean have been with surface ships, for which considerable experience has been accumulated. Recently, new ocean systems have been developed, each requiring specialized material characteristics. These applications relate to offshore drilling and production platforms, surface buoys, instrument platforms, subsurface habitats, submarine vehicles, and ocean instrumentation and research tools. Some of these units are expected to stay at location in the ocean for extended periods of time. The structures of these systems are exposed to wind, waves, and ocean currents together with some thermal gradients and ice. In addition, the ocean environment exerts on the materials chemical, fatigue, stress, and corrosion effects in combination with exposure to barnacles and other living organisms. The materials in the ocean are also exposed to scouring, earthquakes, hurricanes, typhoons, and similar events. As a result, materials are subject to different types of loads and require specific properties to meet their requirements. The materials must have properties which ensure survivability of the ocean structure in case of collisions and accidents during docking and transport, as well as excessive loads during hurricanes and similar events.

Among the applications of material, the most demanding will be for large ocean structures. These structures will have widely varying material requirements depending on the use, intent, design, and application. Among other important factors for ocean materials is a consideration that underwater structures must withstand the hydrostatic pressure which is a function of the depth of submersion and can reach pressures well in excess of 10,000 psi. Buoyant structures require a relatively lighter-weight design, and their material must withstand both the cyclic and dynamic loads imposed by waves, including the special high-impact loads caused by hurricanes, typhoons, and tsunamis.

6-1.1 Surface vessels, platforms, and underwater structures

There is a close relation between the selection of materials and the structures in which they are used. The use of materials in surface vessels is covered in the specifications of the regulatory agencies such as the American Bureau of Shipping (ABS) (1970) and the United States Coast Guard (1968, 1969). While this may limit the selection of materials, it has the advantage that the experience of many users and the statistical result of many load factors due to wind and sea conditions is included in these regulations so that their use will be desirable even in those cases where it is not a requirement. In general, surface ships which have the needed mobility and access to weather forecasting can, on most routes, avoid sea state 6 (Sheets and Boatwright 1970) and higher seas and therefore can be designed for a lower range of loads resulting from the forces and movements associated with the sea.

Platforms and buoys are generally stationary. Therefore, they must be designed for the highest sea state to be expected in the locality of their use. Offshore drilling platforms, many of which are planned for worldwide use, must be designed for maximum sea states.

In the selection of materials, consideration must be given not only to the physical and chemical properties of materials but also to costs, fabrication facilities, and expected maintenance. Consequently, several materials may compete with each other, for instance, forgings, castings and welded parts, steel and nonferrous metals as well as concrete and other nonmetallic materials. The materials presently in use for major ocean engineering components are discussed in subsequent sections.

Underwater vessels are exposed to external pressures and their materials are subjected to compression, bending, and buckling. The load depends on the maximum submergence pressure, and extensive technology has been developed in this field (Krenzke et al. 1965).

6-1.2 Consideration of stress level and fabrication

The fabrication needs for steels and other materials cannot be separated from considerations of the complexities and size of the structure, stress levels, and allowable cost. Therefore, components designed by sophisticated computer and experimental analysis, and for which high fabrication costs are acceptable, cannot be compared on an equal basis with those ocean structures for which simple analysis and fabrication techniques are mandatory because of cost and manufacturing requirements.

Nevertheless, it must be recognized that there is a tendency to use materials of increasing yield strength and operating stress levels as new materials become available. However, for these new materials there must be a corresponding increase in the quality of fabrication practices. The newer high-strength materials show an increase in propensities for cracking, fissuring, and other forms of defects in both the parent material and the weld areas. Such tendencies are penalties stemming from the metallurgical requirements for developing increased strength.

6-1.3 Material technology

Materials used in the ocean must avoid catastrophic failures. Such material requirements are particularly important with regard to avoiding pollution, for which more stringent environmental requirements will be established as time goes on. Therefore, oil drilling or oil transport systems will require materials which must not only meet the usual design requirements, but also withstand such hazards as operator error.

Of the various material characteristics, those relating to yielding and fracture processes are of substantial interest. Corrosion, which has become a specialty in itself, is discussed in the previous chapter. In addition, fatigue and frequency of loading have become of ever-increasing importance as structures in the ocean are being kept on station for longer and longer periods of time. It must also be recognized that the speed of loading owing to wave slap and similar events is considerably higher than "dry" loading speeds. Therefore, material characteristics affected by speed of loading require special attention.

6-2 OCEAN LOADING CONSIDERATIONS

The behavior of ocean materials is closely associated with the configuration of the structure and the associated loading. Many ocean structures, including ships, drilling platforms, and buoys consist of a combination of beams, plates, shells, and reinforcements. In some cases, the beams are constructed as framing systems being built up from simple shapes by welding or similar manufacturing methods to get the optimum configuration for the intended purpose. Steel is still the most commonly used material for many ocean structures, and welding is used extensively. This means that considerable residual stresses are locked in these structures unless special provision is made to completely anneal all welds. In each individual case, there is a maximum load (and stress) which must not be exceeded. In large ships and platforms, high-strength steels are used in a variety of applications to reduce the total weight and come up with an overall economic solution to the design problem. In addition, superstructures, deckhouses, and other appurtenances are frequently built of aluminum or similar lightweight materials in order to locate the overall center of gravity in a desired position.

6-2.1 Primary loads on ocean materials

In the ocean, most structures either float on the surface or are submerged. For floating structures and ships, the load consists generally of longitudinal bending of the main beam, the hull. Submerged structures have to withstand the submergence pressure, which is a direct function of the depth of the ocean. The literature (e.g., Sheets and Boatwright 1970, Myers et al. 1969) presents various means of calculating the needed stresses. In addition to bending and compression, buckling requires thorough analysis for the case of cylinders or shells under compression as well as for the various modes of bending. The above criteria are well described in the literature by Myers et al. (1969) and Comstock (1967).

In many parts of the ocean, there exists a substantial difference between water and air temperatures which results in thermal stresses on the ocean structure. Frequently, temperature changes result in elongations which are partially restrained by the surrounding structure. The thermal stress in the condition of partial restraint in a single direction can be expressed by

$$S_{th} = \mathcal{E}E - \alpha \Delta TE \qquad (6\text{-}1)$$

where S_{th} is the thermal stress, $\mathcal{E}$ is the unit elongation, E is the modulus of elasticity, α is the coefficient of thermal expansion, and ΔT is the change in temperature.

Whereas in small and elastic structures the effect of temperature gradients and thermal stresses is negligible, for large ocean structures having stiffness and restraint, thermal conditions must be carefully analyzed regarding their effect on stress loads and materials.

6-2.2 Environmental loads

The wind and waves of the ocean result in a continuous series of loads and stresses. These forces result in fatigue strength becoming of critical concern for many ocean engineering structures. The fact that in the ocean the loads are frequently imposed on the material in several axes requires a careful analysis of the material and its characteristics to withstand multiaxis loading, in tension, compression, and shear.

The wave forces are likely to excite vibration of the various members in the structure and will result in the accumulation of many cyclic loads within a reasonably short time. The extraordinarily high waves which occur in the ocean during hurricanes frequently result in fast loading cycles.

The occurrences of such high waves and the associated high winds which also cause unusually high loads require a statistical review regarding the number of occurrences of these high loads in order to select the best material for the purpose intended. Generally, wind velocity data and their changes with altitude are available over land areas and, to a more limited degree, over the oceans (Myers et al. 1969).

6-3 DESIGN CONSIDERATIONS

In selecting materials for ocean applications, conventional material characteristics are still important. Primary consideration is given to the yield strength and the ultimate strength of the material. Young's modulus is of considerable interest in analyzing the possible deflections and deformations of structures in the ocean. Poisson's ratio is needed in analyzing the material in cases of multiaxis loading. In addition, performance of the material in fatigue and its fracture resistance have become important material characteristics.

It must be noted that most of the above physical characteristics, as presented in the literature, are based on data taken from standard specimens. These specimens are relatively small and are clearly defined in the specifications

Fig. 6-1 The effect of plate thickness on T-1 steel properties.

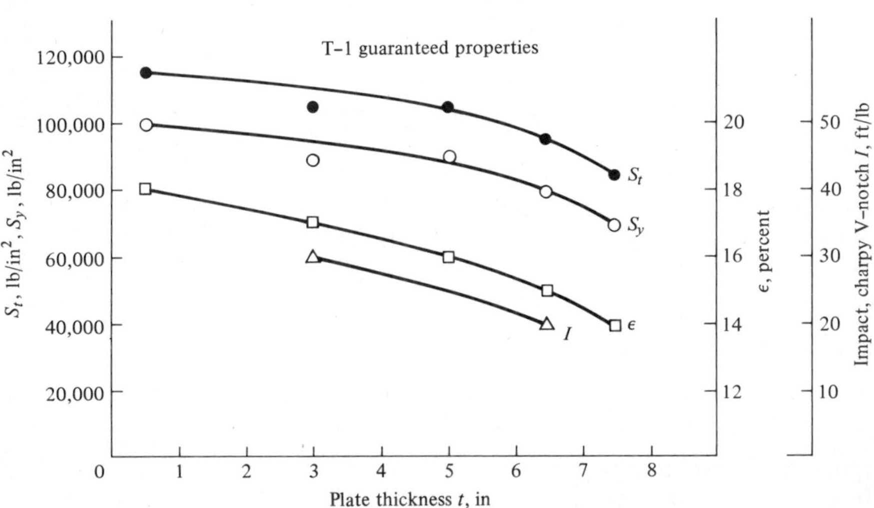

Table 6-1 Qualitative effects of environmental factors on fatigue strength*

	Increase in thickness	*Surface finish*	*Major stress zero to compression*	*Stress intensity*	*Notch effect*	*Total*
HTS	–	–	+	–	–	–––
4340	–	––	+	–	––	–––––
6061 AL	–	–	+	–	–	–––
Titanium	–	–	+	–	–	–––
Magnesium	–	–	+	–	0	––

*Interpretation of fatigue strength variation: (–) small reduction, (––) medium reduction, (+) small increase.

of the regulatory agencies. It is important to remember that metals show a reduction in many of their physical performance characteristics as a function of thickness, as indicated in Fig. 6-1, showing tensile strength S_t, yield strength S_y, impact Charpy V-notch strength I, and elongation & as functions of plate thickness t for T-1 steel.

In addition, surface finish can affect material properties, particularly fatigue characteristics. Forces and loads which result in multiaxis stresses generally reduce the loads which can be imposed on the material. On the other hand, if the stresses imposed by the loads are in compression only, some material characteristics improve. The effect of notches and sudden changes in cross section also affect material properties.

Most fatigue strength data refer to stress values taken on rotating shaft bending tests. Consequently, the stress level under this loading varies from a maximum tension value to an equal value in compression, and simultaneously the stress increases linearly from the center of the shaft to its maximum value on its outer diameter. For material applications in the ocean, this type of loading rarely exists and appropriate considerations of fatigue values under actual loading must be made. As a result of the combined effects described above, certain adjustments to the material characteristics must be made. If they cannot be calculated, a qualitative summary, suggested in Table 6-1, indicates basic changes in allowable stress levels for the various ocean conditions. In many cases, the designer accounts for these effects in terms of a material allowance, increasing the material thickness or selecting an appropriate safety factor. Some allowance for the above conditions is also made in the appropriate specifications of regulatory agencies.

The basic property to be evaluated for a material to serve in the ocean is, in many cases, *toughness*, not strength. This requires the material to have the ability to deform plastically in zones of high stress concentrations and around nonmetallic inclusions. The material must also withstand the loads in the presence of metallurgical notches, as created in welding zones, and in mechanical notches caused by changes in cross section or changes in stiffness. Generally, tests such as the Charpy V-notch impact test and the associated drop weight and explosion bulge test are an indication of material performances related to toughness. From these tests, the notched-to-unnotched tensile strength ratio can be determined, and this is generally used to evaluate the *crack propagation* characteristics of metallic materials. Crack propagation after initiation is another property of importance for materials used in the ocean.

Many material characteristics retain their values in the presence of impurities or crystal defects if the defects are few and constitute a small part of a generally regular material. In this case, the value of the material property is determined by the cumulative effect of all the atoms and molecules. The impurities and imperfections, forming the irregular part of the material, make a unit contribution to the material property of essentially the same value as that of the regular part. Their effect when summed over the entire material is small. Such properties are defined as *structure-insensitive properties*, and the modulus of elasticity E is, for many materials, a structure-insensitive property.

On the other hand, fracture strength, yield strength, plastic deformation, and fatigue strength do not possess the additive features of the constituent atoms. These material properties are very much dependent on the existence of microscopic and macroscopic cracks, dislocations, and voids, so that they strongly depend on anomalies or discontinuities; they are known as *structure-sensitive properties*. Cracks are usually initiated as irregularities at the site of a defect and upon attainment of a certain limited size will progress very rapidly. For those ocean components which are directly exposed to the loads of the waves of the sea or to the pressure of the ocean depths, it is necessary to select materials with properties giving favorably slow crack propagation to avoid catastrophic failures.

6-4 BASIC MATERIAL PROPERTIES

The most important criterion to be considered in selecting a material is its strength. The strength of a material is its capacity to resist the action of all applied forces. However, the strength of a material cannot be represented by a single number because ability to resist loads and stresses depends on the nature of these loads and on the environment in which the material is located. There are enough data available to enable the engineer to calculate with reasonable accuracy the strength of a material under its loading condition provided all environmental conditions are taken into consideration.

For all materials to be used in the ocean a number of mechanical properties are of prime interest. These properties are the tensile strength S_t, modulus of elasticity E, yield strength S_y, and elongation $\mathcal{E}$. These characteristics still form the basis for material selection. However, additional properties such as toughness and fatigue strength must be considered, depending on the application.

The above characteristics are determined from a simple tension test as shown in the stress-strain diagram of Fig. 6-2. A specimen is subjected to a progressively increasing tensile force until it fractures. The normal stress S is equal to the tensile load P divided by the specimen area A.

$$S = \frac{P}{A} \tag{6-2}$$

This value is plotted against the strain

$$\mathcal{E} = \frac{\Delta l}{l} \tag{6-3}$$

where Δl is the stretch and l is the initial length of the specimen.

There is often a linear relation between stress and strain, where the specimen returns to its original dimensions when the load is released. Once the curve departs from linearity, plastic deformation may begin and is usually accompanied by *work hardening*.

Curve A in Fig. 6-2 is typical for mild steel. It also shows the distinct dip indicating that the metal stretches rapidly when a certain stress is reached, even

Fig. 6-2 Three typical stress-strain curves.

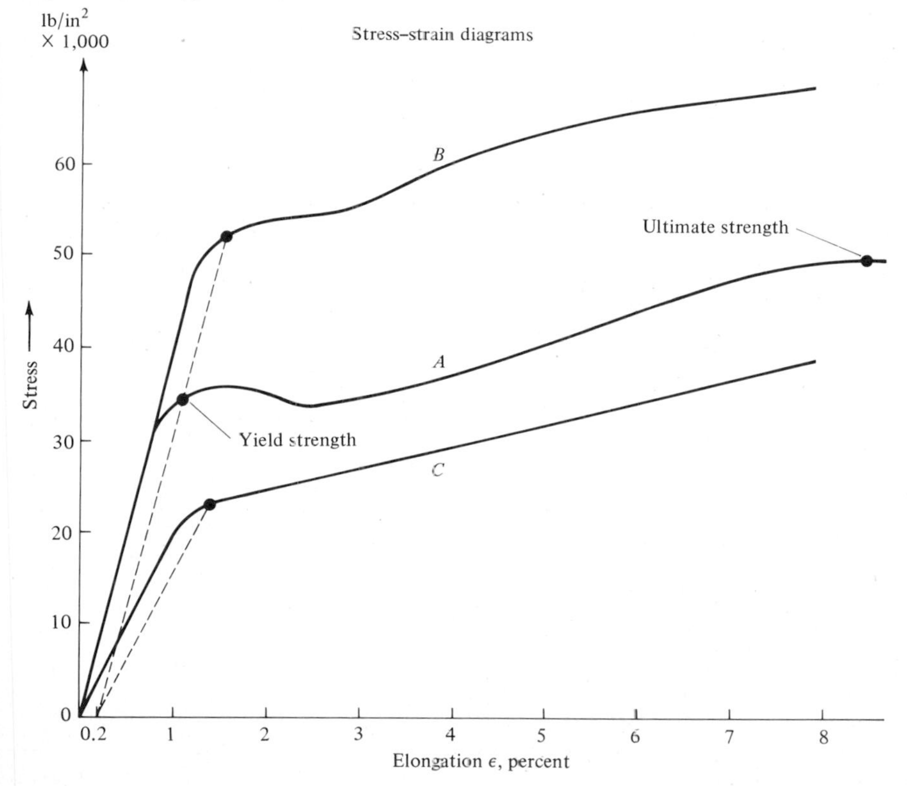

if the load is slightly reduced. Such a material shows a clearly defined yield point. High-strength steels and some nonferrous metals show a characteristic as indicated by curve *B*. In this case, there is no distinct dip in the curve, but there is still a definite change in character. Some aluminum and nickel alloys show a character as expressed in curve *C*. In this case, there is only a gradual curvature and no definite break. The yield strength is generally taken at 0.2 percent offset elongation, unless specifically defined otherwise.

In ocean applications, steel, its various alloys, and some nonferrous metals are the predominant materials. They are all crystalline in nature, and if these materials are loaded within the elastic limit, a temporary deformation of the crystals takes place through displacements of the atoms. Upon removal of the load, the atoms return to their stable position and the crystals recover their original shape. The temporary deformation of the crystals results from an elastic displacement of the atoms in the structure, and the applied loads are not large enough to cause permanent shifting between the atoms.

However, if the material is stressed beyond the elastic limit, it will not return to its original form. It will be subjected to some change in configuration due to yielding or it will fracture. Deformation does not occur unless there are dislocations of certain atoms. Plastic deformation by slip occurs by relative movement of lamellae in the crystal, and the displacement takes place on specific planes which have the largest interplanar spacings by shearing of atomic bonds. The yielding does not occur at a significant rate until the distortion energy has reached a specific value. When the material is subjected to loads in the three major axes, the strength must be calculated according to the maximum distortion energy theory or a similar theory (Timoshenko and Young 1968). It must be remembered that the same material may fracture under a single-axis tension load in a typical ductile fracture pattern whereas it may fracture like a brittle material under multiaxis tension.

In metals, plastic deformation can be extensive, but it is very limited in nonmetals. The characteristics of metals change substantially while being deformed plastically, and there is some strain hardening which increases the strength and usually reduces elongation.

The modulus of elasticity E is another important material characteristic. It is a measurement of stiffness or rigidity of the material. Materials with high moduli deflect less under a given load, and they require a heavier load to produce a given deflection. In some applications greater stiffness may be more important than strength. Such applications exist for ships of relatively large length and narrow beam and also for applications where low-frequency vibrations must be avoided. The natural frequency of vibration is related to bending deflection. The modulus of elasticity has a value of about 30×10^6 psi for steel, 15×10^6 psi for titanium, and 10×10^6 psi for aluminum.

Ductility is another important property of a material. It allows the material to be deformed without rupture. It is generally measured as elongation or as reduction of area in a tensile test. The lack of ductility is commonly called *brittleness*. The areas under the curves in Fig. 6-2 are indications of the energy required to deform the material and, therefore, are one indication of toughness of the material. Therefore, materials of large amounts of elongation and ductility are superior in toughness to brittle materials having low values of elongation.

6-5 MATERIAL FAILURE AND FRACTURE

Fracture in materials is of considerable importance for ocean applications. Fractures in metals which are polycrystalline materials occur as either transgranular or intergranular types, depending on the path of the crack. In the *transgranular fracture* the crack traverses the grains in a line across crystals, whereas in the *intergranular fracture* the separation occurs along the grain boundaries. Intergranular cracks occur frequently in steel at normal temperatures under the simultaneous action of stress and corrosion due to seawater. This fracture is then called *stress corrosion cracking*.

The plastic behavior and the fracture strength of materials, or conglomerates of crystals, are dependent on the collective effects of many dislocations. The theory is presented by Yokobori (1965). Distinct effects are caused by the distribution of dislocations, the existence of solute atoms, and the mutual interaction of atom vacancies and cracks.

Plastic deformation usually starts imperceptibly, and if continued for extended periods of time, it can end with relative suddenness and result in a fracture. Generally, fractures are considered to be *ductile* or *brittle*, and this differentiation is based on either energy absorption, shape, or change of shape accompanying the fracture. The theoretical strength of an ideal crystal has been calculated on the basis of the energy required to form new surfaces at the break. It has been estimated to be up to about 1,000 times the strength which is normally observed. Imperfections and cracks are the causes of this discrepancy. The cracks usually start as microcracks occurring as a result of either solidification or mechanical processing. In addition, an originally sound material can develop cracks on a microscopic scale, and when a high enough stress is supplied, these sources may start a succession of dislocations on favorably located slip planes. It is noted that a grain boundary can intercede with these dislocations and act as a barrier. This will result in succeeding dislocations accumulating against the first one. The resulting accumulation of dislocations creates a stress concentration which eventually will lead to the formation of a microscopic crack. Consequently, it is possible that as a result of extensive dislocations, separation of the material will occur by a slipping apart of the two halves across the plane. In the latter case, the result is a *brittle fracture*, since the two halves are essentially undeformed and the energy absorption is relatively small.

For the case of the *ductile fracture*, it appears that shear strain has a major role, although the physical phenomenon is less well understood. At present it is believed that shear strain alone can cause fracture in compression and torsion.

On the other hand, in a simple tensile test it is impossible to tell whether shear alone is sufficient or whether there is a simultaneous effect of tensile and shear stresses.

It is noted that fracture in metals does not occur instantaneously but usually takes place through the formation of a small crack. This small crack then propagates and grows across the section as time goes on.

Extensive failures occurred in merchant ships which were built in the United States during World War II. About 1,000 ships, or 20 percent of the wartime total, have experienced more than 1,300 failures before the ships were 3 years old. A complete analysis of the causes for these failures was made and the data have been published (Comstock 1967, Parker 1957). These failures did not result from external causes or war damage but must be attributed to fractures caused by the brittle behavior of the steel under conditions of low temperatures and in the presence of geometric notches or welding defects. As a result of these investigations, recommendations were made to reduce all kinds of geometric or metallurgical notches and to improve the toughness of the steel. These recommendations have been included in the specifications of the regulatory agencies.

However, careful design and material selection are still required, particularly for those ships and ocean structures which operate in cold areas where freezing temperatures exist or where there is a possibility of stress concentrations and cold temperatures.

Theories of fracture have been developed assuming that the entire energy of separation is available for the establishment of two new surfaces. If the surface energy per unit area R is the only energy to be expended in creating the new surfaces, then, with the lattice parameter of the crystal a, the theoretical strength S_f is

$$S_f = \sqrt{\frac{ER}{a}} \tag{6-4}$$

The real material strength is considerably less than the theoretical strength. A theory based on an array of flaws in the material calculates the strain energy per unit plate thickness P which results from a crack length ($l = 2c$) in a thin plate under normal stress:

$$P = \frac{\pi c^2 S^2}{E} \tag{6-5}$$

where S is the maximum tensile strength. The surface energy at the crack is:

$$F = 4cR \tag{6-6}$$

For equilibrium the crack size can be calculated for the case of no change in potential energy:

$$\frac{d}{dc}\ 4cR - \frac{\pi c^2 S^2}{E} = 0 \tag{6-7}$$

So,

$$c = \frac{2RE}{\pi S^2} \tag{6-8}$$

The fracture stress for the crack of the size $2c$ is

$$S_C = \sqrt{\frac{2RE}{\pi c}} \tag{6-9}$$

This equation indicates that the stress to propagate the crack decreases as the crack c increases.

6-6 TOUGHNESS AND IMPACT TESTING

Toughness is generally described as the ability to absorb energy before fracture. As such it is represented by the area under the stress-strain curve and consequently involves both strength and ductility. It involves the work expended in deforming the material until fracture.

Toughness is also related to the impact strength although there is no definitive relation between the energy values obtained from static and impact tests. In the impact test there is a multiaxis stress condition in the critical area whereas the static tension test has only single-axis stress conditions.

Brittle materials have low toughness as they have only relatively small values of plastic deformation before fracture. The use of brittle materials in ocean applications should be avoided because fracture can occur suddenly and without any observable deformation.

It must be noted, however, that the same material may behave either as a brittle material or as a ductile material, depending to some extent on the external conditions of loading. Low-carbon steel can show large plastic deformation and elongation in the static tensile test, but it may fracture like a brittle material when tested at lower temperatures or in multiaxis tension or a combination of both.

The factors which influence the fracture and behavior of metals can be divided into two groups. One consists of metallurgical factors such as alloy content, grain size, and prior strain. The other group of factors is associated with operating conditions such as state of stress, strain rate, and temperature. While the first group must be dealt with in the chemical specifications and processing of the material, the second group is controlled by the operating conditions and design of the ocean structure.

A ductile metal tends to behave in a brittle manner when there are predominantly tensile stresses in several axes. Under these conditions, the equivalent yielding strength is increased. However, the fracture strength is not changed. It is also known that a decrease in temperature will generally result in an increase in yield and ultimate strength with a simultaneous decrease in ductility approaching brittle behavior. Some materials show a drastic reduction in elongation if the temperature is reduced below a limiting value. In addition, it is known that for certain materials, high strain rate causes an increase in both yield and ultimate strength, and simultaneously it reduces elongation, resulting in brittle fracture as there is not enough time for plastic flow to occur. Thus, a change from ductile to brittle behavior may occur when operating conditions cause two or three of the above factors to be present.

In order to analyze the combined effect of these factors, impact testing by means of a notched bar is being used. By using a notched bar in this test, there exist predominantly tensile stresses in the several axes near the notch. In the typical Charpy impact test, a bar specimen with a V-notch on one side is supported at each end and is broken by a single blow from a pendulum which falls from a specified height. The tangential velocity of the pendulum shall be not less than 10 ft/s or more than 20 ft/s. The configuration of the V-notch is specified.

After breaking the specimen, the pendulum continues forward until it stops again at a reduced height. The loss in potential energy of the pendulum, corrected for air resistance and friction, represents the fracture energy absorbed by the material. When the specimen is broken, there exists a high stress concentration at the root of the notch. The specimen also has a lateral contraction and, although less highly stressed in the other two directions, tensile stress will be induced near the root. The response of a ductile material to the presence of a notch is generally called *notch sensitivity*. It has been found to be a function of the temperature, strain rate, and the notch configuration. During the test, the notch sensitivity is determined by the energy which is required to fracture the bar. It is known that the fracture energy will decrease with an increase in strain rate and in sharpness of the notch and will decrease with temperature. If such a decrease in fracture energy is small and gradual, then the material has a *low* notch sensitivity. However, when a small change in one of the variables results in a large decrease in fracture energy, then the material is defined as having a *high* notch sensitivity.

The American Society of Testing Materials has standardized the notch bar impact test. The most commonly used specimen is the Charpy notched bar, but many others are in use (Masubuchi 1967). One procedure in evaluating notch sensitivity is to use a notched bar of predetermined configuration, changing a single variable such as temperature. Figure 6-3 shows the Charpy impact resistance for two different types of steel and a nickel alloy. This figure indicates that for low-carbon steel there is a large decrease in fracture energy in a relatively narrow temperature range. This range is generally called the *transition temperature*. Frequently, the transition from ductile to brittle behavior occurs

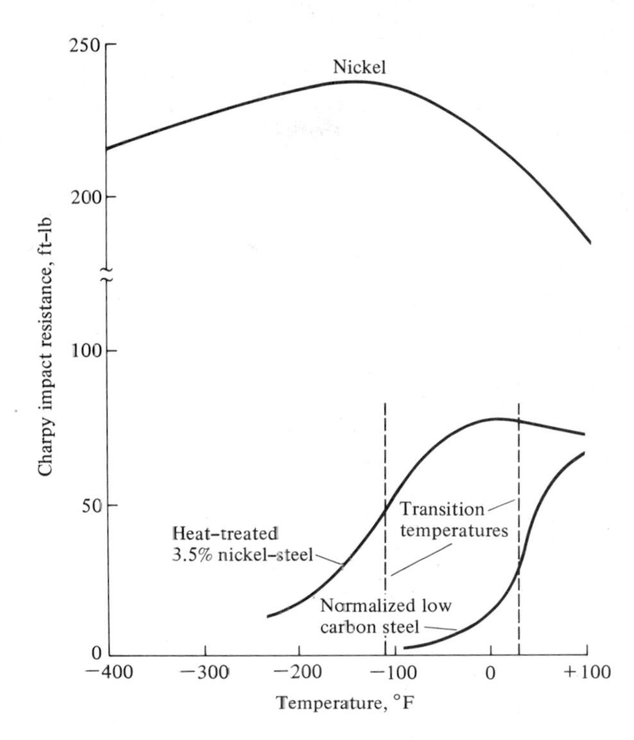

Fig. 6-3 Effect of low temperature on the impact resistance of a face-centered cubic metal (nickel) and two essentially body-centered cubic alloys.

within the range of this transition temperature. In the figure, the low-carbon steel has a higher notch sensitivity than the alloy steel, which has a less steep curve within the lower temperature range. The nickel alloy shows a low notch sensitivity over the entire range of temperatures. This is generally ascribed to the face-centered cubic crystals of the nickel alloy. It is known that steels which have to operate at low temperatures need substantial amounts of nickel alloys to avoid brittle behavior. Beneficial characteristics are also obtained by means of heat treatments resulting in small grain size and complete deoxidation.

6-6.1 Speed of loading

The speed of loading is of particular interest in ocean applications becuase waves hitting ships, buoys, or platforms result in loads which are associated with high strain rates. Considerable data are now available indicating the performance of steels as a function of strain rates of the order of 10^{-6} to 10^{3} in/(in)(s). The static tests are essentially related to strain rates in the lower range of the noted values, and tensile testing machines can be used to obtain data up to 1 in/(in)(s). Above this value impact load tests are made to determine the physical characteristics of the material. Since plastic flow depends largely on the movement of dislocations within the crystals, the speed of loading and associated thermal energy affects change yield strength as well as strain hardening and elongation. When tests are conducted at high strain rates, there is less time for thermal energy to assist plastic flow with the result that high yield and ultimate strength are observed. It is noted that the behavior in compression is quite similar to that in tension.

Consequently, all metals exhibit an increase in both yield and ultimate strength with an increase in strain rate. However, ductile metals also show an increased tendency to fracture in a brittle manner despite the increased strength with high strain rates. The percentage of increase in strength properties is generally greater for metals of low static strength than for those of high static strength. It is noted that there is a small increase in the modulus of elasticity with an increase in strain rate. However, this change is considered negligible in most applications.

Figure 6-4 shows strength values as a function of strain rate on a logarithmic scale indicating substantial improvement in physical characteristics as a result of high strain rate.

In material application in the ocean, it is indeed fortunate that there are improvements in the physical characteristics as a result of high strain rates. Since the maximum load caused by hurricanes and other unusual wave action occurs statistically very seldom, many structures are equipped to withstand these unusual loads because of their improved physical characteristics at high strain rates. It must be remembered, however, that the material selected for these applications must have a satisfactorily high Charpy V-notch energy value within the desired temperature range so that no brittle failure occurs and the yield

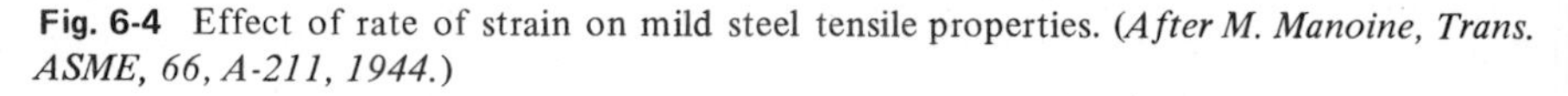

Fig. 6-4 Effect of rate of strain on mild steel tensile properties. (*After M. Manoine, Trans. ASME, 66, A-211, 1944.*)

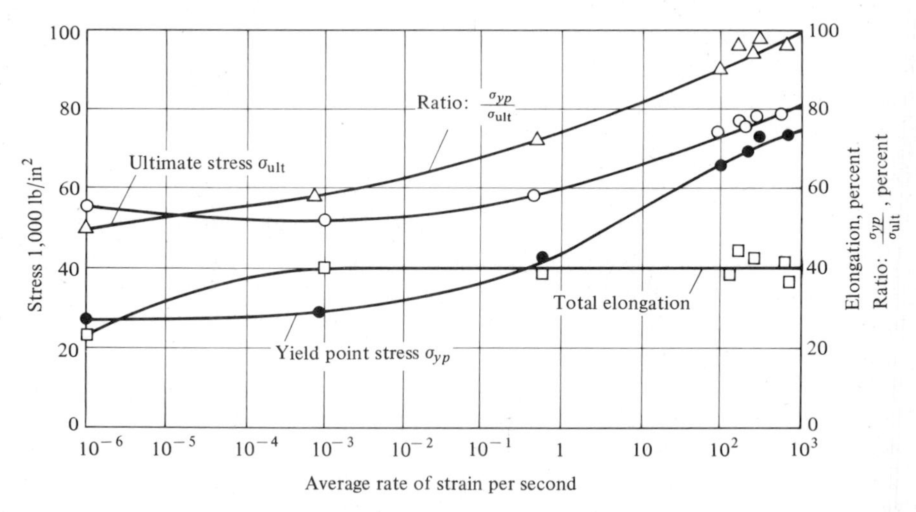

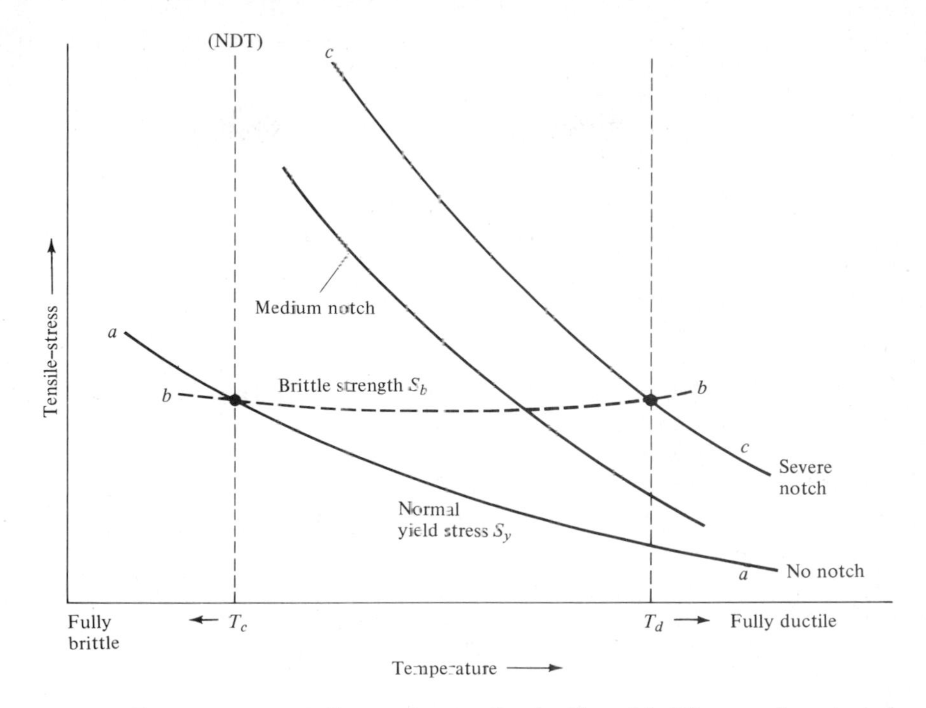

Fig. 6-5 Temperature-stress diagram showing the ductile and brittle zones for a typical steel.

strength rather than the brittle strength is the applicable strength for failure estimation.

6-6.2 Nil ductility temperature and crack propagation

The difference between brittle and ductile fractures can be explained by two kinds of physical characteristics. The brittle strength S_b, producing brittle fracture due to separation, is a material characteristic which is a function of temperature. In addition, the material also has a tensile yield stress S_y, which corresponds to the beginning of sliding. After sliding, if the material is loaded at increased values, failure will occur at the tensile strength S_t. In Fig. 6-5 the values of these two different strengths, S_b and S_y, are represented for a steel as a function of temperature (Timoshenko 1968). It is noted that the resistance to separation S_b changes very little with temperature whereas the yield stress S_y is influenced substantially by temperature, but in addition, its values also are a function of notch configuration and constraints indicating triaxial stresses. There is a point of intersection of these two curves defining the critical temperature, T_c, the *nil ductility temperature* (NDT). Above T_c the resistance to sliding is smaller than that to separation and the specimen will yield plastically, whereas for temperatures lower than T_c, S_b is smaller than S_y and the material will fail with a brittle fracture without plastic deformation. An increase in the speed of loading will increase the resistance toward sliding while the resistance to separation is not changed. As a result, the effect of higher speeds of loading is quite similar to that of a more severe notch and the curve for S_y is moved to the right. The same effect is achieved if the material is loaded by uniform tension in all three directions, resulting in a three-dimensional stress condition. Under these conditions, the critical temperature will move toward higher values when compared to simple tension.

The yield strength is increased by a reduction in temperature while the brittle strength increases only slightly with decreasing temperature. There is a change from ductile to brittle behavior in the material at the transition temperature. Below the temperature T_c the material is completely brittle, whereas above T_d it is fully ductile with or without the presence of a notch. Between these two temperatures, the material is in a notch ductile-brittle zone. The brittle stress curve S_b is above the normal yield stress curve S_y. However, it is also lower than the curve c for the constraint yield stress having the highest notch factor. Consequently, the type of failure depends on the temperature range and the severity of the notch and constraint. Basic information on crack propagation is shown in Fig. 6-6 presenting dynamic tear energy in foot-pounds

Fig. 6-6 Rough correlation between tear energy and yield strength for steel and titanium alloys.

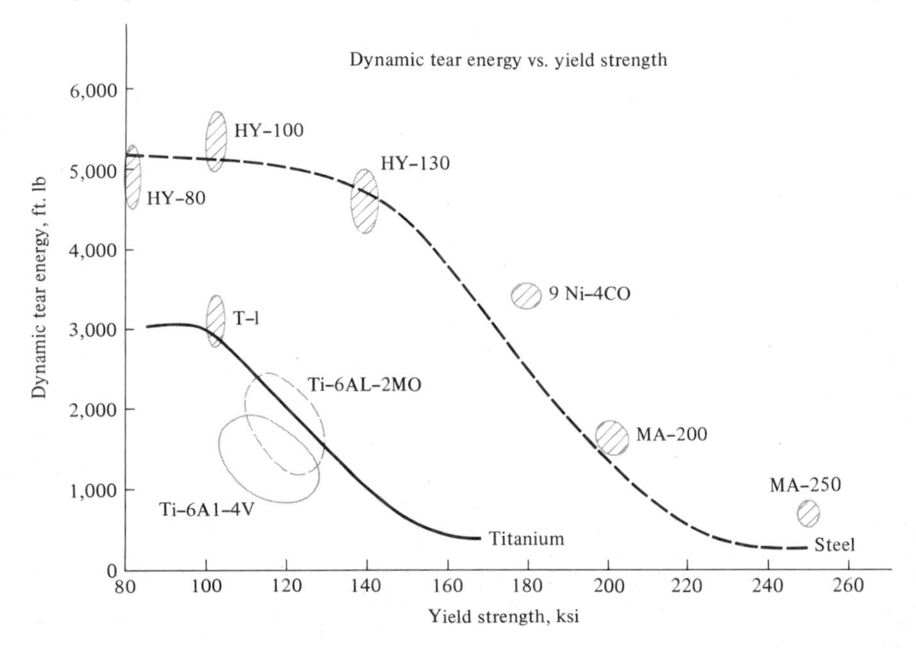

versus yield strength for a number of high-strength materials. Basically the higher-strength steels have a lower dynamic tear energy while materials like HY-80 have a higher tear energy.

6-7 FATIGUE

Owing to the oscillating loads caused by waves and the sea, fatigue is of substantial importance for materials used in the ocean. Fatigue failure occurs because repeated stress in a member exceeds the endurance strength of the material. Frequently, alternating loads are repeated a large number of times and failure can occur as a brittle fracture without any evidence of yielding. Considerable progress has been made in identifying the causes of fatigue failures.

The nature of fatigue is such that prediction of fatigue strength of materials is essentially based on experimental data and its statistical analysis. The fatigue process in ductile metals occurs in four stages. The first stage is strain hardening; second, crack formation; third, crack propagation. This finally results in fracture.

Fatigue cracks generally originate in the surface of a member, probably because surface grains are less restricted by surrounding grains than interior grains are. The joining of microscopic cracks into a visible crack and the propagation of this crack to fracture continues with an increasing number of cycles. The process is slow in the beginning but increases gradually as the size of the crack increases.

Fatigue strength is generally defined by an S-N diagram showing the relation between *fatigue strengths* S_f and the number of cycles N that a specimen can sustain before failure occurs. A considerable quantity of experimental data is available for the fatigue strength of various materials (Lipson and Juvinall 1963).

For a number of materials, the fatigue strength for a rotating beam bending load is presented in Fig. 6-7 as a function of the number of cycles. It is noted that the high-strength materials such as H-11 and 4340 show a considerably larger reduction in strength with increasing number of cycles than the lower-strength materials such as HTS and 6061 aluminum.

Some analytical theories governing fatigue calculations are discussed in Chap. 9.

6-8 LOW-STRENGTH STEELS

The environmental effect of the ocean on materials differs from the usual onshore environment essentially in two areas. In the oceans there is a relative constancy of temperature and composition of the environment. In general, the temperature ranges between the relatively narrow limits of 30°F and 100°F, and the salt content of the water is essentially in the range of 3.2 to 3.5 percent by

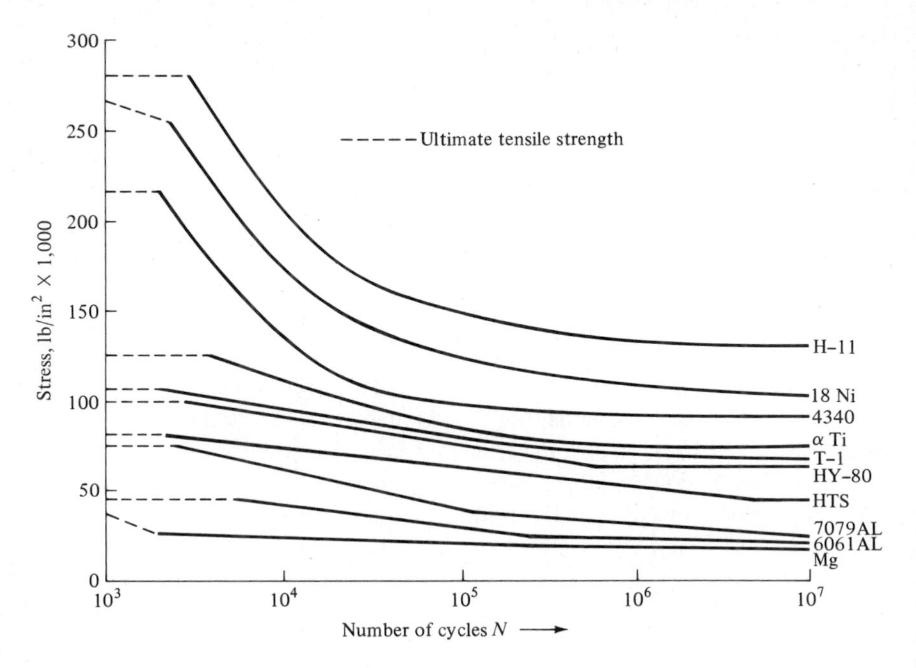

Fig. 6-7 Fatigue strength as a function of number of cycles for a rotating beam bending load.

weight. On the other hand, seawater is relatively corrosive, and the oceans impose a variety of static and dynamic loads of substantial force and magnitude because seawater is relatively incompressible and its density is almost 1,000 times as great as air. As a result, shock and other forces are transmitted through water over great distances. The most desirable characteristic for ocean engineering material is the ability to withstand high forces, fatigue, impact, and simultaneously have ease of fabrication and a competitive price. Steel fulfills many of these requirements and is at present used for the majority of ocean structures.

Low-carbon steel is the most widely used material for hull structures, fittings, tanks, instrument fittings, and buoys. Generally, steels having a yield strength of less than 60,000 psi are considered low-strength and are sometimes called *low-carbon steels.*

Table 6-2 shows the major characteristics of a selected group of low-strength steels. Merchant vessels constructed in the United States must meet the material requirements of the American Bureau of Shipping and the U.S. Coast Guard. In Table 6-2 steels 1 and 4 show ABS-specified steel characteristics. The ABS rules specify chemistry of the steel and process of manufacture.

Steels 2 and 3 in Table 6-2 are widely used for pressure vessels and similar applications. Many Navy and Coast Guard ships have been and still are being

Table 6-2 Low-strength steels

No.	Material and specification	S_y yield strength, psi × 10^3	S_t ultimate strength, psi × 10^3	Elongation, %	Heat treatment	Nil ductility temperature range, °F	Weld-ability	Cost,* cents/lb
1	ABS-class B	32	56	21	None	−20 to +40	Good	13
2	ASTM A-242	46	67	19	†	−20 to +40	Good	15
3	ASTM A-441	46	67	19	†	0 to +70	Good	15
4	ABS-class BH	47	71	19	†	−40 to +40	Good	15
5	MIL-S-16113C HTS	47	80	20	Normalized	−60 to +20	Good	21
6	ASTM-A-302	50	80	15	None	−20 to +50	Special	22
7	CB-C-Mn	60	80	19	Quenched and tempered	−75 to −40	Good	18

*Based on data as of August 1970: an average density for steel is 490 lb/ft³.

†Plates above ¾ in shall be normalized.

built to the specification of number 5. Steels 6 and 7 show further increased yield strength. The steels in Table 6-2, as a group, have improved values of nil ductility temperature (NDT) and notch toughness over ordinary low-carbon steels and simultaneously are easy to form and weld, and are readily available at a low price. These steels also show relatively good fatigue strength as indicated in Fig. 6-7.

6-9 MEDIUM-STRENGTH STEELS

For many applications, such as the icebreakers and buoys in the Arctic, steels of higher strength and simultaneously higher notch toughness are required. This has led to the development of the medium-strength, quenched, and tempered steels, which are described in Table 6-3. Medium-strength steels generally have a yield strength less than 150,000 psi. The quenching prevents the transformation of high-temperature austenite phase into undesirable crystallographic constituents which normally occur with slow cooling. The tempering at a predetermined temperature and duration accounts for the increase in ductility and toughness. The process of quenching and tempering also results in the fine grain which accounts for a substantial part of the improved values in NDT and Charpy V-notch impact values. Some steels such as numbers 1 and 3 in Table 6-3 have almost identical chemistry, and the difference in characteristics is the result of a different tempering temperature and duration. The quenched and tempered steels as a group are sensitive to plate thickness, which affects their physical characteristics. This is due to the larger heat content of the thicker plates during the quenching process, resulting in a slower crystal transformation. As a result, some of the quenched and tempered steels are divided into three thickness groups, each having a slightly different chemistry in order to achieve the desired physical characteristics independent of the thickness. For instance, HY-80 is divided into three plate thicknesses, up to 1¼ in, 1¼ to 3 in, and over 3 in. The increase in plate thickness permits increased values in carbon, nickel, chromium, and molybdenum in the ingot chemistry (Masubuchi 1970). The revision of HY-80 chemistry as a function of plate thickness results in a minimum yield strength of 80,000 psi independent of plate thickness. However, the Charpy V-notch energy requirements result in a reduction from 50 ft·lb at $-120°F$ to 30 ft·lb at $-120°F$ when the plate thickness is over 2 in.

The HY-130 steel, shown in Table 6-3 as number 5, is a recently developed alloy which has a yield strength of 140,000 psi in the plate material. It is designated as HY-130 as the weldments do not reach the 100 percent strength of the parent material. While this steel has substantially improved yield and tensile strength, compared to HY-80, its NDT value is at a higher temperature.

The stainless steel AISI-410 is shown for comparison as its yield, tensile strength, and elongation are similar in range to the rest of the medium-strength steels. However, its NDT value is not available but is expected to be higher than

Table 6-3 Medium-strength steels

No.	Material and specification	S_y yield strength, psi × 10^3	S_t ultimate strength, psi × 10^3	Elongation, %	Heat treatment	Nil ductility temperature range, °F	Weld-ability	Cost,* cents/lb
1	MIL-S-16216 HY-80	80	100	20	Quenched and tempered	−130	†	50
2	ASTM A 543	85	105	16	Quenched and tempered	−120	†	38
3	MIL-S-16216 HY-100	100	120	18	Quenched and tempered	−100	†	50
4	ASTM A 517-67 T-1	100	120	18	Quenched and tempered	−50	†	30
5	HY-130	140	148	16	Quenched and tempered	−100	†	100
6	AISI 410 tempered	100	140	22	Tempered	55‡	Fair	40

*Data are as of August 1970.
†Preheat required.
‡Izod test impact strength in feet per pound.

Table 6-4 Cryogenic steels

No.	Material and specification	S_y yield strength, psi × 10^3	S_t ultimate strength, psi × 10^3	Testing temperature T °F	Elongation, %	Energy Charpy V-notch, ft · lb	Cost,* cents/lb
1	A-553, 9% Ni	108	115	−300	..	....	43
2	A-353, 9% Ni, Grade A	108		−300	..	....	44
3	A 645, 5% Ni	72	100	RT†	29	....	35
		95	141 at −275°F	−275	30	55 at −275°F	
4	A353, 8% Ni, Grade B	95	140	−275	..	....	40
5	A 543	85	105	RT†	16	....	38
		108	140	−175			
6	A-517	100	120	RT†	18	....	330
	T-1	115 at −50°F	135 at −50°F	−50	..	15 at −50°F	
7	A 203	90	115	−80	..	....	22

*Data as of August 1970.

†Room temperature.

HY-80. On the other hand, owing to its simpler processing, the cost of this steel is lower than most of the group. Note that the T-1 steel, with a relatively high value of NDT, also has a comparatively low price.

In recent years, the marine transport of liquid natural gas (LNG) and related low-temperature fluids has become a business which is expected to increase substantially in the future. For marine cryogenic applications, a whole family of special steels has been developed. They are presented in Table 6-4. These steels as a group have above-average values for the Charpy V-notch test, as compared with ordinary steels. The highest value at the lowest temperature can be expected for steels 1 and 2, the low-temperature steels with 9 percent nickel content. Owing to their physical characteristics and their relatively lower price, the steels in Table 6-4 may find in the future increased applications for ocean structures.

6-10 HIGH-STRENGTH STEELS

During the last decade entire families of new steel alloys have been developed for a variety of applications. There is a tendency to produce alloys of ever-increasing strength which simultaneously have also improved characteristics in fatigue and energy absorption characteristics. These improved characteristics are being achieved by reducing impurity limits such as phosphor, sulfur, oxygen, nitrogen, and antimony. Some of these processing techniques resulted in special methods such as vacuum degassing and other special treatments such as electric furnace melting. As a result, there exist today families of *maraging* steels having yield strength of 150,000 to 300,000 lb/in^2 which are relatively ductile and can be heat-treated to a high notch toughness. It can be expected that in the next 10 years the strength of some of the special steels will rise to above 300,000 psi with simultaneous high yield and much toughness. This will permit the design of pressure vessels and other underwater structures of substantially increased performance as compared to the practices and designs which existed in the past.

Table 6-5 lists the two older high-strength steels, 4340 and H-11, with the new maraging steels. The big difference is the Charpy V-notch energy values at low temperatures for the new maraging steels. This improved alloy characteristic is achieved with the new processing techniques and at a substantial increase in cost. Nevertheless, these characteristics give the engineer the tools for improved ocean structures. It can be expected that for these new high-strength steels improved impact characteristics will be developed by means of new processing techniques.

Figure 6-8 shows how the use of steel has changed over the years from 1900 to the present for hydrospace applications. It is interesting to note that until the early 1960s, essentially only three types of steel were used, but at present a large variety of steels is available, each having certain advantages for specific applications. The tendency is to use steels of ever-increasing strength,

Table 6-5 High-strength steels

No.	Material and specification	S_y yield strength, psi × 10^3	S_t ultimate strength, psi × 10^3	Elongation, %	Heat treatment	Charpy V-notch impact, ft · lb	Weld-ability	Cost,* cents/lb
1	4340	220	270	12	Quenched and tempered	11†	‡	70
2	AISI H-11	240	295	12	Quenched and tempered	10 at −200°F	Fair	100
3	Maraging, 18 Ni −200	200	210	15	§	85 at 70°F 45 at −320°F	Good	220
4	Maraging, 18 Ni −250	250	260	12	§	22 at 70°F 14 at −320°F	Good	230
5	Maraging, 18 Ni −300	300	305	12	§	....	....	240
6	HP-9-4-25	220	250	12				

*Data as of August 1970.

†Izod impact strength in feet per pound.

‡Preheat required.

§Anneal and age-harden.

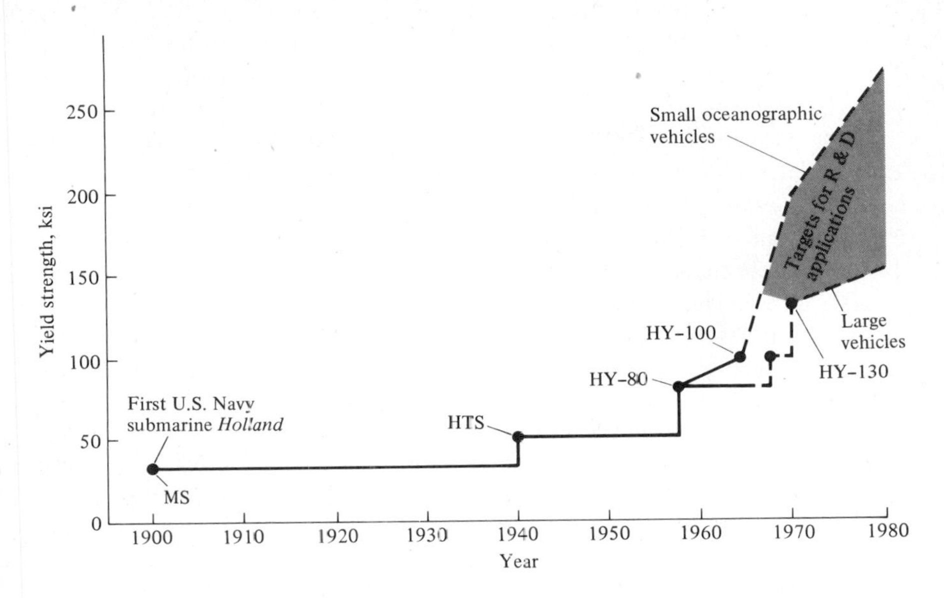

Fig. 6-8 Past and projected requirements for steel hulls for hydrospace.

although NDT and Charpy V-notch energy values are of equal importance. The steels shown in these tables are not a complete listing of the available ferrous materials; there are many more steels obtainable for ocean applications. Tables 6-2, 6-3, and 6-4 represent the basic steel alloys which are in use for ocean applications, and Table 6-5 lists the high-strength steels which will find ever-increasing use during the coming years.

6-11 ALUMINUM

Aluminum and its alloys have found greatly increased use in the ocean in recent years. Small and high-speed boats in particular are frequently constructed of aluminum using the 5456 and 5086 alloys. Hydrofoil craft, surface-effect ships, and the submarine *Aluminaut*, using the 7079 alloy, are examples of aluminum applications in the ocean. In addition, aluminum alloys have been used in merchant ships and naval vessels in superstructures, deckhouses, and the interior to a substantial extent. All-welded aluminum ships are now being built.

The growing use of aluminum can be ascribed to the following characteristics. An aluminum structure is often lighter for the same application than a similar steel structure, since the specific weight of aluminum is 1/2.5. In addition, the cost over the life of the structure is usually lower. The cost of aluminum is higher than steel by a factor of 1.9 when both metals are of the low-strength type for use in railings and other interior outfitting applications. However, if the cost of painting or other surface protection is added to the cost of steel, then the combined cost will exceed that of aluminum by a factor of approximately 1.5 over a 10-year period. Aluminum can perform without protective coatings and is known for its ability to be easily fabricated owing to its high values of ductility. It has long life, as aluminum develops a self-protecting film of oxide on its surface adhering tightly to the base metal.

Aluminum is tough and resilient, and it has satisfactory resistance to impact. It can well withstand the slam action of the waves, owing to the lower modulus of elasticity, of about 10,000,000 lb/in^2 compared to steel of about 30,000,000 psi. This results in increased elastic deflection, and correspondingly more work is absorbed by the material in the elastic range of deformation. For this reason, aluminum has been used quite successfully for small high-speed boats. Aluminum will withstand fairly well the effects of stress concentrations, such as changes in thickness, holes, and notches. In this context, it should be remembered that the value in foot-pounds of the Charpy V-notch impact test gives the energy absorption of the standard Charpy specimen. The ability of the material to withstand stress concentrations is a function of Charpy V-notch test value and the working stress or yield stress of the material, respectively. In other words, aluminum is usually used in larger thickness gauges when compared to steel and the thicker material will absorb more energy. Aluminum has also been used for pressure vessels in liquid natural gas (LNG) transport ships. In these applications, the low-temperature strength of aluminum at $-260°F$ and its relatively low weight make it a very desirable material. The pressure vessels which are mounted in the hull of the ship are usually spheres and can be of substantial dimensions, for instance, 120 ft in diameter with a wall thickness of 2 in. Insulation is provided so that the low temperatures do not reach the ship's hull. Aluminum is resistant to corrosion in air, but when immersed in seawater, aluminum alloys are subjected to corrosion by electrolysis owing to contact of dissimilar metals. However, some of the alloys resist the electrolytic action in seawater to a great extent, and they are in particular use for marine applications. Table 6-6 lists the aluminum alloys which are being used in the ocean. The first six alloys are used extensively in boat construction.

The mechanical properties of aluminum alloys are improved by essentially two processes—heat treatment and strain hardening with thermal stabilization. The heat treatment of aluminum alloys is designated as T and strain hardening as H, each followed by a number designating the detailed treatment. For instance, the T-6 heat treatment consists of solution heat treatment at 850°F with a soaking time being a function of thickness of the material and the chemical composition. The alloy is subsequently cold-water-quenched. The alloys which are strain-hardened are initially annealed at about 700°F and air-cooled before strain hardening and thermal stabilization.

Many aluminum alloys can be welded by the gas metal arc welding process, particularly aluminum alloys 5086, 5083, and 5456. However, careful attention

Table 6-6 Aluminum alloys

No.	*Material*	*Yield strength S_y,* ksi*	*Tensile strength S_t, ksi*	*Elongation, %*	*Charpy V-notch impact, ft · lb*	*Endurance strength S_f (ksi), 5×10^8 cycles*	*Welding*	*Metal treatment*
1	5083	21	42	22	20	19	Yes	
2	5083	33	46	16	15	23	Yes	H-113
3	5086	17	38	22	19	22	Yes	H-112
4	5086	37	47	10	14	16	Yes	H-34
5	5454	36	44	10	15	19	Yes	H-34
6	5456	37	51	16	9	23	Yes	H-321
7	6061	40	45	12	6	14	Yes	T-6
8	7079	68	78	14	3	23	No	T-6

*Minimum value at 0.2 percent permanent set.

must be given to the quality of the weld to avoid porosity and loss of strength in the heat-affected zone.

Note in Table 6-6 that the yield strength of the strain-hardened 5000 series alloys is in the 33- to 37-ksi range with the tensile strength being between 44 and 51 ksi. The 6061 and 7079 alloys are of a different constitution and are heat-treated; the 7079 is a distinctly high-strength material which has been used in the *Aluminaut.* The Charpy V-notch impact values for aluminum are generally much lower than those of the comparable steel alloys. However, since in many applications the aluminum structure contains material thicknesses of the order of two or more times that of steel, the absolute impact value is not too far below that of steel. As the strength level of aluminum alloys increases, the impact energy generally decreases, a characteristic similar to that of steel. However, compared to steel and titanium, the impact energy for aluminum alloys is low, and this must be considered for ocean structures.

Aluminum has a face-centered cubic atomic arrangement and remains ductile even at low temperatures. All the alloys in Table 6-6 are weldable except 7079.

The endurance strength of aluminum alloys is generally low when compared to the tensile strength values. The ratio of endurance strength to tensile strength is about 0.3 for the heat-treated high-strength alloys and is somewhat higher for the lower-strength alloys. The data are essentially based on 5×10^8 cycles for high-strength aluminum alloys. Aluminum alloys require a larger number of cycles to determine endurance strength as compared to steel. The lower-strength materials reach their endurance strength at a relatively lower number of cycles (10^7) whereas the high-strength materials may require 10^9 cycles and possibly more.

The fatigue strength of aluminum is also affected by the size of the specimen or the size of the structure under consideration. Greater thickness and larger size usually have reduced fatigue strength, but tests are required to determine quantitative effects. In regard to fatigue strength, it has been generally found that high-strength aluminum alloys are more sensitive to mean stress values than the lower-strength alloys are. It is also known that in any aluminum alloy, the stress concentrations caused by intermetallic inclusions and by chemical notches caused by corrosion will tend to induce local high stresses. In addition, high mean stresses can also be produced by external loading or by various manufacturing processes. In most of these cases the alloy with a greater ductility is at an advantage because it can minimize local stresses.

The mechanical properties of aluminum as well as its fatigue strength show a wider scatter of data for materials conforming to the same specifications. Consequently, careful analysis and selection of appropriate safety factors is necessary.

Aluminum is being used in ever-larger quantities because it results in a lighter structure and it is easy to fabricate. Since it can be used without protective coatings, it will last a long time, and the total expenditure is relatively low when taken over the life of the structure. In recent years, large quantities of aluminum propellers have been used on small boats. These are usually aluminum castings of moderate strength with paint or other protection. They are usually solution-treated and aged after casting to improve their physical characteristics. The alloys of the 300 type have excellent corrosion resistance. Aluminum propellers have been very successful and have proved to last a long time unless hit by a foreign object.

6-12 TITANIUM

Titanium alloys are used extensively in aircraft structures and in gas turbines as well as jet engines. Recently, titanium has been used in ocean application, particularly in small submersibles, owing to its favorable material characteristics. The submersibles *Alvin, Sea Cliff*, and *Turtle* are using buoyancy spheres fabricated from titanium. Presently, the main sphere of the submersible *Alvin* is to be changed to a titanium sphere.

Titanium has ralatively high strength and low weight resulting in a high strength-to-weight ratio. It has good corrosion resistance and a relatively high modulus of elasticity. It can be used for surfaces which cannot be painted, such as propellers, special valves, and piping. It has good cavitation resistance in both freshwater and seawater and is nonmagnetic. Presently, its price is still relatively high, but a lower price can be expected in the future. Therefore, many new uses for titanium can be expected in ocean applications during the next few years.

Test data indicate that pure titanium has a low value of strength and a relatively high value of ductility. However, small amounts of certain elements such as carbon, hydrogen, nitrogen, and oxygen increase the strength and decrease ductility. The addition of certain metal elements such as aluminum, tin, and vanadium has a similar effect.

Titanium consists of body-centered, cubic lattice crystals at about 1600°F; this crystal configuration is transformed near that temperature into hexagonal crystals which exist at room temperature. The high-temperature phase is called *beta titanium* and the low-temperature phase *alpha titanium.*

Various alloying elements in titanium can raise or lower the transformation temperature as well as increase or decrease the transformation from the beta phase to the alpha phase. Therefore, titanium alloys fall into three groups depending on the microstructure of the crystals, namely, alpha, alpha-beta, and beta. Depending on the type of crystal structure which exists at room temperature, the titanium alloy can be heat-treated. Elements such as iron, chromium, and vanadium have the ability to lower the transition temperature and are important in producing a stable beta phase. On the other hand, elements such as aluminum, oxygen, and tin raise the transformation temperature and result in titanium of the alpha phase. It should be noted that the alpha phase, which is transformed at a lower temperature, results in an alloy with characteristics different from that which is generated at high temperatures. In addition, alloys can be produced which both have crystal elements and are of the alpha-beta phase. If sufficient amounts of the beta stabilizing elements are in the titanium alloy, then the beta phase is stable at room temperature.

Single-phase alloys are usually weldable with good ductility, whereas two-phase alloys are more difficult to weld and the welds are less ductile. However, the two-phase alpha-beta alloys have unusually high strength and, therefore, are attractive for many applications. Two-phase alloys also can be heat-treated since the microstructure can be influenced by proper heat treatment involving quenching and aging.

Owing to the transformation of the crystal structure at a temperature of about 1600°F, the alpha alloys cannot be strengthened by heat treatment and their strength is achieved by cold working and keeping the alloy content within predetermined limits. On the other hand, the beta alloys and the alpha-beta alloys can be heat-treated and, within limits, strength and ductility can be varied, depending on the end use of the material. Titanium alloys also are sensitive to mechanical and metallurgical notches, and impact test values show a relation to ductility and strength similar to that in many high-strength steels. In general terms, the Charpy V-notch impact values show relatively high values for the alloys having a relatively lower yield strength and higher values of elongation, and the Charpy impact values show reduced impact strength for the titanium alloys having high yield strength and lower values of elongation. Titanium alloys, however, do not show a clear transition of fracture toughness with temperature as low alloy steels do. There is a relatively small change in fracture toughness over a considerable range of temperature. There does not exist a typical nil ductility transition temperature value. In titanium, as in many other metal alloys, fracture toughness is a function of impurities, particularly hydrogen, which can result in the precipitation of titanium hydrite in the alloy. At present, titanium alloys have a hydrogen level below 200 ppm. It may be possible to develop new titanium alloys having lower degrees of impurities and higher values of fracture toughness.

Titanium alloys, like many other metals, show a decrease in some of their physical characteristics with an increase in size or thickness. In the alpha alloys, this is caused by a smaller degree of cold working owing to the thickness of the plate. This results in a lower yield strength and tensile strength usually also associated with a lower amount of elongation. In the beta and alpha-beta alloys, the heat treatment becomes more difficult as the thickness of the plates increases because titanium and its alloys have a low thermal conductivity. The slower cooling usually results in a decrease in yield strength, tensile strength, and frequently in toughness, as the plate thickness increases.

Many titanium alloys can be welded with the inert-gas-shielded arc welding process. However, welding is usually more difficult than with either steel or the weldable aluminum alloys. The welding zone must be completely shielded with the inert gas, and temperatures must be carefully controlled in order to avoid porosity and contamination by such elements as oxygen, nitrogen, and hydrogen. Nevertheless, careful processing permits the welding of titanium alloys and results in good-quality joints.

Table 6-7 shows some of the titanium alloys presently in use. Alloys 1 and 2 are of the alpha crystal configuration, alloys 3 and 4 are of the beta configuration, and alloys 5, 6, and 7 are of the alpha-beta configuration. The first two alloys have definitive values of yield strength, tensile strength, and elongation, as they cannot be heat-treated. However, the characteristics of the rest of the alloys can be changed to some degree by a change in heat treatment. The values in Table 6-7 are generally those related to the highest values in yield strength and tensile strength which are available at this time. It is noted that for titanium alloys, the value for the modulus of elasticity does vary over a wider range, about ±10 percent, than that of steel and aluminum alloys. The data also indicate that titanium has relatively high Charpy V-notch impact test values even for the high-strength alloys.

Titanium alloys which have alpha crystal structure of a hexagonal type, when strain-hardened and stretched, can become to some degree unisotropic. This results in both yield strength, tensile strength, and elongation having lower values in the direction orthogonal to the work-hardening direction. Therefore, in the production of sheets and plates, cross rolling is required if uniform physical characteristics are desired. Some of the titanium alloys show a substantial increase in both yield and tensile strength at low temperatures. Alloy 5A is an example, with a yield strength and an ultimate strength of over 200 ksi at a temperature of −320°F. Therefore, this material is well suited for cryogenic applications.

The testing of titanium in fatigue has proved more difficult than might have been expected, even for a relatively new material. The specimens are sensitive to surface effects caused by notch sensitivity and certain difficulties in machining. It also appears that the fatigue strength is affected by the self-heating of specimens due to internal damping characteristics of titanium alloys and by the cyclical stress redistribution effects. Some of these phenomena are not fully understood at this time. Titanium, however, shows a relatively high fatigue strength, both in terms of absolute strength and in relation to the tensile strength. Since titanium has a lower density than steel, this gives titanium alloys an exceptionally high specific fatigue strength. There are many titanium alloys which have a fatigue strength greater than half the tensile strength.

At the present time, titanium alloys have a relatively high notch sensitivity, and therefore surface finish both of specimens for testing and of final

Table 6-7 Titanium alloys

No.	Alloy type	Yield strength S_y, ksi	Tensile strength S_t, ksi	Elongation, %	Modulus of elasticity impact, E $\times 10^6$ psi	Charpy V-notch, ft · lb	Fatigue strength ksi, 10^7 cycles	Crystal structure	Welding	Heat treatment
1	5Al-2.5Sn	117	125	18	16.0	19	93	Alpha	Yes	No
2	8Al-1Mo-1V	150	160	18	18.5	20	92	Alpha	Yes	No
3	1Al-8V-5Fe	221	215	10	16.5	...	60	Beta	No	Yes
4	3Al-13V-11Cr	175	185	8	14.8	8	..	Beta	Yes	Yes
5	6Al-4V	155	170	8	16.5	10–20	92	Alpha-beta	Yes	Yes
5A	6Al-4VELI*	205	220	13	16.4	10	..	Alpha-beta	Some	Yes
6	7Al-4Mo	175	185	10	16.9	18	100	Alpha-beta	Some	Yes
7	4Al-3Mo-1V	167	195	6	16.5	..	124	Alpha-beta	Some	Yes

*Low oxygen content and performance at −320°F.

structures is of importance. It is quite possible that new alloys having less notch sensitivity can be developed. In spite of this notch sensitivity, the specific fatigue strength of notched titanium alloys compares quite favorably with that of any other material because of the inherently high fatigue strength.

6-13 OTHER NONFERROUS METALS

In many ocean systems, such as ships, platforms, and buoys, seawater is used for various purposes, jet propulsion, ballasting, and cooling of engines and condensers. In these seawater systems, heat exchangers, valves, instruments, and piping systems are exposed to a range of pressures, velocities, and temperatures. In these applications, the material selection is made on the basis of the same material characteristics, namely, yield strength, tensile strength, modulus of elasticity, elongation, Charpy V-notch impact test, and fatigue strength. In addition, resistance to corrosion and overall cost become very important. For many of these applications, copper and nickel alloys are the best choice of materials. Table 6-8 shows a typical list of materials.

The cupronickel alloy listed as 1 is the wrought alloy and 1A is the cast alloy. Both are used widely for condenser applications such as tubes, tube sheets, sea chests, and manifolds. These alloys can be hardened by cold working. The material can easily be welded and brazed. However, it may lose some of its strength when it is heated to brazing or welding temperature. Among the nickel-copper alloys, K-monel is one of the most popular types as it has high values of both yield and tensile strength together with excellent elongation. This material can also be precipitation-hardened, giving increased values of yield and tensile strength. As the material is welded, it requires reheat treating to maintain its strength values.

Nickel-copper 400 is another material which has excellent yield strength and tensile strength together with high values of elongation. This alloy receives the higher-strength values by work hardening and, consequently, it will lose some of its strength when heated.

In Table 6-8 alloy 4 and 4A shows the physical characteristics of bronze in both the wrought and cast conditions. Bronze is widely used for valves, pumps, and heat exchangers in applications which do not require high strength. Bronze, even as a casting, has high values of elongation and is used accordingly. Aluminum bronze is characterized by the lower density and simultaneous high yield and tensile strength with good elongation. Table 6-8 also indicates that the modulus of elasticity of this group of metals changes widely, depending on the constituents of the various alloys. In this group of alloys, the values for the Charpy V-notch impact test and fatigue strength are not always available. However, since the above metals are used mostly for internal outfitting

Table 6-8 Other nonferrous metals

No.	*Alloy type*	*Yield strength* S_y, *ksi*	*Tensile strength* S_t, *ksi*	*Elongation,* %
1	Cupronickel 30: Cu 68.9 Ni 30 Mn 0.6 Fe 0.5	20–22 68	44–60 73	45–40 12
1A	Cupronickel: Cu 69.1 Ni 29 Mn 1.5 max Fe 0.9	37	68	28
2	K-500 Monel: Ni 65.00 Cu 29.50 C=0.15 Mn=0.60 Fe=1.00 Al=2.80 Si=0.15 Ti=0.50	40–65 90–120	90–105 130–170	45–25 25–15
3	Nickel-copper 400: Ni 66.00 Cu 31.50 C=0.12 Mn=0.90 Fe 1.35 Si 0.15	25–45 90–130	70–85 100–140	50–35 15–2
4	Bronze: Cu 90 Zn 10	10 54	37–40 61–74	45 5
4A	Bronze: Cu 85 Zn 5 Sn 5 Pb 5	17	37	30
5	Aluminum Bronze: Cu 89 Al 10 Fe 1	27 42	75 85	25 15

*Cold-drawn.

Charpy V-notch impact, ft · lb	Fatigue strength ksi, 10^8 cycles	Metal treatment	Modulus of elasticity, $E \times 10^6$ psi	Note	Density, lb/in³
. .	. .	Annealed	22	Wrought	0.323
		Half hard			
78	18	As cast	21	Cast	0.322
. .		Annealed	26	Wrought	0.306
	45*	Age-hardened			
. .		Annealed	26	Wrought	0.319
	42*	Work-hardened			
. .	. .	Annealed	17	Wrought	0.318
		Hard			
11	11	As cast	13.5	Cast	0.318
16	22	As cast	16	Cast	0.272
20	27	Heat-treated			

of ocean systems, such values are to some degree of less interest than in applications involving the main structure of ocean systems.

Note that the fatigue strength of these alloys, where available, is based on a relatively large number of cycles, namely 10^8. These types of alloys tend to give a relatively greater fatigue strength relative to tensile strength in the annealed condition. Cold working and heat treatment change the tensile strength appreciably more than the fatigue strength. Basically, the ratio of fatigue strength to tensile strength is of the order of 0.40. In the castings, a fine-grain crystal structure will improve fatigue strength.

6-14 FIBERGLASS

The most prominent nonmetallic material for ocean applications is fiberglass-reinforced plastic. A substantial number of small boats and buoys are made of this material, which is, of course, just one of an ever-increasing number of composite materials. They consist of reinforcing materials which are frequently of a fibrous nature and a bonding material. The reinforcing material gives strength to the structure and can consist of glass fibers or carbon graphite, nylon, silica, or metals such as steel, aluminum, boron, and tungsten. The bonding material can be epoxies, polyesters, phenolics, or silicones. The most commonly used material is glass fiber with an epoxy or polyester binder.

Table 6-9 shows the properties of various fiberglass materials (1 through 4). The data presented are typical values, and it must be recognized that typical characteristics of any nonmetallic materials are substantially influenced by the manufacturer's processing methods and permissible tolerances of the base materials. Consequently, the performance characteristics of any of the nonmetallic materials vary to a substantial degree from one supplier to another and from one manufacturer to another. Fiberglass polyester mat is used widely in the production of small boats and buoys. Its great popularity is based on the absence of maintenance and its durability under a variety of operating conditions. Larger fiberglass structures use fiberglass cloth, which has a higher strength, density, and modulus of elasticity, as indicated for number 2. Even greater strength can be achieved with unidirectional cloth and epoxy binder. Typical material characteristics are shown for number 3 in Table 6-9. The tensile strength and compressive strength are presented only for the direction of maximum strength. The strength in the other direction depends on the weave of the fiberglass cloth and the arrangements of laying the various plies of cloth within the structure. In many cases, the strength in the second direction is about half of that in the preferred direction. The second direction also frequently has a lower modulus of elasticity. The highest-strength fiberglass structures consist of fiberglass filament in combination with an epoxy binder. Typical characteristics for this material are shown for number 4. Various pressure vessels of substantial size have been manufactured from fiberglass yarn of high strength. This type of

Table 6-9 Nonmetallic materials

No.	Material	Density, lb/in³	Specific gravity	Tensile strength S_t, ksi	Compressive strength S_c, ksi	Modulus of elasticity, $E \times 10^6$ psi	Fatigue strength, ksi, 10^7 cycles	Fiber content, %
1	Fiberglass polyester mat	0.054	1.49	16.9	27	1.45	. . .	36.1
2	Fiberglass polyester cloth	0.065	1.80	53	42	2.7	12.5	62.0
3	Fiberglass epoxy unidirectional cloth	0.0654	1.81	85	65	4.6	21	62.0
4	Fiberglass epoxy filament	0.0750	2.08	150	120	8.0	. . .	85.0
5	Glass fiber S-glass	0.091	2.52	450	. . .	12.5		
6	Polyester resin	0.044	1.22	9.0	20.0	0.51		
7	Polyurethane foam	0.0046	0.127	0.2	0.25	0.004		
8	Glass (tempered)	0.120	3.32	45	400	12		
9	Wood: Douglas fir	0.0197	0.545	19.5	6.6	2.1		
10	Wood: Spruce	0.0162	0.448	19.5	5.5	1.87	2.6	
11	Wood: Balsa	0.0034	0.094	1.3	1.0	0.320		
						0.018		
12	Concrete: cast in place	0.086	2.38	. . . .	5.5	4.0		
13	Concrete: light-weight	0.061	1.68	. . . .	4.5	2.0		
14	Concrete: pre-stressed	0.090	2.49	. . . .	7.0	4.5		
15	Ferrocement	0.090	2.49	1.0	6.5	4.0		

processing has the highest fiber content, resulting also in a higher density and a higher modulus of elasticity.

The fiberglass-reinforced composites show a high strength-to-density ratio and are, therefore, used in many applications where light weight is of importance. It is possible to manufacture glass filaments with diameters of 0.01 mm. The fiber itself has no compressive strength and is supported laterally by the matrix of the binder material. A typical characteristic of a fiberglass fiber is shown for number 5, the S-glass fiber, which is the most common type of commercial high-strength filament in use today. The fiber requires a special finish to give protection from mechanical damage.

The basic characteristics of polyester resin are shown for number 6 in Table 6-9. This material is relatively lighter than the fiberglass, it has a low modulus of elasticity, and its tensile strength is lower than its compressive strength. For very high-strength fiberglass materials, epoxy resins are preferred because of their slightly improved physical properties and improved resistance to water absorption together with decreased shrinkage during curing.

The fatigue strength of fiberglass laminates is relatively low. In addition, fiberglass, owing to its internal damping characteristics, may heat up when subjected to fast-changing stress cycles. It usually reaches its fatigue strength at 10 million cycles. Note that the ratio of fatigue strength to tensile strength is below 0.25. However, owing to the basic arrangement of the fibers, fiberglass laminates have a very high notch strength. The fiberglass laminates are insensitive to notches and other irregularities in both configuration and stress levels.

Fiberglass is inclined to lose strength by the absorption of water when immersed over long periods of time and by exposure to ultraviolet light. It is, therefore, desirable to give it a protective coating when immersed in the ocean or exposed to sunlight over extended periods of time. The absorption of water usually results in a substantial decrease in strength, particularly compressive strength, which is based on adherence of the resin to the glass fibers. Ultraviolet light, on the other hand, causes brittleness.

Fiberglass also tends to delaminate upon application of heat. The color of protective coatings should be kept in mind to reduce blackbody absorption of radiation from the sun, which can result in relatively high surface temperatures. The working temperature of a material should also be kept in mind if fiberglass is to be used for a pressure vessel, for instance. Most fiberglass resins will burn.

The fiberglass sandwich has been used a great deal where the filler is either end-grain balsa wood or a foam material. It should be noted that certain foams will "melt" when polyester-base resins are applied to them.

The fiberglass laminates will, no doubt, be used in many new applications owing to their excellent behavior and resistance to corrosion. Freedom from notch and fretting effects is also an important feature. One disadvantage is the variability of the material, which is due to the number of processes involved in manufacturing; this can result in delamination if a high level of quality control is not exercised.

6-15 GLASS

In the last few years, a great number of various types of glass have been studied for use as ocean materials. In Table 6-9 number 8 shows typical characteristics for a tempered glass. The primary components in glass are SiO_2, Al_2O_3, and a number of oxides in small quantities, such as CaO, Na_2O. The high-silicate glass has great strength in compression, which is based on the high-energy bonds of its configuration, producing minimum dislocations. The resulting networks are responsible for the high compressive strengths of the silicate glass. This type of glass becomes increasingly resistant to mechanical impacts and underwater shock with increasing depth. This is caused by the increasing ductility of glass with increasing compression stress. Attempts to make this type of glass ductile at lower stress levels have been unsuccessful.

The strength in glass is achieved by a thermochemical treatment which may include tempering and certain phase changes on the surface. While glass shows substantial promise as a material when used in compression, it is difficult to design a structure which has only compression loads and is not loaded in tension, shear, or bending under certain operating conditions. It also has been found difficult to produce large sections of glass without defects. As a matter of fact, the tolerances in physical characteristics of glass and its quality control in processing are at this time not equal to those of metals. Therefore, the manufacture of large pieces of glass to strict physical specifications is expensive when compared to the price of commercial glasses.

6-16 CONCRETE

In many ocean applications, concrete has been in use particularly for foundations, dams, and similar structures. Concrete is known to have excellent compressive strength and has, therefore, been used in many stationary structures. It is resistant to attack by seawater. The typical strength characteristics and weight for concrete are shown in Table 6-9, items 12 through 15. The maximum 28-day compressive strength value of a specimen is usually considerably higher than the listed value for use in large structures. Concrete is to some degree permeable to liquids and gases and requires relatively impermeable aggregates for use in ocean applications. If needed, a protection of the exterior surface can be applied by using corrosion-resistant materials or resistive coatings.

Since concrete has excellent strength in compression and low strength in tension, it is necessary to design concrete structures to have a minimum of tension, bending, and shear stresses. Appropriate reinforcements of the structure, including prestressing or the use of ferrocement, may be desirable. Ferrocement consists of wire mesh or a similar reinforcement that gives the cement considerable stability and permits some tensile stress loads as indicated for number 15.

Recently, this type of material has been used also in the construction of barges, boats, and the General Dynamics "monster buoy." The appropriate reinforcement of the cement with steel wires as in ferrocement and prestressed concrete has found ever-increasing use in the last decade. Some very large structures of prestressed concrete are being built as pressure vessels for LNG storage tanks at LNG terminals and as containing vessels for nuclear reactors. All concretes and ferrocement are sensitive to dynamic loads and have relatively low fatigue strength, but very few data are available. Many concretes are nonisotropic materials, and therefore substantial differences in fatigue strength will result as a function of the type of load. Concrete also suffers some deterioration during freezing and thawing. When a concrete structure is installed near the ocean surface zone, sharp corners may be subject to breaking by abrasion unless protected. For these applications, abrasion-resistant concrete has been developed. Recently, lightweight concrete has found increasing applications, particularly for floating structures. Epoxy resins are being used increasingly as coatings to protect concrete from both erosion and abrasion.

6-17 WOOD

One of the oldest materials used in the ocean is wood. For many years, wood was the only material used in shipbuilding, and it is presently still being used extensively for pilings, docks, and similar applications. Recent developments in the use of cements and resin glues for the production of wood laminates have resulted in many new applications using wood as structural members. In Table 6-9, numbers 9 through 11, typical strength values for wood are presented. The data show both tensile and compressive strength of wood in the direction of the grain. In this direction, wood has much higher strength than across the grain. The tensile strength across the grain is frequently very substantially less, on the order of 40 to 1, whereas the same ratio for compressive strength is on the order of 6 to 1. Wood can have a tendency to creep, and working stress for long-time loading must be conservative. The fatigue strength is about 25 percent of the maximum tensile strength. The strength of wood is also a function of the processing, storage, and its ultimate use. Moisture content can greatly affect the strength of wood. Balsa wood is frequently used as an insulating material or filler material inside metal structures to avoid buckling.

6-18 NEW COMPOSITE MATERIALS

Many new high-performance composites have been proposed for various applications. At this time, boron-epoxy composites have been used in modern aircraft with a density of about 0.082 lb/in^3, a modulus of elasticity of 42 million psi, and a tensile strength of about 225,000 psi. Graphite composites show equally interesting physical characteristics with a density of about 0.065

lb/in^3, a modulus of elasticity of 35 million psi, and an ultimate strength of about 150,000 psi. One of the 12-m America's Cup yachts had the main mast fabricated from high-strength boron composites. The main disadvantages of such high-strength composites is their price. In 1970, the price of graphite composites was about $300 per pound and that of boron composites, $230 per pound. The advocates of high-strength composites predict by 1980 a price of $40 per pound for boron composites and $15 per pound for carbon fiber composites. If this price is realized, there will be many new applications in the ocean owing to the excellent physical characteristics of these new materials.

6-19 BUOYANCY MATERIALS

Buoyancy materials have a specific gravity which is considerably lower than that of water. When integrally included in an underwater structure, they provide buoyancy. Such materials have found applications not only in small submarines but also for oil well drill pipe, deep-sea buoys, and similar applications where the reduction of weight under water is important. The most common buoyancy materials are liquids, such as gasoline with a specific gravity of 0.70, and solids. Among the solids, wood with a specific gravity of 0.50 has been widely used. However, buoyancy materials are now desired which do not absorb water and do not compress over the entire range of water depths where they will be used. For this purpose, *syntactic foams* have been developed. They consist of hollow glass spheres, which have a very high compressive and shear strength, dispersed in a plastic matrix. The most efficient syntactic foams use glass spheres of extremely small diameter, called microballoons, with an epoxy resin binder. This material has a compressive strength of 15,000 psi, a compressive modulus of 550,000 psi, and a tensile strength of 6,000 psi with a tensile modulus of 550,000 psi. The weight is about 40 lb/ft^3 or a specific gravity of 0.64. Therefore, they have a buoyancy of about one-half their own weight in air. They have low water absorption, and the cast material can easily be handled with ordinary woodworking tools to be shaped for its intended use. Syntactic foams are also used as sandwich core materials and for void fillers, replacing the wood which was formerly used.

6-20 COATINGS

Coatings are used extensively in the marine environment to protect surfaces against deterioration from salt spray, barnacles, corrosion, pollution, and all other contaminants of the sea. Fouling increases with water temperature and decreases with lower temperatures and depth in the ocean. Certain new plastic coatings have evidenced antifouling protection for several years and some epoxy coatings have given corrosion protection for several years. The coatings serve as a physical and chemical barrier against the attacks of the sea. In order to apply various chemical coatings successfully, methods of surface preparation are required including chemical cleaning, removal of oil, grease, metal oxides, and other surface contaminants. Frequently, the material requires various prime coats which may consist of phosphates or similar films. The coatings are usually applied in very thin layers only a few thousandths of an inch thick.

Protective coatings are applied in up to five coats, resulting in a film about 0.010 to 0.020 in thick. The most common anticorrosive coatings for long-period underwater exposures are (1) cold-tar epoxy, (2) epoxy, (3) polyurethane, (4) vinyl anticorrosive, and (5) neoprene and similar rubber coatings.

Many underwater surfaces of ships and buoys including tanks, bilges, and voids must be protected with various coatings. For these applications, the coatings consist of a primer coat, followed by one or two anticorrosive coats, followed by an antifouling coat. There are a large number of commercial and military specifications covering treatments and coatings.

Wooden structures need careful protection and require as a first coat a material which will seal the pores of the wood. After the sealing is accomplished, various types of paint and antifouling coatings can be applied. Polyurethane coatings have been used successfully as a protection for wood. Aluminum requires careful cleaning and preparing the surfaces with a deoxidizer. This is then followed by several protective coatings of the epoxy or polyurethane type. In some ocean structures, nonslip coatings are used for walkways and decks. These consist of an aggregate which is mixed with a chemical coating such as epoxy or urethane.

Modern coatings give protection up to 3 years and this duration of protection is about twice as long as could be expected 15 years ago. The technical progress in coatings will permit better protection and longer duration of protection in the coming years. The application and development of coatings has become a specialty and each case may require a different technology (Burns and Brakley 1955).

PROBLEMS

6-1. What is the ratio of yield strength to tensile strength for a low-, medium-, and high-strength steel and for a high-strength aluminum alloy?

6-2. What is the ratio of fatigue strength of 10^6 cycles to tensile strength and yield strength for the same materials as in Prob. 6-1?

6-3. What conclusions can you draw about the content of phosphorus (P) and sulfur (S) in the various steel alloys?

6-4. The beam of an oil drilling rig is loaded by a static load resulting in a maximum stress of $S_1 = 13{,}500$ lb/in^2. During a hurricane, the beam is hit by a wave resulting in an additional load of 10,000 psi which occurs at a strain rate of 10 in/(in)(s). The steel has a yield strength of 27,000 psi and an ultimate tensile strength of 57,000 psi as indicated in Fig. 6-4. Assuming the material performs as indicated in Fig. 6-4 and the load is such that the stresses can be added, what is the yield strength and ultimate tensile strength of the steel when it is loaded at the above strain rate? What is the ratio of static stress to low-strain-rate yield and tensile strength and what is the same ratio at the specified strain rate for the combined stress?

6-5. Which steels can be selected for an arctic application where a Charpy V-notch energy value of 80 ft/lb or higher is needed at 0°F?

6-6. What steels can be used in a liquid natural gas tanker in contact with LNG which has a temperature of −268°F?

6-7. What is the ratio of yield strength to cost for the following steels: ABS Class B, ABS Class BH, HTS, HY-80, HY-100, HY-130, and maraging 200?

6-8. A platform has a diesel engine with a freshwater cooling system and a freshwater-saltwater heat exchange system. Recommend a material for the freshwater-saltwater heat exchanger and the saltwater piping system if light weight and low maintenance are more important than cost. Be sure to utilize Chap. 5 material, as well.

6-9. A barge of 100-ft length and 18-ft beam is to be built. Discuss the use of fiberglass versus ferrocement as the building material and comment on the differences of physical characteristics.

6-10. A fiberglass boat is designed as a sandwich-type construction having an inner and an outer hull with polyurethane foam between the two hulls.

(*a*) Will this boat sink with a large hole in the bottom?

(*b*) If the hull by itself is designed to be buoyant with a specific gravity of 0.80, what is the ratio of fiberglass polyester mat to foam if fiberglass mat is the selected material?

6-11. Compare lightweight concrete and concrete cast in place for use in a habitat 12 ft in diameter, 50 ft long, and 1 ft thick. What are their significant differences in physical characteristics? How much ballast is needed to sink the habitat?

6-12. A steel pipe, 12 in in diameter and 20 ft long, has a thickness of 1 in and is to be used underwater in a drilling rig. What is its weight underwater? A syntactic foam cylinder shall be added on the outside of this pipe to reduce the weight underwater. What is the thickness of this syntactic foam cylinder if the weight underwater is to be reduced by 20 percent?

6-13. A beam is rigidly mounted and stressed to a value of 9,000 psi in tension. The modulus of elasticity E for the material is 11×10^6 psi. The temperature is reduced by 100°F and the stress becomes 12,000 psi. What is the value of the coefficient of thermal expansion α for this material?

6-14. A large number of tie rods are to be used above water for a drill rig, each carrying a pure tensile force of 6 tons. Basing your design on the yield stress with a factor safety of 2, design the cheapest tie rod and give the type of steel, the rod diameter, and the cost per foot.

6-15. A ¼-in-long crack propagates under a stress of 37,000 psi. The material is steel. What is the value of this material's surface energy per unit area R?

6-16. One way of defining toughness is to note the area under the tensile test, stress-strain curve. Assuming that all the low-strength steels in Table 6-2 have a stress-strain curve represented by a line having a slope equal to E up to the yield strength and then another straight line connecting the yield point to the point of ultimate strength and final elongation, derive a general equation for the summed area under these two lines based on E, S_y, S_t, and percent elongation.

(*a*) Which is the "toughest" low-strength steel based on this equation? Which is the least tough? What is the energy absorption per unit volume of material for each?

(*b*) Which is the best steel based on unit energy absorption per dollar?

6-17. Referring to Prob. 6-16, compare the "toughest" low-strength steel with the "toughest" medium-strength and "toughest" high-strength steels, using data from the appropriate tables.

6-18. Using the equation derived in Prob. 6-16, obtain the area under the stress-strain curve for the aluminum alloys in Table 6-6 versus their Charpy V-notch impact values and their ratios of fatigue to tensile strength. Comment on any correlations you detect and attempt to explain the relations by a qualitative explanation.

6-19. The hoop compressive stress of a submerged sphere with "thin" walls is given by $S = (R/2t)P$, where R is the radius, t is the sphere wall thickness, and P is the unbalanced pressure. Using the materials in Table 6-9, design the smallest sphere that will support 1,000 lb at a depth of 500 ft in seawater (density of 64 $\mathrm{lb/ft^3}$). Do not use wood or glass. The sphere has air at 1 atm inside.

In the following problems, select one or more materials for the noted duty. Explain your choice by listing the material characteristics that are important in the noted application and state why other materials were not selected.

6-20. A welding habitat will fit over an underwater pipe at depths of up to 600 ft, and inside a saturated, dry welder will work. The unit must remain pressurized at sea bottom pressure when raised to the surface (so that the welder can be decompressed; see Chap. 8). Lifetime will not be in excess of 1,000 dives, and average time in the ocean will not exceed 1 day per mission. The unit must be hoisted in and out of the water by a barge-mounted crane. Raising and lowering will be relatively slow. Select an appropriate material for the unit.

6-21. A material is wanted for holding a taut-moored buoy in a location having an average wave period of 3 s for at least 1 year. During storms, large, high-speed wave shocks are expected. A composite cable might be possible since maximum shock loading will be at the surface, the lower part of the cable acting as a "spring." Water depth is in excess of 1,000 ft. Select a material.

6-22. A swimmer-assist vehicle needs 100 lb of buoyancy at depths of up to 300 ft. The unit should be as compact as possible and the device may remain wet for weeks at depth outside a saturated habitat. Select the buoyancy material.

6-23. A fish-farming operation requires miles of underwater "fence" that must be as cheap as possible, remain in the water for years, and still receive a minimum of surface preparation. Strength requirements are low and the location is sheltered. The fence will be of slate on 1-in centers and antifouling will be accomplished by raising the material in the air periodically, allowing the growth to dry, then burning off the residue with flamethrowers. Design this mesh.

6-24. A man-carrying research vehicle will dive to the deepest ocean regions. It must be as light as possible to minimize buoyancy systems, will undergo no more than 100 missions, and will be in water no more than 1 day per mission. Maximum emergency rate of ascent or descent is 1,000 ft/min. Select a material.

REFERENCES

American Bureau of Shipping (ABS) (1970): "Rules for Building and Classing Steel Vessels," New York.

Burns, R. M., and W. Brakley (1955): "Protective Coatings for Metals," Reinhold, New York.

Comstock, J. P. (1967): "Principles of Naval Architecture," Society of Naval Architects and Marine Engineers, New York.

Department of Transportation (1968): Marine Engineering Requirements, *Coast Guard Federal Register*, vol. 33, no. 245, Washington.

Krenzke, M., K. Hom, and Proffett, J. (1965): "David Taylor Model Basin Report No. 1985," Carderock, Md.

Lipson, C., and R. Juvinall (1963): "Handbook of Stress and Strength," Macmillan, New York.

Masubuchi, K. (1967): "Materials for Ocean Engineering," M.I.T. Sea Grant Project GH-1, Cambridge, Mass.

Myers, J. J. (ed.) (1969): "Handbook of Ocean and Underwater Engineering," McGraw-Hill, New York.

Parker, E. R. (1957): "Brittle Behavior of Engineering Structures," Wiley, New York.

Sheets, H. E., and V. T. Boatwright (1970): "Hydronautics," chap. VI, Academic, New York and London.

Timoshenko, S., and D. M. Young (1968): "Elements of Strength of Materials," 5th ed., Van Nostrand, New York.

U.S. Coast Guard (1968): "Rules and Regulations for Tank Vessels," CG-123, Washington.

—— (1969): "Rules and Regulations for Cargo and Miscellaneous Vessels," CG-257, Washington.

Yokobori, T. (1965): "The Strength, Fracture and Fatigue of Materials," Erven P. Noordhoff, Groningen, Netherlands.

7

Underwater Sound

F. H. Middleton

7-1 INTRODUCTION

The purpose of this chapter is to give the ocean engineering student a basic introduction to ocean sound theory. It is definitely not intended as an exhaustive treatment, nor could it be in one chapter. Wherever possible, the interested reader will be directed to one of the many good books that have appeared in recent years. The subject of underwater sound cannot be sensibly discussed without some physical and mathematical concepts, but these concepts are not strange to the upper-class engineering undergraduate. With this chapter objective, the physics and mathematics will be kept to a reasonable minimum.

Underwater sound is obviously a physical phenomenon which can be used in many ocean applications besides the most widely known Navy sonar systems. Certainly, the military requirement for better sonar systems has produced the bulk of the progress in the field over the past 30 years. Navy sonar systems take on many different forms, but in all these forms, the system performance is generally limited by some particular property of the ocean, its boundaries, or its inhabitants. These same ocean properties in general must be considered in any nonmilitary application of underwater sound in the field of ocean engineering.

One might divide the present discussion into four parts: sound generation, sound radiation, sound propagation, and sound detection.

Before looking into each of these compartments, it will be profitable to establish the connection between ocean engineering and underwater sound. Perhaps a noninclusive list of different kinds of uses of underwater sound will serve this purpose.

The most simple underwater sound device might be the fathometer, or depth sensor, used on ships of all sizes to determine and record water depth. The underwater telephone used to communicate between surface craft and manned underwater vehicles or divers is another, a fairly simple system. Seismic or subbottom profiling systems, both reflective and refractive, exist in simple and in extremely complex forms. In the category of general communication using underwater sound, there are both simple and refined systems for command, control, and navigation between underwater points. So-called "transceiver beacons" are used in ocean water depths in excess of 20,000 ft for extremely precise open ocean navigation systems. Inexpensive, small, battery-powered acoustic beacons (or "pingers") are commercially available for almost routine attachment to any valuable ocean instrument. In the event of an instrument malfunction, the beacon emits a signal which permits at least locating the instrument for possible recovery.

One rapidly growing application is underwater surveying, or object location performed in many different ways. One special system, called a "side-scan sonar," is suitable for generating an acoustic scan of the sea floor. The resulting display of sequential scans can be of the quality of a crude photographic image of the sea floor. The beauty of a side-scan "picture" is that it can be produced in water that permits essentially no light transmission. Any object, regardless of size, can be detected in the water column, or even under the floor of the sea, if only it has acoustical properties different from the region around it. Large concrete erosion mats can easily be located under considerable sediment cover. Pipelines under the sea floor, cables beneath a river bottom, and of course even fish schools can be located acoustically if the principles of underwater acoustics are understood and properly applied.

In any of the above applications, and many others not mentioned, the same questions must be answered each time by the designer of the system. There is always a question about generating the sound, coupling it to the water, propagating the sound, and detecting a useful echo. To get into these details, it is important to start by understanding the physical phenomenon of underwater sound itself.

7-2 PHYSICS BACKGROUND

As in any other energy-propagating mechanism (such as light transmission), sound propagation can be considered as a wave process or as a ray process. It is a bit more understandable for the beginner to consider these processes one at a time, and so the ray process will be postponed until later in the chapter. To get a feel for underwater sound, it is helpful to define carefully the physical observables, and this is readily done in terms of solutions to the appropriate wave equation. To do a good job of developing the wave equation here would require considerable time and space, and it might be a safe assumption that most engineers are familiar with the wave equation in one form or another. For the reader who is not, there are many excellent developments that are readily available in the book references at the end of the chapter.

In its most simple cartesian form, the wave equation can be written:

$$\frac{\partial^2 p}{\partial x^2} + \frac{\partial^2 p}{\partial y^2} + \frac{\partial^2 p}{\partial z^2} = \frac{1}{c^2} \frac{\partial^2 p}{\partial t^2} \tag{7-1}$$

In words this equation states that "the laplacian of the excess acoustic pressure p is equal to the second partial time derivative of the same pressure divided by the square of the propagation velocity c." It should be mentioned that precisely the same equation is satisfied by other physical observables than the pressure, but more about that later.

There is a whole class of functions which are suitable solutions to this wave equation, but one particular solution will be selected from all these, for simplicity. A plane sinusoidal wave, and particularly one which propagates in the direction of the x axis of the coordinate system, is indeed easy to deal with. The solution might appear as follows:

$$p(x, t) = p_1 e^{j(\omega t - kx)} \tag{7-2}$$

It is clear that this form of excess acoustic pressure corresponds to a simple harmonic plane wave (because p is independent of y and z) and p is clearly propagating in the direction of the positive x axis (because of the negative sign in the exponent).

k = wave number, l/m
x = position along x axis, m
ω = angular frequency, $\omega = 2\pi f$, with f the frequency in hertz
t = time, s
p_1 = arbitrary constant
λ = wavelength

In the case of homogeneous, isotropic medium, wherein the propagation velocity c in meters per second is constant, there are simplifying interrelationships.

$$\omega = 2\pi f \qquad \lambda = \frac{c}{f} \qquad k = \frac{2\pi}{\lambda} = \frac{\omega}{c}$$

By manipulating these quantities, the solution can be written in several equivalent forms.

$$p = p_1 e^{j(\omega t - kx)} = p_1 e^{j\omega(t - x/c)} = p_1 e^{jk(ct - x)} \tag{7-3}$$

That this is truly a solution to the wave equation can be demonstrated by performing the time and space differentiation operations indicated in the wave equation and collecting terms to establish the equality.

Certainly, the restriction to a simple harmonic solution, and a plane wave, propagating in the direction of an axis has simplified the mathematics, but it has caused no loss of generality. This is shown clearly in all the basic book references.

One more point to note in this simple case is that the solution here allows for no change in amplitude as time goes on, or as the position coordinate x increases. There is thus no loss mechanism allowed in the solution. Another way of saying this is that a wave front (surface of constant phase) is not changed in shape as it moves through the medium with increasing time. This is important to note, because in any other than plane wave cases, the conclusion does not hold and losses called "spreading losses" come into play.

Keeping things simple in the discussion (still with no loss mechanisms getting involved or inhomogeneous media to complicate the solution), consider the cylindrical coordinate system. Here, the wave equation takes on a different form, familiar to engineers, as follows:

$$\frac{1}{r}\frac{\partial}{\partial r}\left(r\frac{\partial p}{\partial r}\right) + \frac{1}{r^2}\frac{\partial^2 p}{\partial \theta^2} + \frac{\partial^2 p}{\partial z^2} = \frac{1}{c^2}\frac{\partial^2 p}{\partial t^2} \tag{7-4}$$

If one assumes, as in the cartesian coordinate case, that p does not depend upon two of these coordinates (θ and z), great simplifications in the mathematics result with no loss of generality. If indeed p depends only upon the coordinate r and the time t, then the second and third terms in the equation have no significance. This means we are assuming a circularly symmetric solution and the wave equation simplifies.

$$\frac{1}{r}\frac{\partial}{\partial r}\left(r\frac{\partial p}{\partial r}\right) = \frac{1}{c^2}\frac{\partial^2 p}{\partial t^2} \tag{7-5}$$

As in the plane wave case, a solution $p = (p_1/\sqrt{r})e^{j(\omega t - kr)}$ will satisfy this wave equation, and again it is only one of an infinite number of possible solutions. The important change which appears here is that the amplitude of the excess acoustic pressure decreases as r increases because of the $1/\sqrt{r}$ term that appears. The result is what is commonly referred to as *cylindrical spreading loss.* It is clearly not a loss in the sense of dissipation of energy, but simply follows from a conservation of energy principle. As the wave front propagates outward from $r = 0$, the surface area of a wave front (infinite length in the z dimension) increases directly with r, and the energy density (corresponding to p^2) must decrease correspondingly. This will be discussed in more detail after we define energy and acoustic intensity.

To complete the discussion of simple three-dimensional waves, consider the comparable form of the spherical wave equation.

$$\frac{1}{r^2}\frac{\partial}{\partial r}\left(r\frac{\partial p}{\partial r}\right)=\frac{1}{c^2}\frac{\partial^2 p}{\partial t^2} \tag{7-6}$$

The simplest one of the infinite number of possible solutions can be written:

$$p=\frac{A}{r}\,e^{j(\omega t-kr)} \tag{7-7}$$

where A is an arbitrary constant. The prime point to notice here is the r in the denominator instead of $\sqrt{r}$ as in the cylindrical case. This form of the excess acoustic pressure corresponds to a wave of pressure propagating outward from the point $r = 0$, uniform in all directions (no dependence on θ or ϕ, the azimuth and polar angles). Since the pressure here decreases with increasing r, p^2 decreases as $1/r^2$, just as the surface area of a sphere increases. A given amount of acoustic power is spread over progressively larger areas (as r increases) and if no power is lost or gained in this process, the power density (power per unit area) must decrease exactly as fast as the sphere surface area increases. This spreading process is called *spherical spreading loss.* We see then that in the plane wave, the pressure amplitude is insensitive to distance x; in the cylindrical case, the pressure decreases as $1/\sqrt{r}$; and in the spherical case, the pressure decreases with $1/r$ as r increases. The reader should note that these conclusions would obtain in the ocean only if the seawater was homogeneous (propagation velocity constant) and no boundaries were nearby. We must discuss later the many ways in which the ocean departs from this ideal medium.

7-3 PHYSICAL OBSERVABLES

The excess acoustic pressure p of the last section is the most important of the physical observables that are directly observed by the usual hydrophone placed in a sound field. Another parameter which is derivable from the pressure is the particle velocity associated with a propagating sound wave. Using Newton's second law of motion and Hooke's law, one can arrive at the connection between the pressure p and the particle velocity u. It is $\mathbf{u} = j/\omega\rho$ grad p. This equation looks complicated, but its meaning is quite simple. It says that given any sound field where the observed pressure is p, the particle velocity u is produced by multiplying the gradient of the pressure by the constant $j/\omega\rho$. Here, j is the $\sqrt{-1}$ of complex algebra, ω is the angular frequency in radians per second ($\omega = 2\pi f$, where f is in cycles per second or hertz), and ρ is the density of the medium where the sound wave exists.

The factor j serves only to cause the particle velocity to be advanced in phase by 90° relative to the pressure in the simple harmonic case (except for another segment of phase angle that might be associated with gradient p). The pressure p is a scalar observable, and the gradient of any scalar is a vector observable. It is the space rate of change of the scalar, and the particle velocity is thus a vector. In the ocean, the direction of the particle velocity vector is in the direction in which the wave is propagating. One should note and remember that the particle velocity has absolutely nothing to do with the propagation velocity.

In water, it is common to measure the pressure in units of dynes per square centimeter or microbars. An atmosphere (1 bar) is about 14.7 psi and thus the unit of measure is one-millionth of 1 atm or 14.7×10^{-6} psi. This is the reason for the term "excess acoustic pressure." At a depth of about 33 ft in the ocean, the static pressure or hydrostatic pressure is 2 atm or almost 30 psi or about 2×10^6 μbar. The sound pressure unit is tiny in comparison.

The particle velocity unit is often centimeters per second, meters per second, or any convenient unit of distance per unit of time. In the cgs systems, it would be consistent to use centimeters per second. Using the relation between the pressure and particle velocity in the plane wave case will show that the ratio P/u is ρc (ρ = density and c, propagation velocity). This quantity is a physical characteristic of the medium and is referred to as the specific acoustic impedance. This is about 1.5×10^5 cgs rayls (dyn/cm^2 $\times$ s/cm) in seawater in comparison to 42 rayls (g/cm^2 $\times$ cm/s) = g/cm·s in ordinary air. Water has thus the order of 3,000 times higher impedance than air and this is the basic reason why an ordinary loudspeaker designed to "couple" or radiate into air is a poor underwater acoustic radiator.

An ordinary hydrophone senses the excess acoustic pressure, and if it is calibrated, one knows precisely the number of volts produced by a unit of excess acoustic pressure, or microbar. This is often expressed by the manufacturer in terms of decibels (dB) which will be discussed shortly. For the present discussion, it would be entirely sufficient for the hydrophone calibration lab to stamp its receiver sensitivity on the box in such a form: "1 μbar sound pressure produces 0.1 μV." Then, if the hydrophone were placed in a plane wave whose sound pressure was 100 μbar, the open circuit output voltage on the terminals of the transducer would be 10 μV.

7-4 ACOUSTIC INTENSITY

In a plane wave, the acoustic intensity is proportional to the square of the pressure. Or

$$I = \frac{p^2}{\rho c} = pu = u^2 \rho c \qquad \text{erg/cm}^2 \tag{7-8}$$

The more usual unit of power is the watt, or newton meter per second, very large compared to an erg or dyne centimeters per second. Since 1 N = 10^5 dyn and 1 m = 100 cm, we have 1 N·m/s = 1 W = 10^7 ergs/s. Also:

$$1\text{W/cm}^2 = 10^7 \text{ ergs/cm}^2$$

To utilize sound in the sea, one must be able to transfer acoustic energy from one point to another. This very practical problem usually revolves around, first, the question of what intensity the hydrophone can detect. Then, what intensity or total power must the source be capable of radiating at some distant point to produce this necessary level at the receiver? In addition to the transmitting and receiving response of the two transducers (radiator and hydrophone) the designer must account for all ocean effects. These include spreading (as mentioned earlier), refraction, reflection, absorption, scatter, and directivity of the transducers. All these ocean effects are important and are really what the subject of underwater acoustics is all about.

The quantitative bookkeeping in an underwater sound field problem is greatly simplified by changing multiplication to addition by using logarithms, or decibels. This is quite practical because the ratio of a source intensity to an ordinary receiver intensity in a typical underwater application will be three, four, five, or more orders of magnitude (factors of 10). The decibel is most readily applied to ratios of intensity, and the definition is easily expressed as follows:

$$\text{IL} = 10 \log\left(\frac{I}{I_0}\right) = 10 \log\left(\frac{P}{P_0}\right)^2 = 20 \log\left(\frac{P}{P_0}\right)$$

The symbol IL means the intensity level of I W/cm^2 relative to a reference intensity I_0 W/cm^2. Suppose that we select 1 W/cm^2 as the reference intensity. Then the "level" corresponding to $I = 100$ W/cm^2 is $\text{IL} = 10 \log(\frac{100}{1}) = 10 \times 2 = 20$ dB. The customary notation is IL = 20 dB//1 W/cm^2, where the double slant line symbol means "relative to a reference of . . ." The output of a radiator is conventionally measured at a considerable distance from the face of the radiator, say 100 yd, and then converted to the intensity level at a distance of 1 yd from the source. Since the intensity is inversely proportional to the square of the radius, $I_1 = (100^2/1^2)\, I_{100} = 10^4 I_{100}$, which can be expressed:

$$\text{IL}_{1m} = 10 \log I_1 = 10 \log\,[10^4 I_{100}] = 20 \log 10^2 I_{100}$$

In words, $\text{IL}_{1m} = +40$ dB relative to I_{100m}. Spherical spreading is thus accounted for by applying the operator 20 log (r_2/r_1). To give a simple illustration, suppose a spherical radiator in the ocean (assumed isovelocity) is operated far from the influence of any boundaries, and a calibrated hydrophone placed 100 yd from the source measures a pressure which corresponds to 3 W/cm^2. The symbol SL is used to indicate source level, and in this case:

$$\text{SL} = 10 \log \frac{100^2}{1^2} + 10 \log 3$$

$$= 20 \log \frac{100}{1} + 10 \log 3 = 40 + 4.8 = 44.8 \text{ dB//1 W/cm}^2 \qquad (7\text{-}9)$$

7-5 RADIATORS—SOURCES OF SOUND

There are many, many different kinds of sources of sound that have practical importance in a huge array of applications in ocean engineering and oceanography. The most commonly used high-powered sources include explosives, air guns, sparkers, boomers, hydraulic sirens, and any other device that sets water particles into violent motion. An ordinary hand grenade has been used as a high-amplitude acoustic source in ocean floor penetration for the production of subbottom profiles for geological exploration. Air guns are similar sources in that a chamber is pumped up to 2,000 or 3,000 psi and suddenly opened to the seawater, producing a loud bang with energy sufficient to penetrate well into the sea bottom in water depths greater than 20,000 ft. Such techniques make it possible to detect and display layers or strata of geological importance in the sea floor. Another like means of producing a useful pulse of sound is a "sparker," essentially an underwater spark plug that makes a loud short acoustic bang that is useful. A similar device, called a "boomer," is a large (2- or 3-ft diameter) plate which is suddenly displaced, producing large-amplitude particle motion in seawater. Even hydraulic sirens have been used for this purpose. All these devices can be used for applications where very large source levels are required.

In the general problem of underwater communication such as from a surface ship to a small manned submarine on a deep search mission, a different kind of source is desirable. For voice frequencies, another large array of sources is available. Generally, these sources can be compared to ordinary loudspeakers, except that to match seawater, they are much higher impedance sources. A common source is a *magnetostrictive* transducer. This kind of source is composed of one or more elements which are made from laminated stacks of metallic sheets of material arranged like a transformer core. The material is selected for its ability to expand and contract when a magnetic field is applied. A simple form is illustrated in Fig. 7-1. A wire coil is wound on the core in such a way that when an alternating current is passed through it, the length of the assembly L changes with the current. There are some important details that are really beyond the scope of this discussion, such as the fact that there must be a magnetic bias supplied to procure linear behavior. Let it suffice to say that the length L can be made to change by an amount ℓ, such that ℓ is proportional to the current in the coil. Now if a group of these magnetostrictive elements are cemented in some way to a large diaphragm, the motion of the face of the diaphragm will correspond to the current applied to the array of coils. It is thus possible to make a source which is fairly large in terms of the acoustic wavelength. This is basically the procedure used to produce a radiator that is directive. The radiated acoustic energy can then be directed toward the receiver or the target.

Another class of underwater radiator is the *piezoelectric* or *ferroelectric* transducer. This discussion will not go into detail on these devices any more than

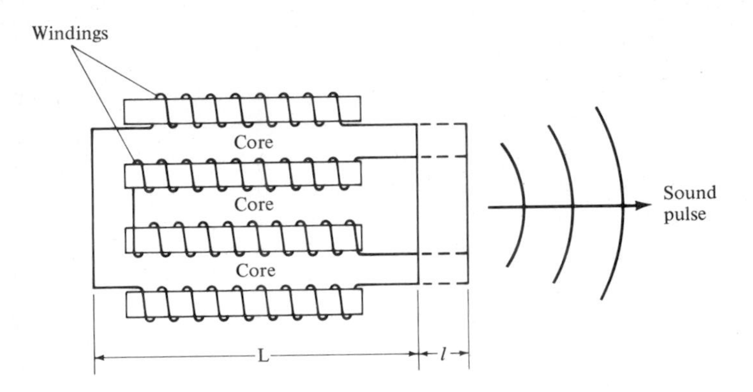

Fig. 7-1 Side view of a magnetostrictive element with distance ℓ, greatly exaggerated.

to say that an electric field applied to this transducer also produces a motion normal to the face of the radiator. Both magnetostrictive and ferroelectric transducers can be used as sound radiators or sound receivers, in general.

There are many sizes and shapes of magnetostrictive and ferroelectric transducers for use in water, but they all have several features in common. They all have a face which moves normally so as to impart a corresponding motion of water particles, or launch an acoustic wave in the water. They are all fairly dense and relatively rigid, and they exhibit resonant properties characteristic of any vibrating structure. An ordinary loudspeaker for use in the air is, by contrast, relatively compliant, of small mass, and it also exhibits resonant properties. The soft, compliant cone results in a small impedance appropriate to coupling to a low-density (low-impedance) material such as air.

7-6 DIRECTIVITY OF A RADIATOR

The transducer designer must be concerned with all the above aspects and others that are equally important. Some applications of underwater sound require high directivity so that the radiated energy is confined to a thin beam. To determine whether or not a transducer is directive, one must compare the dimensions of the radiating surface to the acoustic wavelength in the water. Another way of saying the same thing is that a given radiating face (assumed flat) becomes more directive as the frequency increases (or wavelength decreases).

The practical quantitative manner to express the degree of directivity of a radiator is by means of a *directivity factor* or a *directivity index*. Before defining these useful quantities, consider a simple circular piston radiator such as shown in Figs. 7-2 and 7-3. At very low frequency, the diameter D is very small compared to the wavelength, and the polar radiation pattern is said to be uniform. That is, the radiated intensity is uniform for all angles θ from 0 to 90°.

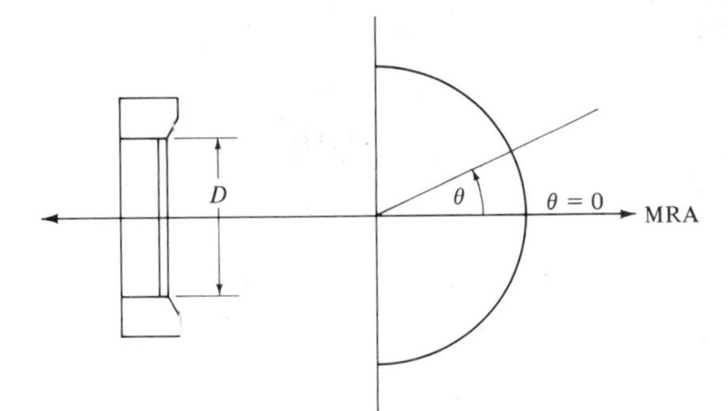

Fig. 7-2 Polar radiation pattern from a radiator when $D << \lambda$.

The angle $\theta = 0°$ is along the axis of symmetry, sometimes called the *maximum reresponse axis* (MRA) of the radiator. If the frequency is increased in such a way that the wavelength becomes very small compared to D, then a radiation pattern might become somewhat as shown in Fig. 7-3.

The latter case is a highly directive transducer. Except for angles close to $\theta = 0$, the radiated intensity is very small. This pattern will be used to illustrate a definition of the directivity factor d. This factor is defined as:

$$d = \frac{\text{total acoustic power of transducer}}{\text{total acoustic power of an omnidirectional transducer}} \tag{7-10}$$

The denominator power is the power that would be radiated by an omnidirectional transducer which produces the same intensity in all directions, and that intensity is the same as the intensity of the MRA of the directional transducer. The directivity index is simply related to d.

$$\text{DI} = 10 \log \frac{1}{d} = -10 \log d \tag{7-11}$$

The narrower the beam (main lobe), the smaller the number d (d is always less than 1 if any directivity exists) and the larger the DI. The DI is directly expressed in decibels. Suppose, for example, that a particular transducer is so directive that it produces only 5 percent as much power as it would have, had its

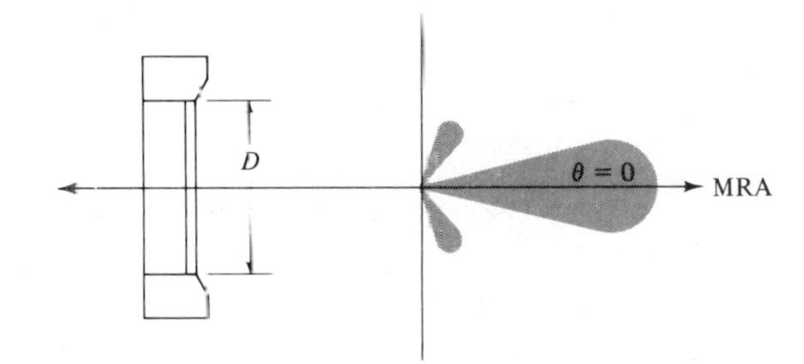

Fig. 7-3 Polar radiation pattern from a radiator when $D >> \lambda$.

MRA intensity been uniformly radiated in all directions, θ. Then $d = \frac{1}{20}$ and DI $= 10 \log 1/d = 10 \log 20 = 10 \times 1.3 = 13$ dB. In the general case, it is necessary to determine the total power by integrating over the pattern. This process can be written:

$$d = \frac{1}{4\pi}\iint \left| \frac{P(\theta,\phi)}{P_0} \right|^2 d\Omega \tag{7-12}$$

Here, $P(\theta,\phi)$ is the radiated excess pressure as a function of the angle θ and the angle ϕ is the second angle in spherical coordinates, $d\Omega$ is the element of solid angle, and P_0 is the reference pressure (on the MRA). The quantity $[P(\theta,\phi)]^2$ is proportional to the intensity radiated, and when multiplied by an element of solid angle, produces an element of radiated power. It can thus be seen that, even though it looks quite different, the result is the same as the verbal definition of directivity factor: a power ratio. At least one of the quantities that is usually specified by a transducer manufacturer is the directivity factor or directivity index. More commonly, a polar plot of the intensity pattern at one or more frequencies is sent along with the other performance specifications. The reader might note that high directivity is not necessarily a good feature. In some applications where one might wish to "flood" a water volume with acoustic power, the ideal transducer might be omnidirectional, or $d = 1$ and DI $= 0$ dB.

7-7 TRANSMITTING SENSITIVITY

The usual sound radiator is driven by an amplifier, and another useful parameter that is needed is the transmitting sensitivity. This parameter indicates the radiated pressure (on the MRA) that is produced by 1 A or 1 V on the terminals of the transducer. Certain standard conditions are specified in making the measurement. It is actually necessary to make the pressure observation far enough away from the radiator so that "far-field" conditions prevail and this pressure is then converted to its equivalent at a standard distance of 1 yd (or 1 m).

In a typical case, a convenient drive amplitude is applied that produces a good response on the calibrated hydrophone and then appropriate scaling is employed. For example, suppose the calibrated hydrophone is placed 10 yd away and on the MRA. If the amplifier output current is 1/10 A, and the hydrophone pressure observed is 20 μbar, the sensitivity is obtained as follows. Assuming spherical spreading, the 20 μbar at 10 yd corresponds to 200 μbar at 1 yd. If the drive current had been 1 A (instead of 1/10 A) the corresponding pressure would have been 2,000 μbar. The result is usually given in terms of decibels or $S_T = 20 \log 2{,}000 = 20 \times 3.3 = 66$ dB//1 μbar at 1 yd 1 A.

The specification sheet for a transducer will necessarily indicate which references are used. From this point on, the user can employ S_T to determine the required drive amplitude to generate a desired sound field in the water.

Radiators, as mentioned earlier, have natural resonances and these are important to the user. If maximum radiated power is the goal, and some flexibility in useful frequency is permissible, then certainly operating at the fundamental resonant frequency will help matters. In high-power applications, the conversion efficiency is important. This simply expresses the ratio of useful radiated total power to the electrical input power in terms of efficiency.

7-8 PROPAGATION

In making use of an acoustic communication system, that is, from a surface ship to a submerged submarine, the major question involves the propagation process. The character of the medium becomes important here, the presence of boundaries, top and bottom, the presence of thermal gradients, etc. The ocean is not nearly homogeneous or isotropic, and the mechanisms that influence sound propagation become dominant. In transmitting an acoustic signal from one point to another, it is essential to know the relative sizes of these propagation effects.

In vertical transmission, the refraction effects are of small consequence. This is because the usual temperature gradients are vertical (except at a place like the edge of the warm Gulf Stream). An acoustic ray from a source at the surface may propagate vertically to a deep receiver without much refraction (ray bending). Even in this configuration though, the passage of a sound wave through a piece of ocean water is complicated. Any small bodies suspended in the water can be excited so as to cause scattering of some of the acoustic energy. That part of the energy that is scattered back toward the source, owing to particles suspended in the water, is called *volume reverberation.*

Some of the energy which is caused to travel from the source to the receiver is lost by frictional processes. These processes are a function of frequency (attenuation) so that the range from source to receiver dictates the appropriate frequency to use. Attenuation is roughly proportional to the square of frequency. Practically speaking, attenuation starts to become large at 10 kHz and gets enormous above 100 kHz. This only means that long-range (miles) propagation is not feasible at high frequencies. A 200-Hz blast from an air gun on the surface of 20,000-ft-deep ocean water for submarine profiling is not seriously impeded by attenuation mechanisms, while employing 10 MHz over more than a few feet is difficult.

The radiated sound from a source suffers spreading loss (regardless of frequency) just due to conservation-of-energy considerations, some scattering loss due to suspended material, and attenuation which is a function of the acoustic frequency. In addition to these mechanisms, which are well understood, the influence of reflection at boundaries and refraction due to thermal gradients must be considered.

7-9 WAVE AND RAY THEORY

To solve an acoustic field problem, there are two general approaches which have merit in their own regimes. One is called *wave theory* and the other, *ray theory.* In a medium where the propagation velocity is essentially constant, the wave theory is highly suitable. This is based upon Eq. (7-1) earlier in this chapter. The form there is a cartesian three-dimensional wave equation.

$$\frac{\partial^2 p}{\partial x^2} + \frac{\partial^2 p}{\partial y^2} + \frac{\partial^2 p}{\partial z^2} = \frac{1}{c^2}\frac{\partial^2 p}{\partial t^2}$$

In this equation, the propagation velocity c is presumed to be independent of position, that is, constant. If the ocean were isothermal, had constant salinity, constant density, etc., the sound field problem would indeed be simple.

As in many physical situations where a second-order linear differential equation with constant coefficients governs the behavior, the solution $[(p(x,y,z,t)]$ is simple. Equation (7-2) is a particular solution to the wave equation which corresponds to a plane wave, one whose amplitude is constant with increasing propagation distance. The key to this simple solution is the property of constant velocity c. This equation is further idealized in that no term is included to account for scattering or attenuation. Indeed, this wave equation only provides an insight for ideal sound field problems. When the velocity of propagation does vary with position in the medium as in the ocean, another approach is warranted. This is the ray theory.

In most places in the ocean, the propagation velocity depends on the position, particularly upon the depth. A useful empirical equation will give the reader a good appreciation of the dependence of sound velocity on temperature, depth, and salinity.

$$c = 4422 + 0.0182y + 11.25T - 0.045T^2 + 4.3(\text{sal} - 34) \qquad (7\text{-}13)$$

where c = sound speed, ft/s
T = temperature, °F
y = depth, ft
sal = salinity, ppt

Inserting all possible ranges of these three ocean parameters into Eq. (7-13) will cause a maximum change in c of only about 4 percent. Indeed, most sound propagation observations will be carried out in an ocean region where the velocity variation is much less than 4 percent.

Even though the velocity change is rather small, the sound field solution is much more difficult because of the dependence of c [in Eq. (7-1)] upon position coordinates. An entire research field is opened up to the propagation specialist, and the associated mathematical techniques are indeed sophisticated. Here too, for the limited objective of this chapter, it should suffice to pick out the simple practical results that are provided by the ray theory. The reader can at least appreciate how a sound field is influenced by an inhomogeneity in the ocean or by an air or a sediment boundary.

The Eikonal equation, which is an approximate wave equation, is shown in Eq. (7-14).

$$\left(\frac{\partial W}{\partial x}\right)^2 + \left(\frac{\partial W}{\partial y}\right)^2 + \left(\frac{\partial W}{\partial z}\right)^2 = \frac{c_0{}^2}{c^2} \tag{7-14}$$

In this equation, W (the Eikonal) depends upon the space coordinates x, y, and z and under the right conditions, if $W(xyz)$ satisfies the Eikonal equation, then $p(x,y,z,t)$ also satisfies the wave equation

$$P = A(x,y,z)e^{j(2\pi/\lambda_0)[C_0 t - \omega(x,y,z)]}$$

where $C_0 = C$ at source depth $\lambda_0 = C_0/f$. It can do this only if c and the space rate of change of c (gradient of c) are extremely small on a distance scale of the order of the sound wavelength. Practically this only means that ray theory is in trouble in high-velocity gradient regions of the ocean such as at the edge of the Gulf Stream or at sharp boundaries where c changes abruptly.

A surface defined by $W(xyz)$ = constant is called a wave front. Any line perpendicular to this surface is called a ray, and the essential utility of ray theory is that it provides a means of tracking a ray over its entire path in the ocean. The purpose here is not an exhaustive treatment of ray theory but rather a discussion of those simple and useful results that provide an explanation for the gross behavior of sound in the sea.

The first powerful simplification results from the assumption that the sound speed will depend only upon the vertical coordinate y. This is not a bad assumption in many practical cases, and the utility of ray theory is just as well demonstrated.

x = horizontal range (distance from source)
y = vertical distance from sea surface (positive downward)
s = distance along ray from source
t = travel time along ray path
θ = angle of ray relative to horizontal (positive upward)
$g = dc/dy$ = sound velocity gradient

After some algebraic manipulation, the simple result is called Snell's law.

$$c = c_v \cos\theta \tag{7-15}$$

where c_v is a constant called the vertex velocity. Equation (7-15) says simply that the cosine of the ray angle is proportional to the sound velocity. When any ray bends up and down in the ocean, its entire behavior is governed by this equation. When $\theta = 0$, the ray is then horizontal, $\cos\theta = 1$, and $c = c_v$. The name *vertex velocity* comes from this characteristic, and it follows that regardless of any other considerations, if and when a particular ray ever becomes horizontal, its c and c_v will be the same. (Remember the original assumption, c depends only on depth y.)

To visualize the meaning of this result, consider a source located at a depth y_0 and at $x = 0$. Each ray starts out at a different initial angle, $\theta_1, \theta_2, \theta_3$. They all come from the same source (same depth y_0, therefore same velocity c, Fig. 7-4). The angles are all different and thus the cosines are also different, and it is apparent that each ray has its own unique vertex velocity.

$$c_{v_1} = \frac{c}{\cos\theta_1} \qquad c_{v_2} = \frac{c}{\cos\theta_2} \qquad c_{v_3} = \frac{c}{\cos\theta_3}$$

Clearly, a value of c_v and a particular ray go together, or one might conclude that the value of c_v completely characterizes a given ray.

Some more manipulations, which will not be presented here, result in a most useful set of relations called the *ray equations*, appropriate in a region where g is constant.

$$\begin{aligned} x &= \frac{c_0}{g\cos\theta_0}(\sin\theta - \sin\theta_0) \\ y - y_0 &= \frac{c_0}{g\cos\theta_0}(\cos\theta - \cos\theta_0) \\ s &= \frac{c_0}{g\cos\theta_0}(\theta - \theta_0) \\ t &= \frac{1}{2g}\left(\ln\frac{1+\sin\theta}{1+\sin\theta} - \ln\frac{1+\sin\theta_0}{1+\sin\theta_0}\right) \end{aligned} \tag{7-16}$$

In these equations, the subscript 0 stands for conditions at the source. These equations are quite adaptable to either analog or digital computer application so that rays may be automatically tracked. In fact, there have been special-purpose computers built commercially which do this and rapidly generate an entire family of ray plots.

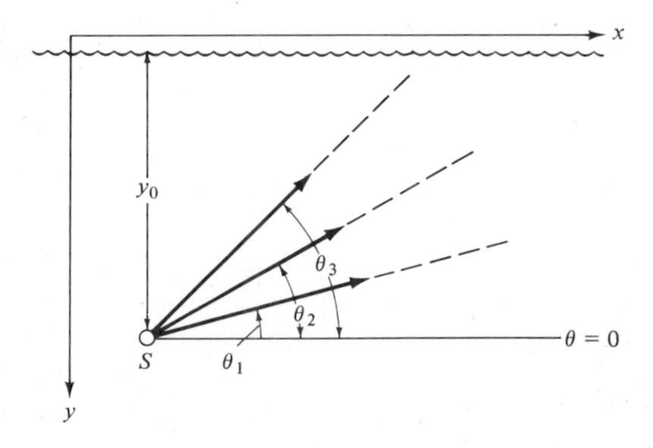

Fig. 7-4 Three sound rays starting from a single point in the ocean.

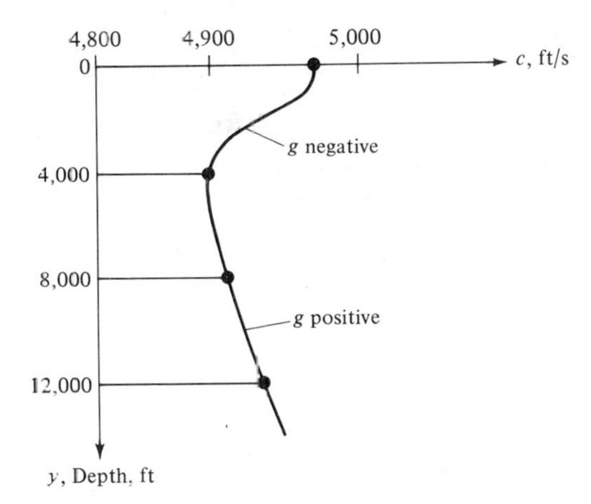

Fig. 7-5 Typical sound velocity profile in the open, deep ocean.

A typical sound velocity profile is shown in Fig. 7-5. In the upper portion, the gradient is negative (velocity decreases as depth increases) and it is obviously positive in the deeper region. In the upper region the temperature change is dominant, and in the lower region the temperature is nearly constant so that the positive gradient is caused by the fact that the depth alone is changing. It can be shown that in a region where the gradient is constant (straight line profile) all rays are circular arcs. The curvature of a ray is proportional to the size of the gradient. If the gradient is positive, the rays curve upward, and, conversely, downward in a negative gradient region. A qualitative example of a set of different gradients is shown in Fig. 7-6. Three different rays are shown, and accurate plots could be drawn by applying the *ray equations* (7-16) to the individual ray initial conditions of angle θ_0 and velocity c_0.

Reference can be made to Fig. 7-6 to point out some other interesting detail. The upper part of the profile that is nearly vertical is often referred to as

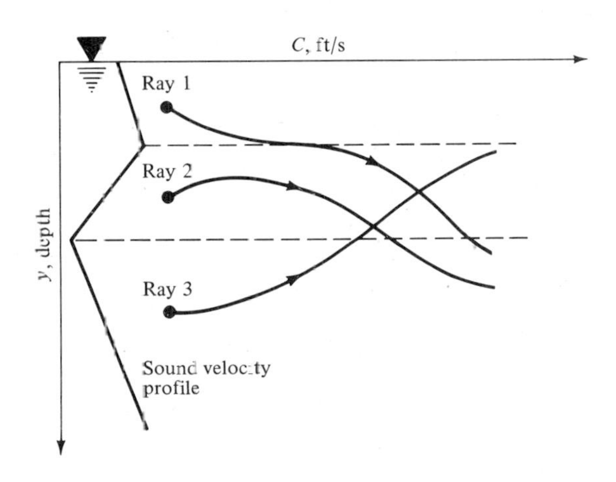

Fig. 7-6 Typical paths of rays starting at different points in the deep ocean.

the mixed layer. This is a variable layer and depends on weather conditions and sea state. The next layer down where the gradient is large and negative is called the main thermocline. The deep region, gradient small and positive, is typical of the deep ocean and is associated with nearly isothermal water. The gradient here is always positive and small, and is due almost completely to the increase in depth. A velocity profile is produced by a device called a sound velocimeter. This is usually an instrument which is lowered on a wire sensing both sound velocity and depth of the instrument. One version is expendable; it is simply dropped to the sea floor, acoustically telemetering its measurements to the surface.

Another form of Snell's law is illustrated by the phenomenon of ray refraction at a flat boundary between two media with different sound velocities. This application is exactly the same as the one in ray optical theory. In Fig. 7-7, a ray is incident on the boundary between regions 1 and 2 at an angle θ_1, relative to the normal. Note that the incidence angle is measured in a manner common to that of physicists. The complement of θ_1 is in more common use in the field of underwater sound; it is called the *grazing angle.*

With the geometry as set up in Fig. 7-7, the angle of refraction θ_2 is obtained from this familiar form of Snell's law. (In this example, it has been assumed that $c_1 > c_2$.)

$$\frac{c_1}{\sin\theta_1} = \frac{c_2}{\sin\theta_2} \tag{7-17}$$

Naturally, if the grazing angles were used in this equation, the sine functions would be replaced by cosines, another common form of Snell's law.

In long-range ocean propagation, a process known as ray convergence and divergence is important. Three rays from the same source are sketched in Fig. 7-8 to illustrate convergence. A source S can produce rays in all directions, but three are sufficient to illustrate the point. A hydrophone moving along at fixed depth below a surface ship traveling between points A and B would experience a

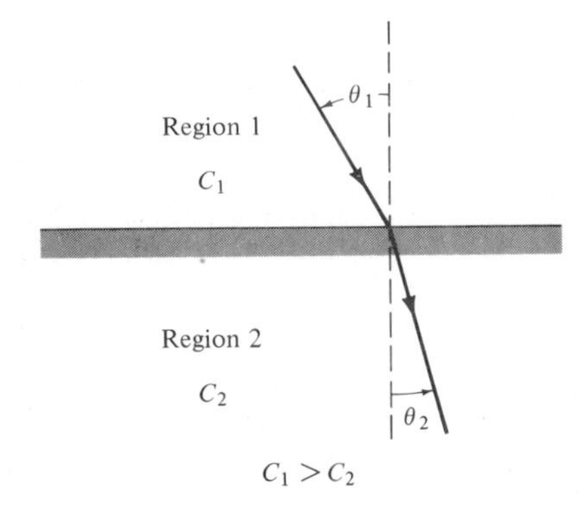

Fig. 7-7 Refraction of a ray entering a region of different sound velocity.

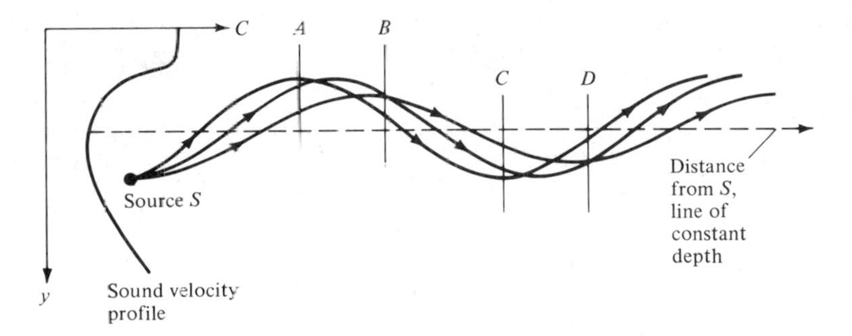

Fig. 7-8 Convergence and divergence of rays from a single source.

strong field because of convergence of rays. The converse is true between points B and C because of ray divergence. The pressure level ratio between a convergent and a divergent zone can easily be 20 to 40 dB or more under the right circumstances.

The profile sketched in Fig. 7-8 illustrates another interesting acoustical feature common to all of the deep ocean. The velocity minimum occurs at a depth of 3,000 to 4,000 ft; this is called the deep sound channel, or SOFAR channel. Above the channel axis the sound velocity gradient is negative so that any ray moving above the axis is bent downward toward the axis. Conversely, a ray moving downward below the axis encounters a positive gradient which curves it upward toward the axis. Thus, all rays are directed toward the axis and the resulting propagation range is phenomenal. A bomb fired at a depth near this SOFAR channel axis can be detected literally thousands of miles in all directions from the shot.

In the above cases, it should be noted that ray bending is caused by the sound velocity gradient. It was assumed that the velocity depended only upon the depth coordinate. For this reason, a sound source carried by a surface ship could launch primarily vertical rays from a directive source aimed downward. Any vertical ray would remain vertical from the source to the bottom and back again. This is how a fathometer works to detect water depth. In shallow coastal waters, a rather high frequency is employed because a directive source is cheap, small, and lightweight. The water column is short enough that fairly high attenuation can be tolerated. For deep water, hundreds of kilohertz cannot be used because of the attenuation that would be encountered. One is forced to a larger, more expensive source to obtain the same directivity and tight beam. The tight beam is essential in precision bathymetry so that only a small spot is *insonified* so as to produce an echo from a point directly beneath the ship. It is obvious that one might be forced to mount the transducer on a stable platform so that ship's roll and pitch can be compensated.

7-10 DETECTION

Underwater sound detection is another very broad topic in itself, but the heart of the subject is the transducer. Much of the discussion on radiators is pertinent here because the majority of transducers work either way. Hydrophones are mainly ferroelectric ceramics or magnetostrictive elements. Several manufacturers have a complete line of hydrophones suitable for nearly any application. The primary specifications furnished by the manufacturer are the frequency response and the receiving sensitivity. As in any other vibrating structure, the size, shape, and material in a hydrophone determine its natural resonant frequency and therefore its frequency response. Generally, the frequency response is displayed in the form of a curve of output voltage as a function of frequency with a constant applied sound pressure level. The receiving sensitivity is given in much the same way as indicated earlier for transmitters. In this case, the reference level is usually volts output per microbar of acoustic pressure.

Another important consideration for the user in selecting a hydrophone is the directivity index. Again here, the same principles apply as in transmitting transducers. The existence of directivity is determined by the size of an active surface of a hydrophone in terms of the wavelength of the impinging sound field. This means of course that a small hydrophone (in wavelengths) is nondirectional. The speed of sound in water is about 5,000 ft/s and so a sound wave at a frequency of 5,000 Hz has a wavelength of 1 ft. Consequently, a hydrophone having a working surface 2 in long will have little directivity at all at low frequency. Depending upon the shape of the active surface, at some high frequency, significant directivity will appear. There are commercially available hydrophones with an active element less than 1 mm (40 mils) in the major dimension. On such a small hydrophone the sensitivity is quite low, but omnidirectionality prevails above 1 MHz.

We have seen that at high frequency, high directivity is rather simple and cheap to produce in a hydrophone. In many ocean engineering and oceanographic applications, where low frequencies are necessary, one is willing to pay the price for sufficient directivity to make a communication system work. This can be accomplished by means of assemblies of nondirectional hydrophones arranged in a line, distributed over a plane, or even throughout a volume of the medium. These devices are called line arrays, area arrays, and volume arrays, respectively. A detailed study of array theory is well beyond the scope of this discussion, but it is not difficult to illustrate the principles.

A pair of identical nondirectional elements (Fig. 7-9) displays a degree of directivity that can be easily determined from Eq. (7-7). It happens to be simpler to develop an expression for the transmitting pattern which is identical to the receiving pattern. This is the total radiated pressure from the two sources (driven in phase) which would be detected by a hydrophone located at a point P. This development is available in every book on underwater acoustics.

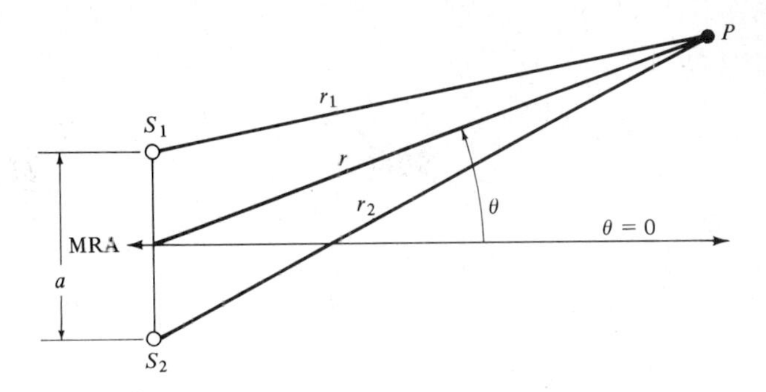

Fig. 7-9 Two sources radiating to a point P.

$$P(\theta) = \frac{A}{r_1} e^{j(\omega t - kr_1)} + \frac{A}{r_2} e^{j(\omega t - kr_2)} \tag{7-18}$$

In Eq. (7-18), A is an arbitrary constant depending on the driving amplitude applied to the transducers. K in the exponent is the wave number ($2\pi/\lambda$). To get a very good approximate result, it is necessary to make one assumption that greatly simplifies the algebra. We make the point P be located a very long distance from the sources so that $r >> a$. Notice that r_1 and r_2 appear in Eq. (7-18) in two different ways, one important and the other relatively unimportant. The difference between r and r_1, r and r_2 becomes unimportant in the denominator as r becomes very large (far field). The radiated pressure from each source is spherically spreading (owing to r in the denominator) and the difference in size between $1/r_1$ and $1/r_2$ is insignificant. On the other hand, the factors kr_1 and kr_2 in the exponents represent phase differences, and these are important.

To follow this phase difference, we write $r_1 = r - a/2 \sin\theta$, and $r_2 = r + a/2 \sin\theta$.

The phase difference between the wave front at S and the origin O is $a/2 \sin\theta$ and it is a phase lead. Between O and S_2 it is also $a/2 \sin\theta$; this time, S_2 lags the point O. The three rays in Fig. 7-10 are assumed parallel, a reasonable assumption if P is a long way from O. Now these expressions for r_1 and r_2 are inserted in Eq. (7-18).

$$p(\theta) = \frac{A}{r}\left(e^{j\omega t - kr + (ka/2)\sin\theta} + e^{j\omega t - kr - (ka/2)\sin\theta}\right)$$

$$= \frac{A}{r} e^{j\omega t - kr}\left(e^{j(ka/2)\sin\theta} + e^{-j(ka/2)\sin\theta}\right) \tag{7-19}$$

$$= \frac{2A^{j(\omega t - kr)}}{r} \cos\psi$$

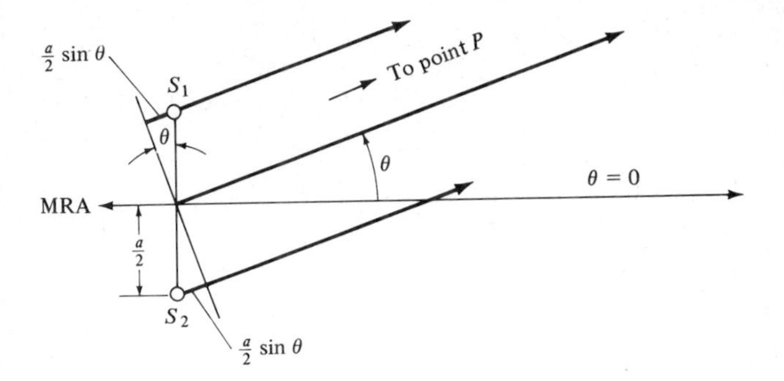

Fig. 7-10 Geometry of two radiating sources in relation to the main axis between them.

where

$$\psi = \frac{ka}{2}\sin\theta = \frac{\pi a}{\lambda}\sin\theta$$

If the two sources were both located at the same point O, the total pressure in the far field would be the coefficient of $\cos\psi$, or $2Ae[j(\omega t - kr)]/r$. The directivity is thus all contained in the $\cos\psi$ part. It is common in directivity expressions to give them as a ratio, the pressure in a direction θ divided by the pressure in a direction $\theta = 0$ (the MRA of the pair of sources). The second form of the equation is simply a trigonometric identity.

$$\frac{P(\theta)}{P_o} = \cos\psi = \frac{\sin 2\psi}{2\sin\psi} \tag{7-20}$$

It would not be suggested here except that the general result for a line array of any number n of elements uniformly spaced a units of length apart and excited in phase is:

$$\frac{P(\theta)}{P_o} = \frac{\sin n\psi}{n\sin\psi} \tag{7-21}$$

To appreciate the utility of the result of Eq. (7-21), we should sketch the pattern that corresponds. To do this, a polar form is most convenient and it is desirable to note that the ψ vs. θ relation contains the all-important separation distance a, in units of λ, the wavelength. Two elements arranged this way constitute a "dipole pair," and a special case results when the distance a is just $\lambda/2$. The resulting pattern is called the "half wavelength dipole" pattern. In this case

$$\psi = \frac{ka}{2}\sin\theta = \frac{2\pi\lambda}{\lambda 4}\sin\theta = \frac{\pi}{2}\sin\theta$$

This is now the fixed connection between ψ and θ. The desired result is the $P(\theta)/P_o$ vs. θ relation. This becomes

$$\frac{P(\theta)}{P_o} = \cos\left(\frac{\pi}{2}\sin\theta\right) \tag{7-22}$$

And a polar plot of this relation is the end result, a beam pattern. When $\theta = 0$, $(\pi/2)\sin\theta = 0$ also, and $\cos\theta = 1$.

This is the maximum response axis. In the direction $\theta = 30^\circ$, $\pi/2 \sin\theta$ is $\pi/4$ and the cosine is 0.707. At 60° the cosine in Eq. (7-22) becomes 0.342 while at 90° it is zero. Note that cosine ψ is an even function so that the pattern is symmetric in the other quadrants.

The result is the familiar "figure 8" pattern of a half-wave dipole pair, a similar result appearing in the electromagnetic antenna literature. A few comments are in order on this result. First, one could at least roughly locate the angle θ to a distant source this way using a dipole pair of hydrophones. One might mount the hydrophones on a stick to maintain the separation distance a at $\lambda/2$. Presumed is a knowledge of the frequency of the source so that λ will be known.

Incidentally, the $\lambda/2$ spacing above does not happen to produce the maximum directivity that is possible with two elements. Another point is apparent from an examination of Fig. 7-11. The maximum is rather broad. That is, it would be difficult to determine the direction $\theta = 0$ to a distant source by manually turning the array over the maximum response direction. On the other hand, once the direction of a distant source was roughly located, it would be advisable to rotate the array by 90° to settle in on the null. The null is very sharp, and the ear would be much more capable of locating a sharp null direction than it would a rather broad maximum. This dipole pair is a perfectly practical and useful arrangement for detecting the direction from which a sound wave is coming.

Equation (7-21) was the result that was presented but not derived for a uniform line array of n elements. With more than two elements in the linear

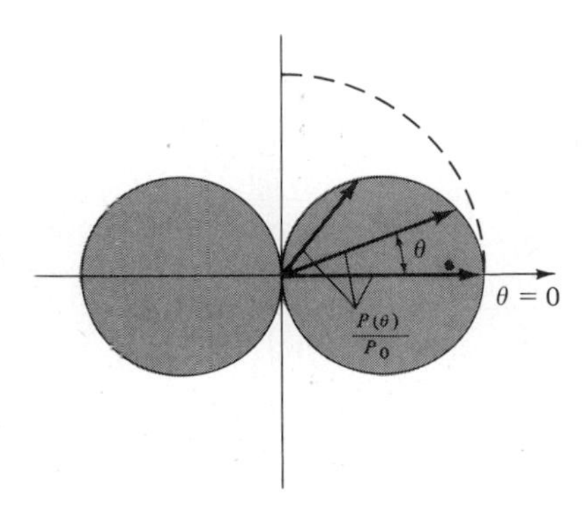

Fig. 7-11 Polar plot of Eq. (7-22) for a dipole pair.

array, minor lobes appear which can be troublesome. In designing a linear array, it is generally desirable to make the major lobe as narrow as possible. Unfortunately, in general, the more narrow the main lobe becomes, the larger in relative amplitude the minor lobes become. This brings up a subject that is also too specialized for this discussion; it is called "array shading." Shading simply means adjusting the excitation of each element in the array to be different, depending upon its position in the array. In general terms, reducing the drive level of the elements toward the ends of the line array will tend to reduce the size of the minor lobes relative to the major lobe. It is possible, in fact, to select an array of n elements and to then specify a desired major lobe width. Then, excitation levels can be selected so as to meet this major lobe requirement with the bonus condition that the minor lobes are as small as possible and all of the same size. This design technique is called the *Chebyschev* distribution. The details of this and several other similar processes are discussed at length in the references and in texts on the subject of electromagnetic antenna theory.

The pattern shown in Fig. 7-12 should be viewed as a cross section of a three-dimensional pattern. It is called a broadside pattern because the MRA is at right angles ($\theta = 0°$) to the axis of the array. Looked at along the axis of the array, the pattern would be circular since the line of the array is an axis of circular symmetry. That all the elements are excited in phase is the reason that the MRA is in the direction shown in Fig. 7-12.

One additional important feature of linear arrays should be mentioned for completeness. One can insert a bit of phase shift (or time delay) in the excitation of an element relative to its neighbors. The beneficial effect is that the main lobe (such as in Fig. 7-12) can be tilted all the way up or down to 90° from the broadside direction. When the extreme of this "steering" is applied so that the MRA (main lobe axis) is in line with the array axis, the resulting array is called "end-fire." Steering is important in an array which is physically large and fixed. Electrical steering can thus be useful when it is not possible to rotate the array

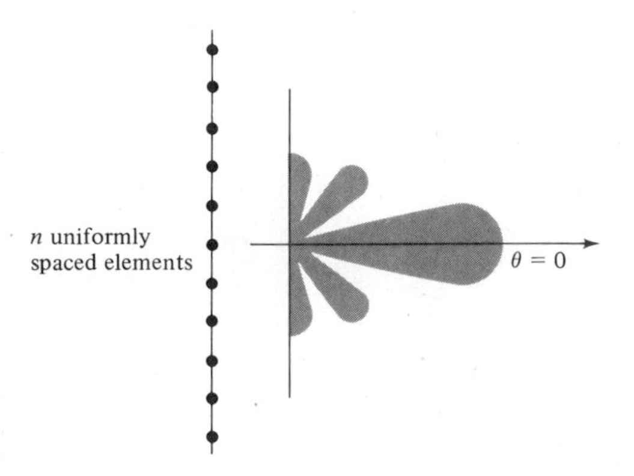

Fig. 7-12 A typical polar plot for a broadside array.

physically. Arrays of up to a mile or two in length are used in some subbottom profiling applications, as in offshore oil surveying.

Before leaving the subject of arrays, it should be mentioned that beyond a linear array, one can assemble an area array of elements and even a volume array. As the name implies, an area array can be formed by distributing isotropic hydrophone elements over a surface, perhaps flat. A simple case for discussion would be a rectangle $A \times B$ in dimensions. If both A and B are large in terms of the wavelength of interest, a large amount of directivity can be obtained.

For the purpose of the present discussion, we shall omit the derivation but show the result and explain it. Suppose there are n elements, uniformly spaced along the A side of the rectangle. In the direction B there might be m uniformly spaced elements. The total number of elements is thus $n \times m$, or one might observe that the rectangular array is constructed by assembling m line arrays, each with n elements. The individual directivity function (pattern function) for a single line array of n elements was given earlier as Eq. (7-21).

$$\frac{P(\theta)}{P_o} = \frac{\sin n\,\psi}{n \sin \psi}$$

where

$$\psi = \frac{\pi a}{\lambda} \sin \theta$$

Figure 7-12 was a rough sketch of a typical line array pattern. The angle θ is measured in the plane of Fig. 7-12, $\theta = 0$ being the "broadside" direction, and $\theta = 90°$ being the end-fire direction. In that case, the angle ϕ might be measured in a plane through $\theta = 0°$ and normal to the array axis. The pattern in Fig. 7-12 is intuitively independent of the angle ϕ as Eq. (7-21) shows because the pattern would be exactly the same at any angle ϕ, since the array axis is an axis of perfect circular symmetry.

In the rectangular array case, this is not true and the shape of the pattern will indeed be a function of the angle ϕ as well as θ. The result is still quite simple and appears as in Eq. (7-23).

$$\frac{P(\theta, \phi)}{P_o} = \frac{\sin \psi_A}{\psi A} \frac{\sin \psi_B}{\psi B}$$

where (7-23)

$$\psi_A = \frac{\pi A}{\lambda} \sin \theta \cos \phi \qquad \psi_B = \frac{\pi B}{\lambda} \sin \theta \sin \phi$$

It should be noted that each factor in Eq. (7-23) is identical to Eq. (7-21). That is, the present result is the product of two line array pattern functions, one A in length and the other B in length. The fact that the two line arrays are at right angles to each other is the reason for the complication of ψ_A and ψ_B functions over the Eq. (7-21) case.

This result is important in itself but it also illustrates an array principle called the "product theorem." Equation (7-21) stayed quite simple for one important reason. Each of the n elements was nondirectional or isotropic. Suppose this was not the case. All is not lost, because if the n elements are identical, even though not isotropic, the resulting pattern function can be obtained by multiplying the "pattern function" by the "array function." Equation (7-21) is the array function in this case. Let each element have a pattern function $P_1(\theta, \phi)/P_0$. The total directivity function of such an array of directive elements would be

$$\frac{P(\theta, \phi)}{P_o} = \frac{P_1(\theta, \phi)}{P_o} \frac{\sin n\,\psi}{n \sin \psi}$$

This is the meaning of the product theorem.

Without going into many details, the reader will not be surprised to learn that a volume array may be handled by simple extensions of the above discussion. Volume arrays (and area arrays) enjoy the features of shading and steering, and they are indeed used in some highly capable radiating and listening systems. In such cases, the array design, shading computations, and steering time delays are readily programmed on a computer. There is a large amount of literature on the subject of array design and utilization, and so we may terminate this discussion with the hope that the reader has at least an intuitive feel for how one may obtain and use directivity by means of arrays.

There is a practical subject for ocean engineers that should not be overlooked in a discussion such as this. Reference is made to Eq. (7-17) and Fig. 7-7. There we were considering a plane wave striking a boundary between two different media at an angle relative to the normal. In one very important ocean engineering application, namely, reflection profiling, the incidence will be normal. That is, the sound rays strike a boundary at right angles. This is true in the usual fathometer, a rather high-frequency, tight-beam ocean bottom detector.

In such a case, a rather simple and important relation prevails. A pressure reflection coefficient (only valid at normal incidence) is exactly analogous to the voltage reflection coefficient on an electromagnetic transmission line. The quantity $\rho_1 c_1$ is called the plane wave specific acoustic impedance of region 1. Likewise, $\rho_2 c_2$ is the impedance of region 2, wherein ρ_2 is the density and c_2 is the propagation velocity of the same region. The reflection coefficient gives the ratio of the reflected pressure amplitude to the incident pressure amplitude.

$$R_P = \frac{P_1^-}{P_1^+} = \frac{\rho_2 c_2 - \rho_1 c_1}{\rho_2 c_2 + \rho_1 c_1} \tag{7-24}$$

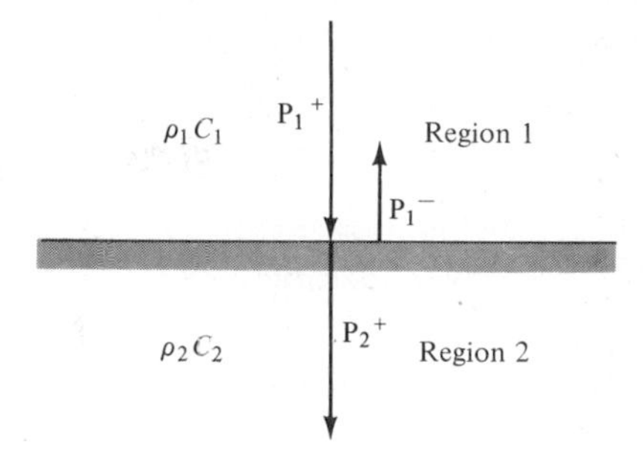

Fig. 7-13 Nomenclature of a vertical ray striking a bottom having properties different from water. See Eq. (7-24).

The interesting point here, for either subbottom profiling or for bathymetry, is that there must be a difference in the acoustic properties of media at a boundary in order to produce an echo. The symbol P_1^- in Eq. (7-24) indicates the reflected pressure amplitude and P_1^+ the incident amplitude.

In the fathometer application, all that is necessary is that the sea floor exhibit an impedance significantly different from that of the overlying seawater. Generally this condition is met but not in all regions of the sea floor. Indeed, sediments generally have an impedance considerably higher than that of seawater so that strong echoes are produced. There are regions, however, where the impedance is only slightly different, and sometimes even smaller than that of seawater. In such cases, the echo is very weak. In subbottom profiling, the various sediment strata do have different impedances so that the boundary between adjacent layers can produce a respectable echo such that the profile is distinct.

The same relation, Eq. (7-24), applies to reflection at the air-sea interface resulting from a plane wave propagating upward in the sea. In this case, $\rho_2 c_2$ may be considered as corresponding to air, and $\rho_1 c_1$ to the water. In mks units, it was pointed out earlier that the impedance of seawater is about 3,000 times as large as that of air. The important point here is that the reflection coefficient becomes nearly -1. The negative sign means that the phase of the reflected plane wave pressure is exactly out of phase with the incident pressure. The sum of the incident pressure and the reflected pressure is almost zero. For this reason, such a reflecting boundary is called a "pressure release" boundary. The opposite extreme is something like a flat rock sea floor, where any particle velocity associated with an incident wave from water must nearly vanish. In this case, the pressure nearly doubles because the pressure reflection coefficient approaches $+1$. Thus, the sum of the incident pressure and the reflected pressure is about twice the incident pressure. The term "rigid boundary" is often employed here.

From the previous discussion related to plane waves and flat boundaries, it is clear that if the boundary is not flat, the conclusions will be different. A rippled boundary produced by water current over the boundary (even in deep water) can produce a complicated reflection picture. If the size of the ripple structure, particularly the peak-to-peak distance of the ripples, is comparable to the acoustic wavelength, the reflected wave pattern is anything but the one discussed above for flat boundaries. Lord Rayleigh discussed this case at length in his book "Theory of Sound."

In addition to locating the sea floor, another highly important ocean engineering application is bottom search or object location. The picture should be at least qualitatively clear by this point in the discussion. An object must first of all have acoustical properties different from the water and from the sediment upon which it sits. Any object has a property called the "target strength" which gives a measure of the strength of the echo which it can produce. It depends upon the impedance of the object and its size and shape. Here too the size is determined on the basis of the wavelength of the sound wave employed. Generally, the best reflector is an object like an air-filled balloon. A solid metal object in seawater can also be a good reflector, but not as good as an air-filled volume of the same size.

It is of some practical importance to an ocean engineer to put a number on the ability of an underwater object to produce an echo. This is contentionally accomplished in one of two ways; a target strength or a scattering cross section. These two quantities are naturally related and the interested reader can readily find the relations in the references. It will suffice for the present purpose to give a verbal and a mathematical definition of the target strength of an object.

Let us use a fish as an example and suppose he is swimming in deep water, far from any boundaries such as the sea surface or bottom. Suppose a plane sound pulse is passed through the water region where the fish is located. First, the excess acoustic pressure in the plane wave is measured or determined in some other way. As the sound wave packet passes by the fish, his swim bladder or other parts of his body will be set in motion and this motion produces a radiated sound field of its own. In general, some radiation will be produced in all directions from the fish. For purposes of determining the target strength, however, the one direction back toward the source of the acoustic pulse is singled out. One has only to measure the pressure amplitude associated with the echo at a fairly large distance from the fish. The only other piece of necessary data is the distance from the fish to the observation point. One assumes spherical spreading from the fish and converts the observed pressure to the corresponding pressure at a standard reference distance of 1 yd or 1 m. If we denote this last pressure mentioned as p_1 and the incident pulse pressure p_0, then the target strength T is given.

$$T = 20 \log \frac{p_1}{p_0} = 10 \log \frac{I_1}{I_0} \tag{7-25}$$

Now we see that a given fish has a given target strength which is ideally independent of frequency. The literature contains tabulated values of target strength for various fish, and it should be noted that those species with large swim bladders have correspondingly high target strengths. Much of the body of most fish is a poor reflector of acoustic energy simply because it has a specific acoustic impedance much like that of water.

The usual search sonar is called *active* when the system supplies its own acoustic energy. When the radiator and the receiving hydrophone are mounted on the same platform or ship, the system is called *monostatic*. In contrast, if the radiator is on one ship and the receiver on another, the term is *bistatic*. A most important search tool is a *side-scan sonar*, so called because it produces an acoustic scan out to the side of the vessel. It has been used in shallow water mounted on a surface vessel or in the deepest parts of the ocean, where it is carried by a towed housing or a manned submersible. In either application the result is a fair approximation of a line-by-line optical photograph of the sea floor. See Fig. 7-14.

The same highly directive transducer is used as the radiator and receiver. Frequencies generally employed are several hundred kilohertz since the desired range is perhaps 50 to a few hundred feet out to the side of the transducer. The transducer is often only 6 in to 1 ft in maximum dimension since the wavelength is so small. Still, the main lobe can be about a degree wide between the 3-dB points. The transducer is oriented so that this tight beam is aimed out to the side (often both sides) of the ship track. The acoustic rays strike the bottom at a

Fig. 7-14 A side-scan sonar picture of a mushroom anchor (bowl downward, shank upward) connected to chain running from upper left to right center. Note the sound "shadow" cast by the shank (white line pointing upward). (*Photo. F. H. Middleton.*)

small grazing angle. Any bottom protrusion or object that protrudes above the bottom will produce an echo or a pattern of echoes that results in a line-scan picture. The final display form can be generated on the face of an oscilloscope screen, on a TV screen, or on a facsimile recorder paper.

Ocean engineers and ocean scientists are finding good uses for more and more of the smaller acoustic devices called beacons and transceivers. A beacon can be quite small, simple, and inexpensive. One type is about the size of a small flashlight, and it has its own batteries and activating devices. The main use is to attach it to the outside of an expensive instrument that might be lost in a cable failure or some such accident. The beacon either can be operated all the time or can be started by any number of devices such as time delay devices, corrosive links, or pressure switches. The beacon radiates high-frequency pulses so that one can readily home on the radiator to accomplish an instrument recovery.

Beacons or transceivers are used in increasing numbers for precise navigation in both deep and shallow water. The transceiver is so named because it is capable of transmission and reception. In a typical open ocean navigation application, the surface vessel can drop several transceivers spaced a mile or two apart on a triangle buoyed a small height above the sea floor. Each one can be interrogated or called by its own coded acoustical pulse from the surface ship. It is arranged so that it will respond by transmitting another coded pulse, perhaps at a different frequency so that the round trip can be accurately timed. A rather simple program on a minicomputer on the surface ship can produce a fix with a maximum error of the order of 1 m.

Many instruments and devices that have not been mentioned in this discussion are commercially available. Space does not permit covering all of them, even if there were not new applications coming along continuously. It is hoped that enough different acoustic systems have been mentioned so that the reader has developed a practical feel for the important considerations that go into using underwater sound. In many cases, it will be necessary to consult the more complete discussions in the references, and the reader is encouraged to do this. A reasonable engineering background is all that is necessary to follow up specific details in this field because many excellent basic books have appeared in recent years.

PROBLEMS

7-1. A calibrated hydrophone is lowered from a surface ship into a sound field that is known to be a plane sinusoidal wave. According to the manufacturer's sensitivity data, the excess acoustic pressure appears to be 200 μbar.

(*a*) What is the pressure in newtons per square meter?

(*b*) What is the magnitude of the particle velocity in meters per second?

(*c*) What is the intensity in watts per square centimeter?

7-2. The directivity factor was defined as a ratio of total powers.

$$d = \frac{1}{4}\iint \left[\frac{P(\theta, \phi)}{P_0}\right]^2 d\Omega$$

For a two-element array with spacing a, the result of the integration process can be shown to be:

$$d = \frac{1}{2}\left(1 + \frac{\sin(2\pi a/\lambda)}{2\pi a/\lambda}\right)$$

When a source is highly directive, this factor d is much less than 1, and so $D = 10 \log 1/d$ is large. For this two-spot (dipole) source, what is the spacing a, in wavelengths λ, that will produce the largest directivity index? What is the maximum directivity index?

7-3. Some interesting insight can come from simple manipulation of Eq. (7-13), the rough empirical sound velocity equation. Find the derivative of the velocity with respect to depth $g = dc/dy$. From the result, determine the size of the temperature gradient which will produce isovelocity conditions (i.e., $g = 0$ when $T = 50°\text{F}$; when $T = 30°\text{F}$).

7-4. A section of the ocean has a positive constant velocity gradient of 0.01 ft/(s)(ft). If a sound ray is launched at 10° below the horizontal, at what range will it reach the same depth as the source? Repeat for a ray 20° below the horizontal.

7-5. A sound velocity profile is as shown with a negative gradient region above a positive gradient region, much as in the case of a SOFAR channel in the deep ocean. An acoustic source S is situated as shown, and a ray starts out from the source horizontally at a depth of 4,000 ft. Use the ray equations to determine:

(*a*) The range to point A.

(*b*) The angle of the ray at point A, θ_A.

(*c*) The range to the vertex point B in the upper region.

(*d*) The depth of the vertex point B.

(*e*) The range X_C.

(*f*) The travel time t_c from the source to the point C.

(*g*) Suppose a ray is launched horizontally directly above the source. How long will it take to cover the shorter straight path to point C?

(*h*) Why does this shorter ray arrive 0.1 s later than the larger ray above?

Fig. P7-5

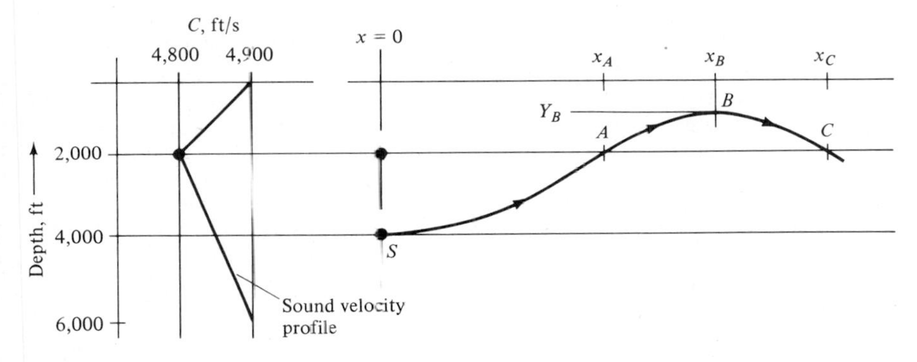

7-6. The pattern function for the two-point array with in-phase excitation is:

$$\frac{P(\theta)}{P_0} = \cos \psi$$

where

$$\psi = \frac{\pi a}{\lambda} \sin \theta \quad \frac{P(\theta)}{P_0} = \cos\left(\frac{\pi a}{\lambda} \sin \theta\right)$$

(*a*) Suppose $a = \lambda/4$ and make a cartesian plot of $P(\theta)/P_0$ as a function of θ for $-\pi/2 < \theta < +\pi/2$.

(*b*) Convert this to a polar plot.

(*c*) Suppose $a = \lambda/2$ and make a cartesian plot of $P(\theta)/P_0$.

(*d*) Convert this to a polar plot.

7-7. An object is suspended in the ocean far from any boundaries and it is struck by a plane wave acoustic pressure pulse of 100 dyn/cm^2. The echo from the object is detected by a calibrated hydrophone placed on a line toward the source 20 m from the object. The excess acoustic pressure detected by the hydrophone is 0.1 dyn/cm^2. What is the target strength of this object? What is the spreading loss (spherical) on the echo from the object to the hydrophone in decibels?

7-8. An omnidirectional acoustic source and a target are located 2,000 m apart in isovelocity water well removed from all boundaries. The source produces 1-kW sinusoidal acoustic pulses and the target has a strength of 0 dB. At the operating frequency, the water produces an attenuation of 2 dB/km.

(*a*) What is the source level in decibels per microbars per meters?

(*b*) What is the PL at the target?

(*c*) What is the PL of the echo at the source location?

(*d*) What is the corresponding pressure?

(*e*) What is the corresponding intensity in watts per square centimeter?

7-9. A soils lab determines that a core has the following properties: wet specific gravity of 1.35 and sound velocity of 1,610 m/s. The water alone has a specific gravity of 1.025 and sound velocity, *in situ*, of 1,531 m/s. What is the reflectivity of this bottom? How many decibels loss is suffered by a sound wave striking the bottom from a vertical direction?

7-10. We have a receiver with sensitivity of -80 dB//V/μbar and a transmitter which provides a sound pulse of 100 dB//1 μbar, 1 yd. For the bottom material described in Prob. 7-9, how deep can we transmit a pulse from a surface vessel and detect a 1 = μV return with the noted hydrophone? Assume no energy dissipation.

7-11. For the system described in Prob. 7-10, energy attenuation by water is found to be 2 dB/km. What does the bottom-detection range now become?

7-12. A number of studies have suggested that the human ear in water suffers a loss of about 50 dB compared to air, owing to the poor coupling between the ear and sound waves in water. An older diver has a 10-dB loss at 1,000 Hz referenced to 0.0002 μbar. How strong a sound pulse must be generated (referenced to 0.0002 μbar) such that he will hear it 40 ft from a nondirectional transducer? (The noted diver loss is an air loss due to ear damage from pressure accidents.)

REFERENCES

Bartleberger, C. (1965): "Lecture Notes on Underwater Acoustics," Defense Documentation Center (AD 468 869), Washington.

Kinsler, L., and A. Frey (1962): "Fundamentals of Acoustics," Wiley, New York.

Lindsey, R. B. (1960): "Mechanical Radiation," McGraw-Hill, New York.
Officer, C. B. (1958): "Introduction to the Theory of Sound Transmission," McGraw-Hill, New York.
Rayleigh, J. (1945): "Theory of Sound," vols. I and II, Dover, New York.
Urick, R. (1967): "Principles of Underwater Sound for Engineers," McGraw-Hill, New York.

8

Underwater Life Support and Diving

Hilbert Schenck, Jr.

8-1 INTRODUCTION

Diving and life support theory forms an offshoot from the more traditional areas of ocean engineering in that it is closely allied with bioengineering subjects as well. The ocean engineer who is not primarily involved in life support research, development, or immediate use is still likely to be concerned with the subject in regard to its applications and possibilities. He usually wants to know the answer to questions such as "How deep?" "How much work?" "How many men needed?" "How long can they stay down?" and so on. Some underwater tasks can be accomplished only with divers. In other cases, the diver competes with alternate systems involving remote handling, deck-operated TV and cameras, midget submarines, bells, dry habitats, and so on.

In this chapter we will introduce the basic physiological factors that set limits on both wet diving and 1-atm enclosed situations (such as a deep-diving research submarine). The Haldane theory of stage decompression will be derived, both because of its historic importance and because extensions of this theory still govern short-term diving operations in deep water. We will then describe the various types of diving equipment and derive the air-demand equations for these devices. Heat transfer both to and from the body will be reviewed. Finally, we will discuss the various environmental problems that beset the short-term and saturated diver including propulsion, buoyancy and body-work considerations, and other important limits that prevent the diver from achieving the same work rate that might be expected dry on the surface.

8-2 BASIC PHYSIOLOGICAL PARAMETERS

Since the bones, fat, muscles, and blood of the human body are relatively incompressible, they do not suffer damage when subjected to elevated surrounding hydrostatic pressure, provided that any inner, gas-filled spaces can equalize their pressure with that of the environment. This is accomplished in the lungs and throat either by providing respirable gas at local pressure or, in the case of a free (nonequipment) dive, by collapse of the chest cavity at a rate that maintains lung and surrounding pressures equal. In the case of the inner ear, gas must flow through the Eustachian tube from the throat to equalize pressure. Otherwise, the eardrum will suffer inward rupture. Similarly, sinuses must have access for gas flow in and out if serious tissue destruction is to be prevented.

The most important physiological data needed by engineers involved in ocean life support problems deal with respirable gas demand. Figure 8-1 shows the general shape of the volume-time curve of an average adult lung (after Miles 1966) for two cases, (*a*) when the body is relatively quiescent and (*b*) when metabolic demands require absolute maximum ventilation, called the *vital capacity*.

The absolute volume of gas pumped by a diver's lungs per minute (called the *minute volume*) is equal to the product of the breathing or *respiration rate*, breaths per minute, and the *tidal volume*. As a general rule of thumb, experienced divers meet respiratory needs by breathing slower and deeper whereas the novice usually obtains the same minute volume by breathing faster and shallower. Figure 8-1 defines other lung parameters of interest: the *inspiratory reserve, expiratory reserve,* and *residual volume*. Table 8-1 gives reference values for "small," "average," and "large" lungs.

A respiration rate of 10 to 12 breaths per minute is typical for a "resting" condition; an absolute maximum of 120 breaths per minute might be possible for very brief periods. Sustained breathing requirements for bodily output work of the order of 1/3 hp are found to be in the minute-volume region of 2.5 ft^3, or 70 l. Some tests are run at what is called the *maximum breathing capacity*, using forced breathing timed by a metronome; this is the region of 5.0 ft^3/min, or 140 l/min.

Figure 8-1 shows that the breathing cycle is roughly sinusoidal in shape but with a longer "tail" produced by the spring of the chest wall. To a first approximation, we can assume a perfect sinusoid having the equation:

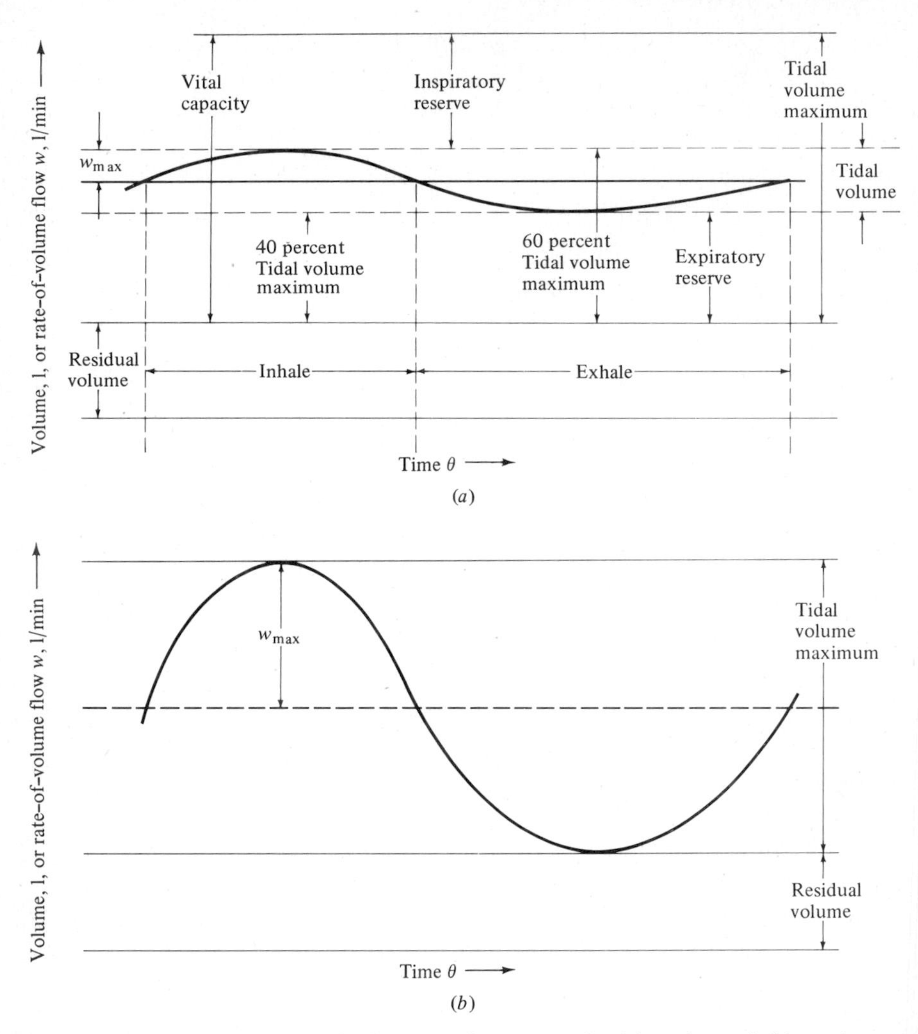

Fig. 8-1 Volume-time or rate-of-volume vs. time curves for (*a*) resting and (*b*) maximum ventilation. Note that lung volume and rate of volume are out of phase by 90°. The time of one cycle is not necessarily the same in the two cases.

$$\dot{w} = \dot{w}_{\max} \sin \frac{\theta}{\theta_{\max}} \tag{8-1}$$

From the properties of the sine curve we know that the average flow over an inhale or exhale period is

$$\dot{w}_{\text{av}} = 0.637 w_{\max} \tag{8-2}$$

Although the exhale is actually a little longer, we can assume the exhale and inhale periods to be equal so that the average gas *demand* (say from a scuba tank to a diver) over the complete cycle is given by

$$\dot{w}_{\text{av,cycle}} = 0.318 w_{\text{max}} \tag{8-3}$$

or, allowing for the longer "tail" of the cycle shown in Fig. 8-1, an approximate equation might be

$$\dot{w}_{\text{av,cycle}} \cong 0.3 w_{\text{max}} \tag{8-4}$$

The meaning of these equations to designers of respirable gas flow apparatus is the following: peak flows in the apparatus will be over *three times as great* as the average flow based on minute volume. Thus passages must be sized to accept the peak flow with minimum pressure drop.

In normal air at 1 atm pressure, the oxygen partial pressure is approximately 3.0 psi with nitrogen making up the remaining 11.7 psi. When this mixture is taken into the lungs, some oxygen is removed and carbon dioxide replaces it. In addition, the lung may evaporate some moisture into the air. Under normal conditions, the exhaled air contains CO_2 at a partial pressure of almost exactly 40 mm Hg (or 0.77 psi) pressure while the oxygen partial pressure has fallen from about 3.0 psi to 2.3 psi or less. This exhaled air still contains sufficient oxygen to supply the metabolic requirements of a man. However, it also contains CO_2 at a pressure of 40 mm Hg, which is the equilibrium CO_2 pressure in the normal lung. If this "used" air is now rebreathed, there will be no CO_2 "washout" from the lung since the CO_2 partial pressure in the lung and in the air is equal. This means the body will begin to accumulate CO_2 or undergo *acidosis*. Since the breathing cycle is triggered, in part, by the partial pressure of CO_2 in the lungs and blood, the breather will begin to breathe more deeply, to gasp, and, eventually, to become unconscious.

Diving and respiratory apparatus must therefore accomplish two basic jobs on the air it will process: the oxygen partial pressure should be maintained at a value of at least 3 psi, and the CO_2 partial pressure must be as low as possible, but definitely less than 40 mm Hg. For short-term situations, CO_2 partial pressures approaching 40 mm Hg may be tolerated, although headaches and disorientation can result. For long-term breathing, a CO_2 partial of less than 7.6 mm Hg (1 percent of sea-level total pressure) should be sought.

This means that when air is supplied from a tank, as in aqualung apparatus (demand scuba), considerable oxygen is "wasted" because the air is used once and then "thrown away." The rebreathing type of gear is much more economical

Table 8-1 Typical lung volumes

	Vital capacity		*Residual volume*	
	Liters	*ft³*	*Liters*	*ft³*
Small	3.0	0.105	1.0	0.035
Medium	4.5	0.158	1.5	0.053
Large	6.0	0.21	2.0	0.070

in gas demand, since ideally it must only add new oxygen, but we pay for this economy in greater gear complexity.

When oxygen pressure drops to below its normal 3-psi value, the possibility of *anoxia* or oxygen want exists. Whereas excessive CO_2 accumulation produces gasping and other warning signs, oxygen starvation occurs in a stealthy manner and unconsciousness may occur without warning. Oxygen uptakes of swimmers and divers have been measured under various conditions. Table 8-2 suggests typical values (from Miles 1966).

Notice that there is no absolute relation between oxygen uptake, minute volume, respiration rate, and tidal volume. While these all tend to rise together, as the diver begins to exert, their interrelations depend on the individual man and his response to stress.

Whereas oxygen partial pressures less than 3.0 psi may produce anoxia, oxygen pressures in excess of 2 atm may produce *oxygen poisoning*. This disorder, whose cause is still somewhat obscure, results in seizures and convulsions that can be deadly to divers at work. Since normal air at a total pressure of 144 psia will contain oxygen at a partial pressure of about 29 psi, oxygen poisoning sets this limit (equal to about 300 ft of seawater plus 1 atm from the air) on compressed-air diving. However, even this limit cannot be reached under normal conditions since the average diver will suffer *nitrogen narcosis* at this depth. When normal air is breathed at pressures in excess of about 5.5 atm (equivalent to 150 ft of seawater plus 1 atm from the air), the diver may become euphoric and intoxicated to the extent of taking his air-supply hose out of his mouth. Actually, narcosis due to nitrogen starts at shallower depths than 150 ft, and sets depth limits on air diving. Helium does not produce narcosis and is now used as an inert-gas diluent in all deep-diving activities.

Summarizing, dives to depths of less than 150 ft on air should give no drastic difficulties, unless the user is particularly prone to narcosis. Dives in excess of 150 ft should be performed with a helium-oxygen mixture, with the oxygen partial pressure in the mixture held well below 29.4 psi.

Table 8-2 Typical oxygen demand values

Condition	*Oxygen uptake (at standard temperature and pressure)*
Divers walking in mud	2.35 l/min
Swimming with fins, 1½ mi/h	3.16 l/min (mean value), 4.15 l/min (maximum value)
Underwater swimmers, 0.7 to 0.9 mi/h	1.3 to 1.9 l/min
Resting, quiescent	0.3 to 0.4 l/min

8-3 BOYLE'S LAW AND DIVING

The inverse relation between total pressure and volume is basic to all diving computations. Boyle's law, in diving terms, is often written as

$$V_0(33) = V_d(D + 33) \qquad (8\text{-}5)$$

where V_d is the volume of gas at seawater depth D, in feet, and V_0 is the volume of the same mass of gas at the sea surface. For freshwater, the constant, 33 ft, changes to 34 ft. The application of Boyle's law to the previous material can best be shown in two short examples:

Example 8-1 A skin diver with an "average" lung takes a full breath and swims downward in seawater. How deep can he safely go before crushing his rib cage?

Solution Referring to Table 8-1, we note that the average lung has a vital capacity (maximum breath intake) of 4.5 l. When all breath is driven out, the minimum volume of the collapsed lung is given as 1.5 l. From Fig. 8-1 we see that a fully inflated lung at the surface will have (1.5 + 4.5) or 6.0 l of air. At sea level, the pressure in feet of seawater is 33 ft, and D_{max} is our unknown depth at which the 6.0-l lung will be squeezed to the minimum allowable value of 1.5 l. Thus, Eq. (8-5) becomes

$$6.0 \text{ l } (33 \text{ ft}) = 1.5 \text{ l } (D_{max} + 33 \text{ ft})$$

so D_{max} is 99 ft. Actually, a somewhat deeper dive is possible for persons with reasonably springy chests. However, a dive too far below the 100 ft mark could produce chest collapse and internal bleeding, a so-called *lung squeeze*.

Example 8-2 A standard aqualung tank contains 70 ft³ of air at a reference pressure and temperature of 1 atm and 60°F. A comfortable (suited) diver with a "large" lung is resting at sea level, and at 66 ft in seawater. How long does he have in 60°F water at each location?

Solution From Fig. 8-1 we note that the tidal volume for a resting person is about 20 percent of his vital capacity or, from Table 8-1 and for a large lung, (0.2×0.21 ft³) or 0.042 ft³ per breath. We suggested a resting person might take 10 to 12 breaths per minute. Taking 12 breaths, we obtain a minute volume of 12×0.042 or 0.504 ft³/min. At sea level and 60°F, the capacity of the tank is 70 ft³ of gas, so the number of minutes available at the standard condition is 70/0.504 or 140 min. The actual time available is slightly less, since when the tank blows down to 14.7 psi internal pressure (from a starting pressure of about 2,200 psia), a small amount of gas will remain inside. At 66 ft under water, the pressure is equivalent to 99 ft of seawater or three times atmospheric. Gas from the tank will thus emerge at a greater density and Eq. (8-5) yields (for the 70 ft³ at standard pressure)

$$70(33) = V_d(66 + 33)$$

V_d at 66 ft is now 70/3 or 23.4 ft³ and the time available is reduced to 23.4/0.504 or 46.6 min. Again, the true time will be somewhat less since now the flow will stop when the tank interior pressure falls to 3 atm.

8-4 INERT GAS SATURATION AND DECOMPRESSION

The problem of nitrogen narcosis can be prevented by (1) holding a depth limit of 125 to 150 ft or (2) using helium as an inert diluent. However, any inert gas in a breathing mixture will gradually dissolve in the body's tissues, and once absorbed, it may produce bubbles and blood stoppage when the tissue is suddenly brought to a lower pressure. This is the *bends* or *decompression sickness*, and it is the primary danger in deep diving as well as a major factor in the engineering costs of diving activity.

Oxygen does not dissolve to any extent in simple solution in blood or tissue. The presence of an inert gas in the lungs (usually nitrogen or helium) means that the lung blood will saturate immediately with this gas to whatever partial pressure it may have in the lung, value P_e. This saturated blood now flows through a section of tissue having a lower partial pressure of dissolved inert gas of value P. If we assume that the area of contact between the blood in the vessel is A, and the resistance to transfer across the vessel membrane is R, then the process by which inert gas flows from the blood to the tissue is given by the simple differential equation

$$\frac{dP}{dT} = \left(\frac{A}{R}\right)(P_e - P) \tag{8-6}$$

where dP/dT is the rate of rise of tissue, inert-gas pressure with time, and $P_e - P$ is the inert-gas pressure difference between blood and tissue.

Let us postulate the simple case of a diver having his tissues saturated at 80 percent of 1 atm of nitrogen, P_{at}, suddenly going to a higher bottom pressure, still breathing normal air. His blood is now instantly saturated to this new, higher nitrogen pressure P_e which is 80 percent of the bottom pressure. We can integrate time from zero to T, the time spent breathing the nitrogen pressure of P_e on the bottom. Equation (8-6) becomes

$$P = P_e - \frac{P_e - P_{at}}{e^{AT/R}} \tag{8-7}$$

Clearly, the tissue nitrogen pressure P will approach the new higher nitrogen pressure P_e following this exponential law. A/R is not easily predicted. Dr. J. Haldane, the English physiologist who first successfully carried out this type of

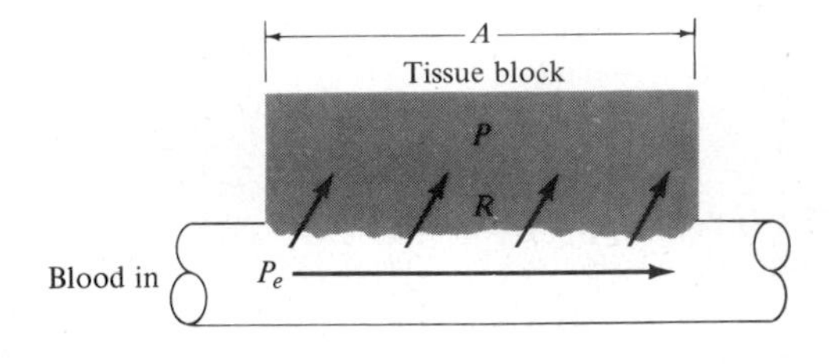

Fig. 8-2 Idealized tissue with nitrogen diffusing from a blood vessel.

study, suggested defining the tissue characteristic A/R in terms of the time it took a given tissue to *half-saturate*. That is, we write Eq. (8-6) in the following way:

$$\int_0^{P_e/2} \frac{dP}{P_e - P} = \frac{A}{R} \int_0^{T_{1/2}} dt$$

integrate, and solve for A/R getting

$$\frac{A}{R} = \frac{\log_e 2}{T_{1/2}} \tag{8-8}$$

Equation (8-8) can now define a tissue by the amount of time it takes to reach half the bottom partial pressure of nitrogen, providing the tissue started its test with zero nitrogen pressure.

The insertion of Eq. (8-8) in Eq. (8-7) will give us a way of estimating the inert gas pressure rise in any tissue for which we may know the half-saturation time. However, the problem is to bring the diver back to the surface without inert-gas bubbles forming in his tissues. From the previous arguments, it is apparent that the desaturation equation will be

$$-\frac{dP'}{dT'} = \frac{A}{R}(P' - P'_e) \tag{8-9}$$

The prime symbols now define the *ascent* parameters. P' is the tissue inert-gas pressure at any point in the ascent, and P'_e is the blood inert-gas pressure at the same point in the ascent. The ascent desaturation equation (8-9) shows that inert-gas elimination can be maximized by making the difference $(P' - P'_e)$ as large as possible. This, in turn, dictates as fast a rise as possible. Haldane learned, however, that a tissue that was brought to a surrounding pressure less than half its internal, dissolved, inert-gas pressure would develop bubbles. He suggested that, with a normal-air dive, the tissue inert-gas pressure P' and the surrounding nitrogen pressure P'_e should be related by a simple proportion

$$P' = nP'_e \tag{8-10}$$

where n of 2.0 was found to be safe. If we now put Eq. (8-10) in Eq. (8-9) and integrate from P–found by Eq. (8-7)–to P' and from 0 to T', we obtain a safe ascent curve for a given tissue:

$$T' = -\frac{n}{n-1}\frac{R}{A}\log_e \frac{nP'_e}{P} \tag{8-11}$$

Example 8-3 Haldane postulated that the longest half-time for any tissue for relatively short dives was about 75 min. Assume a dive to 200 ft for 1 h using normal air. What will be typical tissue nitrogen pressures at the end of the dive?

Solution Consider a tissue with $T_{½} = 40$ min. From Eq. (8-8) $A/R = 0.692/40 = 0.0173$ (1/min). Now Eq. (8-7) gives

$$P = 0.8(233) - \frac{0.8(233) - 0.8(33)}{e^{0.0173 \times 60}} = 128 \text{ ft}$$

Similar computations for other tissues give P(10 min) = 182.5 ft, P(20 min) = 165 ft, P(75 min) = 93 ft. The 10-min tissue is almost completely saturated at sea bottom nitrogen pressure (185 ft), while the 75-min tissue is only about half-saturated. Remember that all tissues started with some nitrogen (26 ft), since the diver is living at this pressure when not diving.

Example 8-4 Assuming an ascent is to be made after this 1-h period with n equal to 2.0, what ascent times are specified by the several tissues?

Solution Again taking the 40-min tissue as an example, we note that its nitrogen pressure P is 128 ft. P_e is surface nitrogen partial pressure, or 26 ft. Then Eq. (8-11) becomes

$$T' = -\frac{2}{2-1}\frac{1}{0.0173}\ln\frac{52}{128} = 103 \text{ min}$$

Notice that an ascent based on this tissue will start with an immediate rise to a surrounding nitrogen pressure of 128/2 or 64 ft. If nitrogen is 80 percent of the mixture, this is a total pressure of 64/0.8 or 80 ft and an actual water depth of (80 − 33) or 47 ft. Such a schedule would be dangerouw for the 10-min tissue, which cannot have a surrounding nitrogen pressure less than (182.5/2) or 91.3 ft at the start of the ascent. This is a true depth of 91.3/(0.8 − 33) or 81 ft. The time to desaturate the 10-min tissue is only 40 min. Figure 8-3 shows the saturation and desaturation for the four tissues. Clearly, an ascent based on the slow tissues will permit an initial reduction in pressure that is too great, while following the fast tissues will not

Fig. 8-3 Tissue saturation and desaturation for 1-h dive to 200 ft. Solid lines are tissue pressures, dotted are safe ascent schedules for the noted tissues. The stepped schedule is the U.S. Navy decompression schedule for the whole body for this dive.

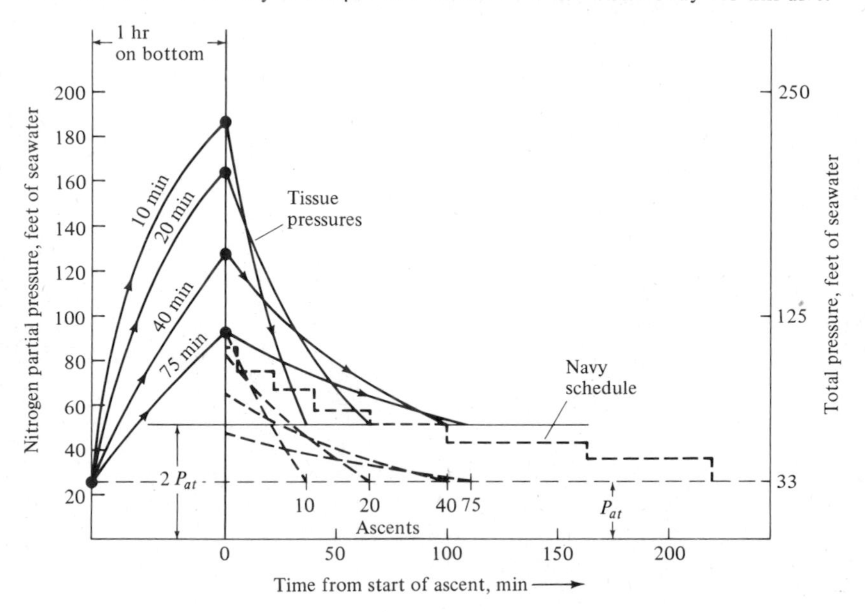

provide sufficient time to get nitrogen out of the slow ones. Thus a complete *decompression schedule* (the stepped line in Fig. 8-3) must be computed with its early part based on fast tissues and its later part based on slow tissues. In fact, the early part of such a decompression schedule may see the slow tissues still gaining some nitrogen. The theoretical schedule is a continuous curve. The stepped approximation shown in the figure was adopted because it is more easily specified and followed. There is one obvious way to improve this situation. If we can introduce pure oxygen where it will be safe (when a total pressure of 2 atm is reached), then we effectively reduce P'_e in Eq. (8-9) to zero. That is, the nitrogen in the tissues "sees" blood that is completely free of dissolved nitrogen. Integration of Eq. (8-9) now gives

$$T'_{ox} = \frac{A}{R} \ln \frac{P}{P''} \tag{8-12}$$

T'_{ox} is the time spent under oxygen and P'' is the allowed nitrogen pressure at the end of the rise.

Example 8-5 How long would the 75-min tissue in Example 8-4 take to desaturate under pure oxygen?

Solution We found that this tissue had 93 ft of nitrogen, which is equivalent to a total pressure of (93/0.8) or 116 ft. We can go to half this, or 58 ft without bubble formation. Since 2 atm (66 ft) is safe with oxygen, we can use oxygen throughout a desaturation period with this tissue controlling. Then Eq. (8-12) gives, with P'' equal to 52 ft, the allowed end nitrogen pressure:

$$T'_{ox} = \frac{75}{0.692} \log_e \frac{93}{52} = 64 \text{ min}$$

Figure 8-3 shows that this same tissue would require 115 min with air decompression.

When helium-oxygen "air" is used, various constants undergo changes. n is less than 2.0, the half-time of various tissues changes, and in general decompression after short, deep dives with helium it is not shorter than after equivalent air dives. However, a mixed-gas dive is easily adapted to pure oxygen above the 2-atm level. In fact, most helium, short-term diving utilizes oxygen decompression starting at a total pressure of 93 ft (almost 3 atm). Divers that convulse readily because of oxygen sensitivity are rapidly weeded out of this type of diving.

Figure 8-3 shows that a deep, rather brief dive requires huge decompression times (almost 4 h in this case). The economic answer is to *saturate* the diver, that is, leave him down for several days or until his job is over, then decompress him at once. Unfortunately the Haldane theory that has been reviewed here falls apart with whole body saturation. The saturated body cannot tolerate an immediate rise to half its former pressure; sometimes a rise of 5 percent of depth will produce bends. Furthermore, computations carried out assuming a maximum tissue half-time of 75 min are drastically in error. For example, if we solve Eq. (8-11) for a fully-saturated 75-min tissue at 300 ft under water, we obtain a rise time of about 6 h. Actually, a saturated test (dry) dive at the Experimental Diving Unit in Washington, D.C., to 300 ft required a uniform ascent from 300 ft to sea level at the rate of 11 min/ft, the entire ascent requiring over 2 days. Apparently, long-saturating tissues never noted by Haldane are governing saturated diving decompression.

It is the bends and their prevention that control much deepwater and offshore engineering work today. A great deal remains to be done both in obtaining better understanding of the phenomenon and in optimizing the diving system for maximization of useful bottom activity.

8-5 GAS REQUIREMENTS FOR DIVING

Diving apparatus can be divided into three basic types: *demand* gear such as the aqualung, *flushing* gear such as the traditional "hard hat" equipment supplied with air from the surface, and *recirculating* gear, in which CO_2 is chemically removed and oxygen is added. Each type has its own theoretical gas supply requirements, as will be shown.

A. Demand equipment We have already illustrated the method of estimating the gas requirements with compressed air and demand regulators in Example 8-2. Since the regulator in demand scuba equipment supplies a volume of air equal to the tidal volume of the diver, and then discharges the exhaled air overboard, the air requirements are simply given by (see Fig. 8-4*a*):

$$\text{Sea-level demand gas per minute} = \text{minute volume}\ \frac{33}{D + 33} \tag{8-13}$$

where D is the depth at which the diver is working. Demand unit computations are usually based on sea-level volumes. That is, the standard demand scuba tank contains 70 ft^3 of gas at standard pressure and temperature so that Eq. (8-13) will tell us how many standard cubic feet are used per minute.

B. Flushing equipment Most diving over the past 100 years has been carried out by a diver in a rigid helmet, an attached rubber suit, and a constant air supply from the surface supplied by hand- or motor-driven pumps. Figure 8-4*b* shows the general nomenclature of such a flushing system. We saw in Sec. 8-2 that dangerous CO_2 accumulation occurred more rapidly than dangerous oxygen decrease. Thus, if we can maintain the CO_2 level in the flushed system of Fig. 8-4*b* at a safe level, we can be certain that sufficient oxygen will be present. As the figure suggests, we will make a volume balance on the CO_2. V_{CO_2} will be the partial volume of CO_2 at any time in the system having a fixed total volume of V_t. Then the rate of change of CO_2 volume with respect to time, dV_{CO_2}/dT, will equal the rate of CO_2 production, $\dot{v}_{CO_2}$, minus the rate at which CO_2 is flushed from the system. This, in turn, will be the flow rate of air out, $\dot{v}_{air}$ times the partial volume fraction of CO_2 (V_{CO_2}/V_t). Or:

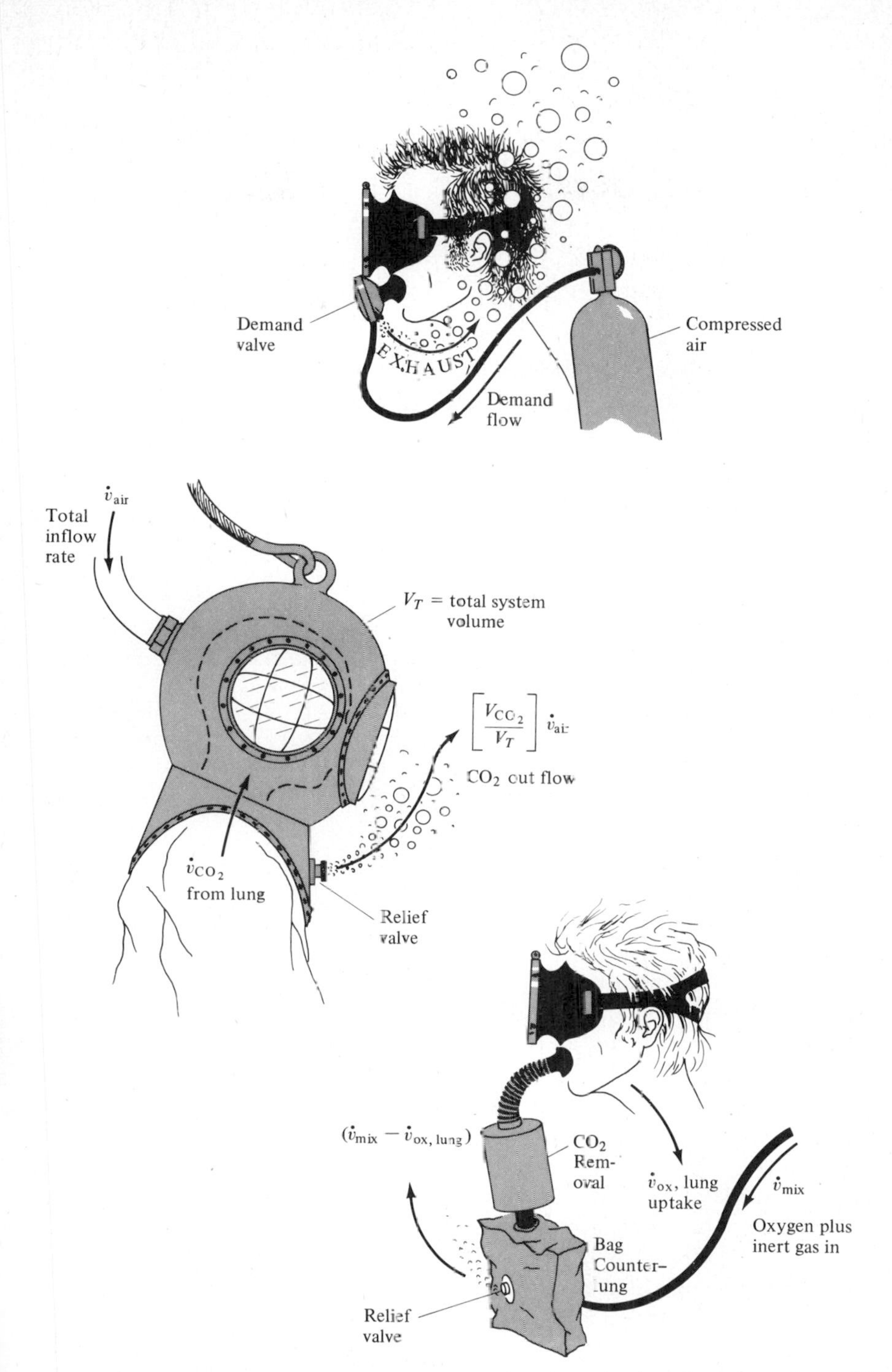

Fig. 8-4 (*a*) Demand scuba air requirements are based on depth and minute volume; (*b*) flushed diving gear; (*c*) recirculating diving gear.

$$\frac{dV_{CO_2}}{dT} = \dot{\nu}_{CO_2} - \frac{V_{CO_2}}{V_t}\dot{\nu}_{air} \tag{8-14}$$

This can be separated for integration

$$V_t \int_0^{V_{CO_2}} \frac{dV_{CO_2}}{V_t\dot{\nu}_{CO_2} - V_{CO_2}\dot{\nu}_{air}} = \int_0^T dT$$

We will assume that the inflow rate of air, $\dot{\nu}_{air}$, and the production rate of CO_2, $\dot{\nu}_{CO_2}$, remain constant. We are assuming that the CO_2 level is zero at time zero (when the dive starts). Then integration gives:

$$V_{CO_2} = \dot{\nu}_{CO_2}\frac{V_t}{\dot{\nu}_{air}}(1 - e^{-T(\dot{\nu}_{air}/V_t)}) \tag{8-15}$$

It is much more convenient to change Eq. (8-15) to a partial pressure form using the identity

$$\frac{V_{CO_2}}{V_t} = \frac{P_{CO_2}}{P_t}$$

which is simply a version of the well-known law (the ratio of partial to total volume equals the ratio of partial to total pressure). Then Eq. (8-15) becomes

$$\frac{P_{CO_2}}{P_t} = \left(\frac{\dot{\nu}_{CO_2}}{\dot{\nu}_{air}}\right)(1 - e^{-T(\dot{\nu}_{air}/V_t)}) \tag{8-16}$$

Although this derivation is based on the assumption that both CO_2 production and inlet air flow are constant with time, it is a simple matter to reintegrate Eq. (8-14) from limits V_{CO_2} to V'_{CO_2} and T to T' where V'_{CO_2} and T' are new conditions after a change in $\dot{\nu}_{air}$ or $\dot{\nu}_{CO_2}$ has occurred.

Figure 8-5 shows the prediction of Eq. (8-14) during flushing experiments on the University of Rhode Island *Portalab* habitat. The upper curve shows a reduction in flushing rate at about 95 min from diver entry and a further reduction at 155 min, then finally a return to full flow 175 min after the start of the test. The theory curve "fits" the data points best with a CO_2 production (sea-level value) of 0.362 l/(man)(min). The lower curve is a similar experiment with a single occupant, a value of 0.408 l/(man)(min) proving better in this case since the single diver is more active while taking data. Data scatter is typical for the small color-tube systems used to obtain CO_2 percentage.

Now Eq. (8-16) is in two parts, a "steady-state part" and a "transient part." When enough time has passed so that the exponent $T(\dot{\nu}_{air}/V_t)$ has a value of 3.0 or more, Eq. (8-16) reduces to

$$\frac{P_{CO_2}}{P_t} = \frac{\dot{\nu}_{CO_2}}{\dot{\nu}_{air}} \qquad T \text{ large} \tag{8-17}$$

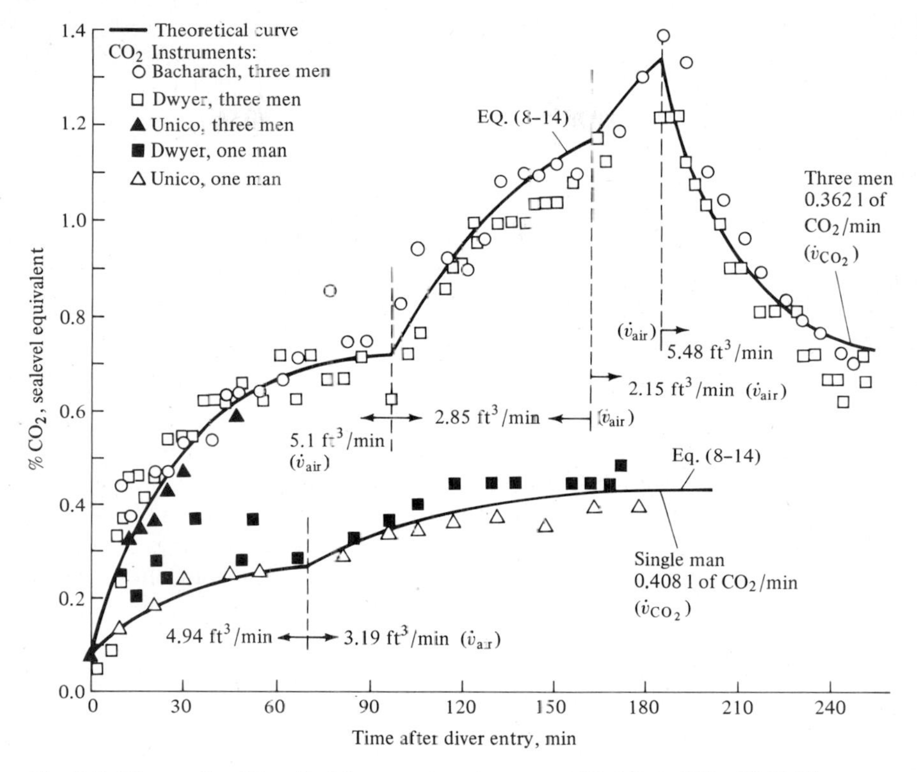

Fig. 8-5 The results of two flushing experiments on a small habitat of V_t of 155 ft³.

Clearly, the term $(V_t/\dot{v}_{air})$ is a *time constant* of the system. When the time constant is very large, as might be the case in a flushed habitat (Fig. 8-5) or diving bell (V_t big), the time to reach an equilibrium CO_2 partial pressure will be long.

Example 8-6 An underwater habitat is to contain six diver-scientists at a depth of 250 ft. Its total internal volume is 2,500 ft³. Assuming that the men will be usually resting inside the unit and that helium "air" is to be circulated through a CO_2 scrubber at some rate $\dot{v}_{air}$ and reintroduced into the unit, estimate the gas rate. How long will it take to reach equilibrium?

Solution Since habitats are designed for long-term use, we must reduce CO_2 equilibrium pressure to less than 1 percent of 1 atm as noted in Sec. 8-2. Let us choose half this value, 0.005 atm or 0.073 psi for our desired CO_2 pressure. P_t at this depth is 8.53 atm or 81 psia. We note in Table 8-2 that a good value for resting oxygen uptake is 0.35 l/min at standard conditions. Since the reaction equation converting O_2 to CO_2 is $O_2 + C = CO_2$, we might assume that one volume of oxygen would yield one volume of CO_2. This is not quite correct, however, since the *respiration quotient* (the ratio of CO_2 output to O_2 input) is not unity but about 0.85. Thus for each 0.35 l of O_2 uptake, the breather puts out 0.35 × 0.85 or 0.296 l of CO_2. There are six men, and so this is an output of 0.296 × 6 or 1.78 standard l/min. At 8.53 atm Boyle's law gives a volumetric output of (1.78/8.53) or 0.21 l/min of CO_2 at depth. Equation (8-17) gives

$$\dot{\nu}_{\text{air}} = 0.21 \frac{8.53}{0.005} = 357 \text{ l/min}$$

Since 1 ft³ is 28.3 l, this is a CO_2 scrubber flow of 12.6 ft³/min. The time constant is $V_t/\dot{\nu}_{\text{air}}$ or 2,500/12.6 or 198 min. Taking three time constants as a rough measure of time to equilibrium, we have a transient period of about 10 h for this layout. If the men are to do anything more than sleep and play checkers, this design flow may be a little low (see Fig. 8-5). If light work is involved, a flow of approximately double would be needed to maintain the CO_2 partial pressure at 0.005 atm.

C. Recirculating equipment Recirculators come in various forms. In the most modern and complex types, electronic oxygen detectors automatically sense partial pressure and supply either additional oxygen or inert diluent to maintain the desired level. Figure 8-4*c* shows simpler equipment in which a fixed inflow of oxygen plus inert diluent enters the breathing unit. The total volumetric inflow is $\dot{\nu}_{\text{mix}}$ and the rate of oxygen partial volume inflow is $\dot{\nu}_{\text{ox,in}}$. Some of this oxygen is taken up by the diver's metabolic needs, $\dot{\nu}_{\text{ox,lung}}$, and the remainder escapes into the water. The system contains a CO_2 scrubber, so there is no CO_2 present. Thus the total outflow is the inflow less the oxygen uptake or $(\dot{\nu}_{\text{mix}} - \dot{\nu}_{\text{ox,lung}})$. The net loss of oxygen due to gas outflow is then the total outflow rate times the ratio of system oxygen volume V_{ox} to system total volume V_t. The rate at which system oxygen volume changes with time (dV_{ox}/dT) is then equal to the entering oxygen rate minus the amount metabolized plus the amount flushed, or

$$\frac{dV_{\text{ox}}}{dT} = \dot{\nu}_{\text{ox,in}} - \dot{\nu}_{\text{ox,lung}} - \frac{V_{\text{ox}}}{V_t}(\dot{\nu}_{\text{mix}} - \dot{\nu}_{\text{ox,lung}})$$

This can be separated and integrated over the limits $V_{\text{ox},1}$ to $V_{\text{ox},2}$ and 0 to T. The result is

$$\frac{V_{\text{ox},2}}{V_2} = \frac{\dot{\nu}_{\text{ox,in}} - \dot{\nu}_{\text{ox,lung}}}{\dot{\nu}_{\text{mix}} - \dot{\nu}_{\text{ox,lung}}} - \frac{\dot{\nu}_{\text{ox,in}} - \dot{\nu}_{\text{ox,lung}}(\dot{\nu}_{\text{mix}} - 1) - \dot{\nu}_{\text{mix}} V_{\text{ox},1}}{\dot{\nu}_{\text{mix}} - \dot{\nu}_{\text{ox,lung}}} \exp\left(-T \frac{\dot{\nu}_{\text{mix}} - \dot{\nu}_{\text{ox,lung}}}{V_t}\right) \quad (8\text{-}18)$$

Again we have an equation with a steady-state part and a transient part. Since mixed gas recirculators are usually personal devices with small volumes, the time constant $[V_t/(\dot{\nu}_{\text{mix}} - \dot{\nu}_{\text{ox,lung}})]$ is usually short. Thus we can write Eq. (8-18) in its steady-state form, noting as before that $P_{\text{ox}}/P_t = V_{\text{ox},2}/V_t$.

$$\frac{P_{\text{ox}}}{P_t} = \frac{R_{\text{ox}}\dot{\nu}_{\text{mix}} - \dot{\nu}_{\text{ox,lung}}}{\dot{\nu}_{\text{mix}} - \dot{\nu}_{\text{ox,lung}}} \quad (8\text{-}19)$$

R_{ox} is now a new term for the partial-volume ratio of oxygen in the entering gas so that $R_{ox} = \dot{v}_{ox,in}/\dot{v}_{mix}$. There are two kinds of application for this sort of rebreather: in one the diver returns to surface at the end of his dive, and in the other he is saturated and returns to a habitat or transfer capsule at bottom pressure. Taking the case of surface return first, it is essential that the P_{ox}/P_t ratio in Eq. (8-19) never drop below 0.2. At, say, a total pressure of 4 atm, this ratio could be safely set at 0.05, and the man would still be getting surface equivalent oxygen partial pressure. If, however, he had to surface suddenly, the oxygen pressure would reach a value of only one-quarter of the normal value and anoxia could occur before the man got out of the water. Putting 0.2 in Eq. (8-19) and solving for $\dot{v}_{mix}$ gives

$$\dot{v}_{mix} = \frac{0.8\dot{v}_{ox,lung}}{R_{ox} - 0.2} \tag{8-20}$$

The next question in the surface-return case is the proper value for R_{ox}. Here oxygen poisoning is the controlling factor. If the mixture is too rich in oxygen, and the diver goes too deep, he may convulse. If he should rest at depth, his uptake would be small and his diving unit oxygen partial pressure would approach that of the entering gas. Taking a 2-atm (66-ft) limit on oxygen partial pressure, we can set R_{ox} from the equation

$$R_{ox} = \frac{66}{D_{max} + 33} \tag{8-21}$$

where D_{max} is the anticipated maximum depth at which the unit will be used.

Example 8-7 Design the gas mixture and flow for a recirculating surface-return diving unit with a depth limitation of 100 ft.

Solution Using Eq. (8-21), we obtain a maximum safe R_{ox} value of 66/133 or about 0.5. A typical design oxygen uptake for normal work is 2 l/min. However, Table 8-2 suggests that values as high as 4 standard l/min are possible. Then Eq. (8-20) can be solved at sea-level conditions:

$$\dot{v}_{mix} = \frac{0.8(4)}{0.5 - 0.2} = 10.7 \text{ l/min}$$

Generally, the unit would have a constant mass flow so that both $\dot{v}_{mix}$ and $\dot{v}_{ox}$ would decrease with depth, but the actual gas use would be independent of depth of operation. Even though we have based this design on very high uptakes, notice that the unit used only a little more than 1/3 ft^3/min, much less than is required from a demand unit by a man resting at sea level (see Example 8-2).

When a saturated diver is using a recirculator of this type, the situation changes. He cannot return to the surface, no matter what his situation, since he would be prone to bends. The ratio P_{ox}/P_t needs only to be whatever is suitable and safe for the depth of operation.

Example 8-8 A saturated habitat at 610 ft will supply helium-oxygen "air" to a diver through an umbilical hose. Oxygen pressure will be at 150 percent of surface value. Specify the gas flow to a recirculator.

Solution Oxygen pressure will be 10.4 ft of seawater with helium making up the remainder of the mixture, so R_{ox} is (10.4/643). Since the diver will be working hard in some cases, we might choose a sea-level oxygen uptake from Table 8-2 of 4 l/min. At 643 ft of total pressure this becomes (33/643) × 4, or 0.205 l/min. Normal oxygen partial pressure is 6.7 ft of seawater. Then Eq. (8-19) gives

$$\frac{6.7}{643} = \frac{(10.4/643)\dot{\nu}_{mix} - 0.205}{\dot{\nu}_{mix} - 0.205}$$

and $\dot{\nu}_{mix}$ is 30 l/min. Such a flow of over 1 ft³/min of dense helium is clearly excessive from an expense standpoint, unless a double hose layout is possible with the exhaust brought back to the habitat. The difficulty can be eased by enriching the habitat umbilical flow with added oxygen. A perfectly safe value would be an oxygen partial pressure in the hose of 50 ft. Then the term 10.4/643 becomes (50/643), and the umbilical flow is 3 l/min. This is still expensive to throw away, but with such a small flow, the use of return hoses of small diameter may become practical.

8-6 THERMAL PROTECTION OF DIVERS

To a first approximation, the body of a diver plus any protective thermal covering can be treated as a simple, one-dimensional thermal circuit, as suggested in Fig. 8-6. The adjustable tissue thermal resistance $1/KS$ is the inverse of the product of the mean thermal conductivity K of the tissues (mainly fat and muscle) and a tissue "shape factor" S. This shape factor is inversely related to the

Fig. 8-6 (*a*) Thermal circuit of an unsuited diver; (*b*) thermal circuit of a diver plus full protective suit; (*c*) thermal circuit of diver plus full suit with heating elements on inner face delivering Q Btu/h.

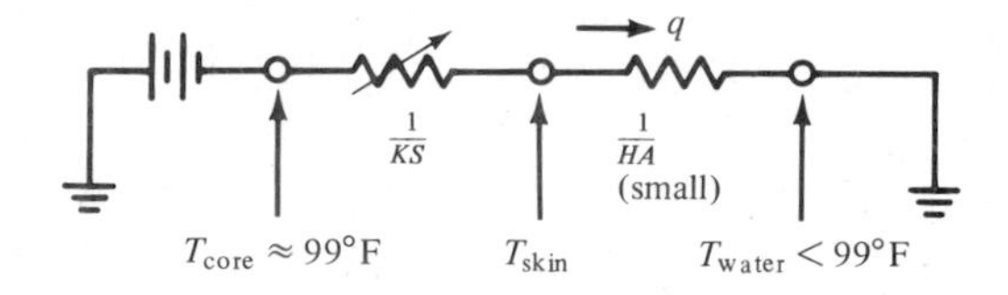

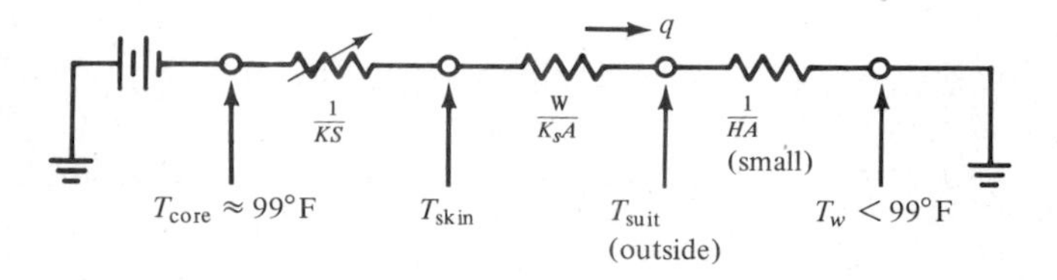

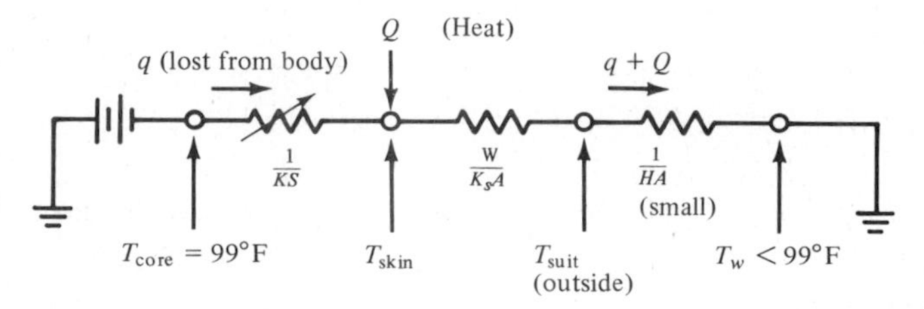

Table 8-3 Thermal parameters of human subjects in water

Subject	*KS, Btu/(h)(°F)*	*Water temperature (°F)*	*Heat loss, Btu/h [used in Eq.* (8-22)]
Obese man resting	9.0	50	440
Obese man resting*	26.1	61	990
Obese man resting	41.0	91	330
Obese man resting	146.0	97	294
Obese man swimming*	59.5	61	2260
Thin man resting*	53.5	61	2040
Thin man resting	41.0	70	1190
Thin man resting	41.0	75.5	960
Thin man resting	41.0	91	330
Thin man swimming*	105.0	61	3960

*The starred and nonstarred data are from different subjects and under different conditions.

mean geometrical heat flow path length between the main core of the body (at temperature T_{core}) and the skin surface, and has typical units of inches or centimeters. S is a complex and variable quantity which is controlled by skin temperature, core temperature, and other mental and physical factors. In water approaching the core temperature of 98.6°F, the blood vessels open and blood flows close to the surface (*vasodilation*) and S becomes large. In cold water, the vessels close and keep more of the blood in the body core away from the cold so that heat is conserved (called *vasoconstriction*) and S is small.

A diver with no suit has two resistances between his core temperature (which the body attempts to maintain at slightly less than 99°F at all costs) and the surrounding water at T_w. These are $1/KS$ and $1/HA$, where H is the heat-transfer coefficient between skin and water and A is the body surface area (about 21 ft^2 for an average adult male). This is the circuit shown in Fig. 8-6*a*. If we assume the value of H is large [30 to 100 Btu/(h)(°F)(ft^2)] then the $1/HA$ resistance is minor and can be ignored to a first approximation. The simplified heat flow equation for the unsuited diver is then

$$q = KS(T_{core} - T_w) \qquad (8\text{-}22)$$

A set of consistent data on heat loss experiments on fat and thin subjects can be found in a paper by Beckman (Second Symposium on Underwater Physiology, 1963). We can use Eq. (8-22) to convert Beckman's data to the KS form (assuming A is 21 ft^2, $T_{core} = 99$°F).

Table 8-3 reveals a number of important points about the way the body controls its thermal output and maintains its core temperature:

1. The fat man has greater variation of his shape factor than the thin man. Between 50°F and 91°F his KS rises from 9.0 to 41.0. The thin man has a constant KS of 41.0 from 70°F to 91°F.

2. Exercise produces vasodilation since muscles must be supplied with blood. The effect is far more drastic as far as increasing KS and heat loss is concerned with the thin man than the fat man.
3. Thermal protection for the fat man needs to be far less extensive at any given water condition than for the thin man. Protective suits of varying thickness should be available. Thermal control requirements vary greatly from one man to the next.

Figure 8-6*b* shows the body thermal circuit with a protective suit in place. Again, we can omit the $1/HA$ resistance, which is tiny compared to that of the suit. The heat loss equation is now:

$$q = \frac{(99 - T_w)}{(1/KS) + (W/K_s A)} \tag{8-23}$$

where w is the suit thickness, K_s the suit thermal conductivity, and A about 21 ft^2 for the average adult. Table 8-4 contains useful conductivity values taken from several end-chapter references, included for design purposes.

In cold, deep water, the use of auxiliary heating is often considered. Heating wires bonded to the inside of a wet suit and supplied by a battery or

Table 8-4 Thermal conductivities

Substance	*K, Btu/(h)(°F)(ft)*
Dry wool clothing	0.023–0.054
Foamed neoprene, 15.6% solid material (7.9 lb/ft^3)	0.031
Solid neoprene	0.108
Natural rubber	0.08
Natural foam rubber	0.025
Butyl rubber	0.05
Rubber-impregnated cloth	0.108
Polyvinyl chloride (PVC)	0.07–0.10
Cotton fabric	0.046
Cast epoxies	0.1–0.8
Silica glass	0.8
Pyrex glass (140 lb/ft^3)	0.65
Corkboard (7 lb/ft^3)	0.0216
Fine glass wool (1 lb/ft^3)	0.02
Ice	0.128
Carbon dioxide	0.0079
Helium	0.0870
Hydrogen	0.084
Oxygen	0.0145
Nitrogen	0.0148
Freon 13	0.00633
Human fat	0.097
Human muscle	0.166

small hot-water tubes is a typical method used at present. Figure 8-6c shows the thermal circuit of this type of system. Notice that the external heating source is not intended to supply heat to the body. In fact, all its heat flows out to the water. The intent is rather to increase the temperature level at the skin, thereby making the diver more comfortable and reducing heat loss across the $1/KS$ resistance. Solving the Fig. 8-6c circuit for T_{skin} (with $1/HA$ assumed small as before), we obtain

$$T_{skin} = \frac{99KS + (K_sA/W)T_w + Q}{[KS + (K_sA/W)]} \tag{8-24}$$

where Q is the heat input in typical units of Btu/per hour.

Example 8-9 A thin man is in 45° water. A ⅛-in-thick foamed-neoprene suit (density of 7.9 lb/ft³) will be used, with or without extra electric heating inserted between the skin and the inner face of the suit. Design the system.

Solution Make a trial first with the suit alone. From Table 8-3, KS for a thin man in the comfort zone is 41.0 Btu/(h)(°F), and so $1/KS$ is 0.0245 (h)(°F)/Btu. K_s from Table 8-4 is 0.031 Btu/(h)(°F)(ft) and W/K_sA is (0.0208/0.031 × 21) or 0.032 (h)(°F)/Btu. Then Eq. (8-23) gives

$$q = \frac{(99 - 45)}{0.0245 + 0.032} = 960 \text{ Btu/h}$$

This computation assumed the skin temperature was above 70°F; otherwise the KS chosen from Table 8-3 is not correct Knowing the flow, we can solve for the temperature drop across the $1/KS$ resistance, $99 - T_{skin} = 960 \times 0.0245 =$ and $T_{skin} = 75.5$°F. This is reasonable for the chosen KS for a thin man. This loss, however, is considerable for, say, a 1-h dive. Let us now add electric heating in the amount of 400 Btu/h, which is possible with belt-mounted batteries. Solving Eq. (8-24) will give a new T_{skin}:

$$T_{skin} = \frac{(99 \times 41.0) + (31.3 \times 45) + 400}{41.0 + 31.3} = 81.0°\text{F}$$

This is still within the chosen KS range. The heat loss from the body is now (99 − 81)/0.0245 or 740 Btu/h, only 220 Btu/h less than with no electric heating. The extra heating would be more efficiently utilized with a thicker suit, so that the difference between skin and water would become greater with a given heater input. For example, a 3/16-in-thick heated suit, with all else the same, would give a T_{skin} of 87°F and a loss of 490 Btu/h.

Data from the Sealab II material (Project Sealab Report, 1967), on how the heat should be distributed over the body, are given in Table 8-5. The table shows clearly the importance of protecting the hands and feet in very cold water, even if such protection may impede the diver's activities to some extent.

8-7 WORK AND MOTION IN THE WATER

The work required to propel a swimming or towed diver can be found from the usual drag equation (with D the drag in pounds-force)

$$D = (C_d A)\rho \frac{V^2}{2g_c} \tag{8-25}$$

where ρ is the density of the water, V is the diver's velocity, g_c is 32.2 lb/(lb_m)(ft)(s^2), and $C_d A$ is the drag coefficient times its related area or the *drag area*. Hoerner (1965) suggests a drag area of 1.2 ft^2 for a man moving feetfirst and of 2 to 3 ft^2 for a man in a ball shape (i.e., doing a "cannonball" into the water). A study of the ama divers of Korea and Japan (National Academy of Sciences 1965) reports drag forces when the divers are towed headfirst by boats which can be fitted by Eq. (8-25) with an AC_d of 0.74 ft^2 at speeds over 1.0 m/s and an AC_d of 0.92 ft^2 at lower speeds. However, this same study suggests feetfirst AC_d values based on weighted ascents by divers of 1.2 ft^2, in good agreement with Hoerner.

Since power is simply the drag force times the velocity, we can compute the minimum energy costs of swimming from the above data. A strong swimmer might manage 50 yd in 30 s, or a velocity of 5 ft/s. With a headfirst AC_d value of 0.74, Eq. (8-25) predicts a drag of 25 lb_f and a power of 125 ft·lb_f/s or 0.228 hp. Foot flippers will not increase this speed appreciably. Instead, they allow the body to exert force on the water more effectively and thus make the energy transfer from muscle to water more efficient.

The general reduction in activity and increase in muscular requirements noted in the above computations applies to most bodily movements in the water. Various studies on diver work rates and efficiency have revealed the severe degree of handicapping produced by ocean environmental factors. Anderson (1969) showed that scuba swimming with a 9-lb weight resulted in double the nonweighted air demand plus a speed loss of 20 percent. Baddeley (1971) describes a loss of speed of divers attempting a manual dexterity test of 35 percent and suggests that this decrement in performance is mainly due to "anxiety." Streimer (1968) compared oxygen uptake of divers working in dry and wet conditions on various tasks and found an increase in oxygen consumption ranging from 10 to 20 percent more for a given job. Table 8-6 from

Table 8-5 Heat loss distribution from the body

Body region	*Heat loss, Btu/h*	*Body area, in^2*	*Btu/(in^2)(h)*
Head	85	282	0.305
Arm (each)	51	140	0.36
Chest	102	264	0.385
Upper back	102	357	0.287
Abdomen	102	186	0.545
Lower back	85	279	0.3
Leg (each)	69	411	0.165
Hand (each)	102	54	1.8
Foot (each)	135	93	1.44

Table 8-6 Wet and dry human performances

Type of test	*Dry land, ft · lb (mean)*	*Sealab (wet), ft · lb (mean)*	*Number of tests*
Pull	236	200	15 dry, 58 wet
Lift	626	606	15 dry, 60 wet

the Sealab II studies shows the reductions in body forces produced by immersion (differences are statistically significant). Overall decrease in assembly times on manual dexterity tests, owing to wet compared to dry tests, was 37 percent in Sealab II. A single group assembly task (involving several divers working together) took twice as long wet as dry, but arithmetic tests done at 1 atm and in Sealab (dry) at a total pressure of 238 ft showed no difference. Weltman and Crooks (1970) reported on two divers working together in a swimming pool to assemble a flanged pipe structure, first in warm water with a surface-supplied diving mask and then in colder water with full scuba, with the latter test taking twice as much time as the former.

The point of these and many other experiments is important for the underwater engineer. Tools, methods, and timing of underwater jobs must be tested under conditions as close to the real thing as possible. Land-based time estimates may be as much as 100 percent too optimistic and dry force requirements may not be met by scuba divers.

PROBLEMS

8-1. A submarine crew is at 200 ft in a compartment 8 ft in diameter by 20 ft long. Fifteen men are present and are at a resting state when the sub is disabled. O_2 renewal and CO_2 scrubbing are not possible. Once the boat is flooded up to local pressure, the men can leave by ducking under a trunk and rising in the water column up through the hatch. Assume it will take 1 min for each man to escape, and the CO_2 partial pressure must not exceed 65 mm Hg, before the last man leaves. How long can the man in charge wait before flooding up and starting the escapes?

8-2. In the case of Example 8-1, the compartment is to be flooded up and then flushed with air from a surface compressor, the men remaining in the boat. We wish to maintain the chamber air at 0.044 psi CO_2 partial pressure. What rate of air must enter the boat? How long will it take before the chamber approaches equilibrium?

8-3. A standard demand scuba air tank contains 70 ft^3 of air at standard (14.7 psia and 60°F) conditions. How much does this air weigh in a fully charged tank?

8-4. A habitat at 610 ft will contain oxygen at 150 percent of surface pressure with the remainder in the mix being helium. If the unit is a cylinder 51 ft long and 12 ft in diameter, what is the weight of the helium-oxygen atmosphere in the unit?

8-5. Divers operating on umbilicals from a 610-ft habitat (see Prob. 8-4) have two backpack 11.5-l tanks charged with 1.6 percent oxygen (by pressure) with the remainder helium. If the umbilical fails, how long do they have to return to the habitat if:

(*a*) They use the gear as a semiclosed circuit unit with CO_2 removal?

(*b*) The CO_2 removal system malfunctions also and they use the unit as a demand scuba?

8-6. Suppose we have a semiclosed circuit, CO_2 removal system in which the *volumetric* inlet flow is constant at all depths. Is this better or worse than an identical unit in which the *mass* inlet flow is constant at all depths? Plot the ratio of the oxygen partial pressure to the total pressure for both cases with a percentage of half oxygen, half nitrogen at inlet.

$$\dot{v}_{ox,lung} = 0.07 \text{ ft}^3/\text{min}$$

Discuss the advantages of the two systems.

8-7. Design the following emergency system which will bring a saturated diver from 300 ft to sea level at a uniform decompression rate of 10 min/ft. The man will breathe from a mixed-gas recirculator using a mask and will be resting. He will start with 5 percent (by pressure) oxygen, 95 percent helium. At 200 ft he will shift to 20 percent oxygen and 80 percent helium. We will assume that the inlet gas flow can be adjusted constantly as required and that P_{ox} has to be only 3 psi at all points of the ascent. Compute the total amount of gas (both mixtures) to accomplish the ascent.

8-8. A mixed-gas recirculating rig with a mixture of 40 percent oxygen and 60 percent nitrogen is at 150 ft of seawater. The diver is exerting moderately and his metabolism requires 1 l/min of sea-level oxygen. The inlet flow is 2 l/min of sea-level gas. What is the partial pressure of oxygen in his unit?

8-9. Using the data from Tables 8-3 and 8-4, compute an "equivalent fat thickness" for the fat and thin man, assuming in each case that a layer of fat is the only resistance between the body core and the water.

8-10. Suppose it is impossible to place electrical heating elements next to the skin. Instead, we bond them into the *center* of the neoprene suit. What is the equation for T_{skin} in this case? What will the values in Example 8-9 become, with all else the same?

8-11. On the basis of material presented in this chapter, design a survival suit for the following situation: An escapee from a vessel or submarine will wait on the surface for rescue. Water temperature is 29°F, and the swimmer must have 6 h of survival time without excessive heat loss. Batteries in the amount of 2000 Btu total output can be included and the swimmer will not need to do any work. Air can be supplied through a snorkel arrangement.

8-12. A thin human man is in water at 61°F with no suit. The heat transfer coefficient between skin and water (*H*) is 30 Btu/(h)(°F)(ft^2). The heat loss from the man is measured and found to be 3960 Btu/h. What is the man's *KS* at this condition? (Note that it is not the same as in Table 8-3, which assumes *H* is infinite.) What is the man's mean outer skin temperature? How thick a layer of neoprene foam with *K* of 0.031 Btu/(h)(°F)(ft) should be added to cut this loss to 1000 Btu/h, assuming exterior *H* and *KS* are not changed by this addition?

8-13. The mean thermal conductivity of a foamed material is the weighted mean of the conductivities of the solid material and the trapped gas with the relative volumes used to weight the two. Obtain a general formula based on this weighting method and check the mean value of *K* for the foamed neoprene in Table 8-4, assuming air or nitrogen is in the void bubbles. Now assume the same material has CO_2 in the voids. What is the new *K* (mean)? How much thinner (percentage) could we make a suit with CO_2 in the voids than one with air?

8-14. The thrust of an ideal propeller is given by $F = (\rho A/2g_c)(V^2_{out} - V^2_{in})$, where A is the area swept by the disk, V_{out} is the exit water velocity, V_{in} is the forward speed of the propeller, and ρ is the density of water. Suppose we design a hand- or leg-driven propeller arrangement to push a man at a speed of 1.5 mi/h. What thrust and what V_{out} will be needed for a 6-in-diameter propeller? Compute the ideal work needed to run it from the steady-flow energy equation. What is the efficiency of this perfect propeller in moving the man at this speed? Do we want a large or small prop for diver propulsion?

8-15. In the case of the dive in Example 8-4, for the 40-min tissue, assume that this controls the first stop location during decompression. What is the maximum safe oxygen percentage that could be used at this stop? Assuming this new oxygen fraction was used, how long would the 40-min tissue take to desaturate to the value that would permit surfacing?

8-16. If the 20-min tissue controls ascents from 70 ft of seawater, how long can we stay at this depth and return immediately to the surface?

8-17. We wish to design a recirculating diving rig for surface return for use at a maximum depth of 200 ft and a maximum sea-level oxygen demand of 3 l/min.

(*a*) What is an appropriate helium-oxygen mixture?

(*b*) What is the sea-level equivalent flow of gas from the bottle?

(*c*) How long could a bottle holding 4 ft^3 of mix at standard conditions last at 150 ft?

REFERENCES

Anderson, G. G. (1969): Measurement of Scuba Diver Performance in the Open Ocean Environment, *Am. Soc. Mech. Engrs. Paper 69-UNT-2.*

Baddeley, A. (1971): Diver Performance, in J. Woods and J. Lythgoe (eds.), "Underwater Science," Oxford University Press, New York.

Bennett, P., and D. Elliott (1969): "The Physiology and Medicine of Diving," Williams & Wilkins, Baltimore.

Hoerner, S. F. (1965): "Fluid-Dynamic Drag," published by the author, Midland Park, N.J.

Lambertson, J. (1963): "Proceedings of the Second Symposium on Underwater Physiology," National Academy of Sciences, Pub. no. 1181, Washington.

Miles, S. (1966): "Underwater Medicine," Lippincott, Philadelphia.

National Academy of Sciences (1965): "Physiology of Breath-hold Diving," Pub. no. 1341, Washington.

Pauli, A., and Claper, G. (ed.) (1967): Project Sealab Report, *O.N.R. Rept.* ACR-124, Washington.

Streimer, I. (1968): How Effective Are Underwater Workmen? *Ocean Ind.*, Oct., pp. 75–76.

Weltman, G., and T. Crooks (1970): Human Factors Influencing Underwater Performance, in "Equipment for the Working Diver," Marine Technology Society 1970 Symposium, Washington.

9

Thermal and Structural Limitations on Ocean Systems

Hilbert Schenck, Jr.

9-1 INTRODUCTION

The previous eight chapters have dealt primarily with specialized technological information related uniquely to the ocean environment. Much of this material will be new to the average undergraduate engineer, whatever his traditional specialty or background, although some of the soils and materials information is closely derived from "dry" engineering. In this chapter we will call upon the familiar disciplines of thermodynamics and structural analysis to provide ourselves with certain special equations that will set limitations on most, if not all, ocean systems. The two most characteristic environmental problems that uniquely govern devices to be used in the ocean are (1) the remoteness of the system from shore "support" and (2) the high external pressures possible in the deep, open oceans. In general we must supply energy in a closed, water-resistant container whether our system is a sound transducer, free-falling corer, research submarine, or wave probe.

PART 1 ENERGY SOURCES

9-2 AVAILABILITY AND WORK PRODUCTION

Most instruments and systems used in the ocean require energy input to operate. Cable-suspended devices may depend on batteries or boat-mounted generators for work production. Free-falling devices such as gravity corers use the energy generated by motion in the gravity field for their operation. Some few instruments utilize the large pressure difference between the surface and the deep ocean. Many underwater and floating systems obtain their energy from some sort of thermodynamic process, the burning of fuel and the expansion of a compressed gas being the most common. Both these processes are governed and limited by the first and second laws of thermodynamics, which have basic application to ocean systems especially where odd or "exotic" energy sources may often be under consideration.

The thermodynamic efficiency of any heat engine is set by the maximum and minimum temperatures between which it operates. This efficiency may be defined as

$$\text{Efficiency} = \frac{\text{work out}}{\text{heat in}} \tag{9-1}$$

and it expresses how well thermal energy produced by chemical, nuclear, solar, or other sources can be changed into work. The second law shows

$$\text{Maximum efficiency} = \frac{T_h - T_c}{T_h} \tag{9-2}$$

where T_h is the maximum temperature of the cycle or energy sources and T_c is the minimum temperature of the cycle or its surroundings. This is the so-called *Carnot efficiency* based on heat flows.

The second law efficiency is a useful limiting equation but is too general for all the possible cases of work production, particularly in the ocean environment. The key to *all* such processes is the so-called *availability*, which tells us the absolute maximum work we can obtain from any working substance, whatever its type or composition. For example, it is evident that if we have compressed air at sea bottom temperature and a pressure of several thousand pounds per square inch, it cannot be used in any purely thermal cycle since Eq. (9-2) predicts a zero efficiency. Practically, we know the air is very useful for work production in all manner of air tools. The maximum amount of work we can expect from this air is measured by its availability A and the measure of how well the tool converts this high-pressure-air energy to work is measured by the *effectiveness* E defined from

$$E = \frac{\text{actual work out}}{\Delta A} \tag{9-3}$$

where we are talking about a *work-producing device*.

There are two general equations for A, depending on whether our work process is to be a closed system or a steady flow case. For the *closed system*

$$\Delta A_c = (u_2 - u_1) + P_0(v_2 - v_1) - T_0(s_2 - s_1) \tag{9-4}$$

where u is the internal energy, P_0 is the *dead state pressure* (i.e., the pressure of the surroundings), T_0 is the surrounding or dead state temperature, v is the specific volume of the working material, and s is the entropy of the working material. The subscripts 1 and 2 refer to the states of the working material before and after the work-producing process.

For the *open system* availability:

$$\Delta A_0 = (h_2 - h_1) - T_0(s_2 - s_1) + \frac{V_1^2}{2g_c} + \frac{g}{g_c}(Z_2 - Z_1) \tag{9-5}$$

where h is the enthalpy, V_1 the initial velocity of the flowing material, and Z the height above a datum of the flowing or working material. Both these equations are derived in Keenan (1941), Kenyon and Schenck (1964), Obert (1948), and many other thermal texts.

In addition, we will need equations for $s_2 - s_1$ for a perfect gas. [*Note*: All temperatures in Eqs. (9-4), (9-5), and (9-6) are absolute.]

$$s_2 - s_1 = C_p\left(\ln\frac{T_2}{T_1}\right) - R\left(\ln\frac{P_2}{P_1}\right) \tag{9-6}$$

and for $u_2 - u_1$ for a perfect gas

$$u_2 - u_1 = C_v(T_2 - T_1) \tag{9-7}$$

and for $h_2 - h_1$ for a perfect gas

$$h_2 - h_1 = C_p(T_2 - T_1) \tag{9-8}$$

In the case of steam or other substance for which tables are available, the various state values in Eqs. (9-4) and (9-5) must be computed or found from the tabulated entries. These two equations are most important in thermodynamics when it comes to studies of new cycles, energy sources, and energy-storage schemes.

The applications of the previous equations to work-producing or work-absorbing systems suggests that there are at least three levels of abstraction to any "feasibility study" of such systems. *Level 1* involves the basic second law considerations just discussed. At this level, we find the absolute maximum (and always unattainable) work output assuming only that the basic laws of thermodynamics prevail. At *level 2*, we consider the question of the thermodynamic mechanism whereby the work can be actually attained. We still assume "perfect frictionless processes," but we now attempt to define specific, realizable processes. At *level 3*, we further compromise the chosen processes by including all manner of practical heat-transfer rates, turbulence, etc. The following example will indicate these levels.

Consider a compressed-air container at 200 atm and an absolute temperature of 520°R. We wish to utilize this air for propulsion purposes on the surface, probably by driving a turbine connected to a propeller. What is the maximum work attainable from each pound of gas?

Level 1. The gas is assumed at the surrounding water temperature 520°R which is then T_0. We will work with a dead state pressure of P_0 of 1 atm, realizing that at depths greater than the sea surface, degradation in performance must occur. Then Eq. (9-6) gives

$$s_2 - s_1 = C_p \ln (1.0) - R \ln \tfrac{1}{200}$$

With ln (1.0) of zero and R of 53.3/778 (Btu/(lb$_m$)(°R)) we obtain an overall entropy change of +0.365 Btu/(lb$_m$)(°R). Note that we are assuming at level 1 that the gas goes all the way to the dead state. This will be the case of maximum work. Then Eq. (9-5) (since we anticipate a turbine, we also anticipate this is an open system availability problem) gives

$$A = -520 \times 0.365 = -189 \text{ Btu/lb}$$

which is the maximum work permitted the system using a steady flow arrangement in a universe in which the second law is operative.

Level 2. How will we really make this power device work? The usual assumption about air-powered devices involves an adiabatic turbine operating reversibility. The work obtainable from such a device connected to a discharging reservoir is, for the perfect gas (see Obert 1948),

$$W_{rev} = \left[C_v(T_1 - T_2) - RT_2 - \frac{P_2}{P_1} T_1 \right] m_1 \tag{9-9}$$

where m_1 is the mass of the gas in the tank before exhaustion and will be 1 lb$_m$ in this example. With the isentropic assumption,

$$\frac{T_1}{T_2} = \left(\frac{P_1}{P_2} \right)^{(k-1)/k} \tag{9-10}$$

For air the exponent is 0.286, and we obtain 4.6 for T_1/T_2. T_2 is 113°R.

Then Eq. (9-9) predicts a work output of 51.4 Btu/lb. The effectiveness of the system (without even considering whether a reversible adiabatic expansion to 113°R is possible) is then only 51.4/189 or 27 percent.

A reasonable question is, what happened to the rest of the available energy? Part of it remains in the gas at 113°R and 1 atm but cannot now be taken out by a turbine since the pressure difference across the turbine would be zero. Using Eqs. (9-5) and (9-6), we can compute how much available energy is left at the low-temperature 1-atm condition. From Eq. (9-6)

$$\Delta S = 0.24 \ln (4.6) - R \ln (1.0) = 0.367 \text{ Btu/(°R)(lb)}$$

where we assume we go to the dead state temperature, 520°R from 113°R. Then

$$A_1 - A_2 = 0.24(407) - 520(0.367) = -91.5 \text{ Btu/lb}$$

This residual availability could, possibly, be realized by some sort of heat engine operating between the temperatures 520°R and 113°R. Adding 91.5 and 51.4, we obtain 142.9 Btu/lb out of a maximum of 189 Btu/lb. Where is the remainder now? The answer lies in the assumptions used in obtaining Eq. (9-9). When a rigid bottle "blows down" through a work-producing device, the available energy is partitioned. Some of it (51.4 Btu/lb in this case) appears as turbine work. The rest is lost as the gas remaining in the bottle expands and its pressure drops. That is, only the first small amount of gas in the bottle is available at 200 atm. Once that bit has gone, the pressure across the turbine is less, and the next increment of gas does less work. Gas expanding within the bottle, instead of within the turbine, loses its ability to do work (its available energy) with no effect on the universe, except for the adiabatic temperature decrease shown by Eq. (9-10).

Clearly, a more efficient manner of using the air would be to "shrink" the bottle to smaller volumes during air discharge, thereby holding the residual gas at full pressure. In the limit, a steady-flow, constant-pressure situation would be approached and the usual thermodynamic expression for steady-flow, adiabatic work would be realized:

$$W = \frac{kR(T_2 - T_1)}{1 - k} \tag{9-11}$$

where k is the ratio of the specific heats (about 1.4 for air). Using 113 and 520°R in Eq. (9-11), we obtain a work output of 97.5 Btu/lb. Adding this to the remaining flow (cold gas) available energy (91.5), we obtain a total of 189 Btu/lb, the total possible as predicted originally. Thus, if we could discharge the bottle in a constant-pressure manner, we would gain 91.5 − 51.4 or 40.1 Btu/lb additional work compared with allowing the gas to discharge and drop its pressure. Summarizing our level 2 computations (all in Btu per pound of gas at 200 atm):

Total available energy	189.0
Work from isentropic turbine and rigid bottle discharge	51.4
Available energy left after discharge	97.5
Work from "shrinking bottle" isentropic discharge	91.5

Effectiveness with normal bottle	27%
Effectiveness with "shrinking bottle"	48.5%

Level 3. Level 3 considerations are beyond our abilities here because they require an actual design to be worked up. Certainly we cannot expect the bottle to achieve a low temperature of 113°R. Heat will flow in and out, frictional forces will use energy, and any turbine will have an isentropic efficiency of less than 100 percent. Actually, the present state of the art would probably limit us to half or less of the level 2 predictions.

One problem with mobile underwater systems is the storage of sufficient heat-producing material to provide sufficient range. One possible (?) scheme is to store thermal energy in a hot block of metal which then generates steam to operate a power plant. Let us apply second law ideas to this system. Consider the material to be iron at 2500°R, C_p (and C_v) of 0.16 Btu/(lb)(°R), and density of 491 lb/ft^3. Taking the dead state (surrounding water) to be 60°F (520°R), we can compute the best Carnot efficiency of an engine working between 2500°R and 520°R from Eq. (9-2):

$$\text{Efficiency} = \frac{2{,}500 - 520}{2{,}500} = 0.79$$

This is, practically, a meaningless computation. The 79 percent efficiency will exist only for the first bit of heat that flows from the hot iron block. After that, the iron temperature will be less and the efficiency will drop. Instead of the Carnot efficiency, we need the effectiveness. From Eq. (9-6)

$$s_2 - s_1 = -C_p \ln \frac{2{,}500}{520} - R(\ln 1.0) = -0.254 \text{ Btu/(lb)(°R)}$$

Then Eq. (9-4) or (9-5) (they give the same results with an incompressible system like this) yields

$$A_2 - A_1 = -0.16(1{,}980) + 520 \times 0.254 = 185 \text{ Btu/lb}$$

which is the maximum work attainable per pound of heated iron. For comparison, it will take 0.16(1,980) or 317 Btu/lb to heat the iron from the dead state to the 2500°R condition.

This calculation is, in general, the same sort of estimate that one would make with a chemical propulsion system, except that in such a case it is necessary to obtain the *equilibrium flame temperature* and *combustion pressure* first, using standard charts or methods of chemical thermodynamics. These are rather too extensive to be considered here. Once the flame temperature and pressure are obtained, the same computation as above will tell us the absolute maximum work possible from the energy generation process. If the hot gases are used to heat steam, only the temperature of the flame is of interest and the computations of available energy should be *run on the steam*.

Note that when energy is stored in a purely mechanical system (such as a rotating flywheel), the available energy is equal to the total energy of the system since all of it can be realized as work.

We noted above that the Carnot efficiency based on the maximum temperature of our hot block of iron was almost meaningless. What, then, is a reasonable efficiency? One possibility is suggested, namely, the ratio of the possible work out $(A_2 - A_1)$ to the heat input.

$$\text{Thermal storage efficiency} = \tfrac{185}{317} = 58.5\%$$

Such a parameter assumes that the cost of energy storage is of primary importance to us. Practically, in underwater propulsion, the cost of heating the block to 2500°R is negligible compared to the wages of the men in the vehicle, the vehicle fixed and operating costs, etc. A more meaningful performance parameter might be

$$U/w\text{, vehicle storage parameter} = \frac{\text{Btu/lb}}{\text{volume storage of 1 lb}} \tag{9-12}$$

That is, *U/w* VSP is the best possible work out of 1 lb divided by the amount of volume needed to store 1 lb. This parameter would recognize that in an underwater device, large volumes mean large fluid drag forces on the vehicle. Compare this with the usual rocket vehicle fuel performance parameter, called the specific impulse (I_{sp}), defined as the thrust divided by the mass flow of the fuel per second. Logically, a high value of thrust and a low mass of fuel flowing are "good" since the rocket must carry its fuel as extra weight up to some distance.

Any underwater parameter based on thrust and weight flows would make no sense since *U/w* vehicles do not necessarily need reaction engines (which are primarily "sized" by thrust figures), nor is a "heavy fuel" bad underwater, where the water buoyant forces will easily support any reasonable propulsive system.

Taking our above noted parameter based on energy and volume, we note that 1 lb of hot iron could generate an absolute maximum of work of 185 Btu/lb and that 1 lb of iron occupies $\frac{1}{491}$ ft^3. Thus from Eq. (9-12)

$$U/w \text{ VSP} = 185 \times 491 = 91{,}500 \text{ Btu/ft}^3$$

For comparison, how does our compressed-gas storage system compare? Taking the air at 200 atm, we noted that the normal bottle discharge scheme gave 51.4 Btu/lb of air. From the perfect gas law

$$v_{\text{air}} = \frac{RT}{P} = \frac{53.3 \times 520}{14.7 \times 200 \times 144} = 0.00135 \text{ ft}^3\text{/lb}$$

our *U/w* VSP parameter is now $51.4/0.00135 = 38{,}200$ Btu/ft^3. Thus the compressed-air scheme will involve us in 91,500/38,200 or 2.4 times more fuel-storage volume. In a long-range vehicle, fuel storage might easily be half the volume of the craft, and thus the total volume will be substantially greater and the work needed to propel it considerably greater also.

These examples illustrate an important point in the thermal sciences. Each technology requires a special examination and, usually, a special set of performance parameters.

9-3 ENERGY SOURCES: STORAGE BATTERIES

Battery-electric power is a primary source of energy for most ocean instruments and many ocean systems. In general, a given battery is characterized by three quantities: its terminal voltage, its internal resistance, and its "suggested" discharge rate. When a battery is connected to its load, as shown in Fig. 9-1*a*, current flows around the circuit. The actual terminal voltage depends on this current flow and the loss of voltage internally produced by the internal resistance. In the Fig. 9-1*a* system, a nominal 6-V battery with a 5-Ω external load and a 1-Ω internal resistance will produce only 5 V at its terminals, the other 1 V being developed across the internal resistance of the cells.

Fig. 9-1 (*a*) Single battery, 5-Ω load, and 1-Ω internal resistance. (*b*) Parallel batteries of identical characteristics. (*c*) Parallel batteries with different internal resistances.

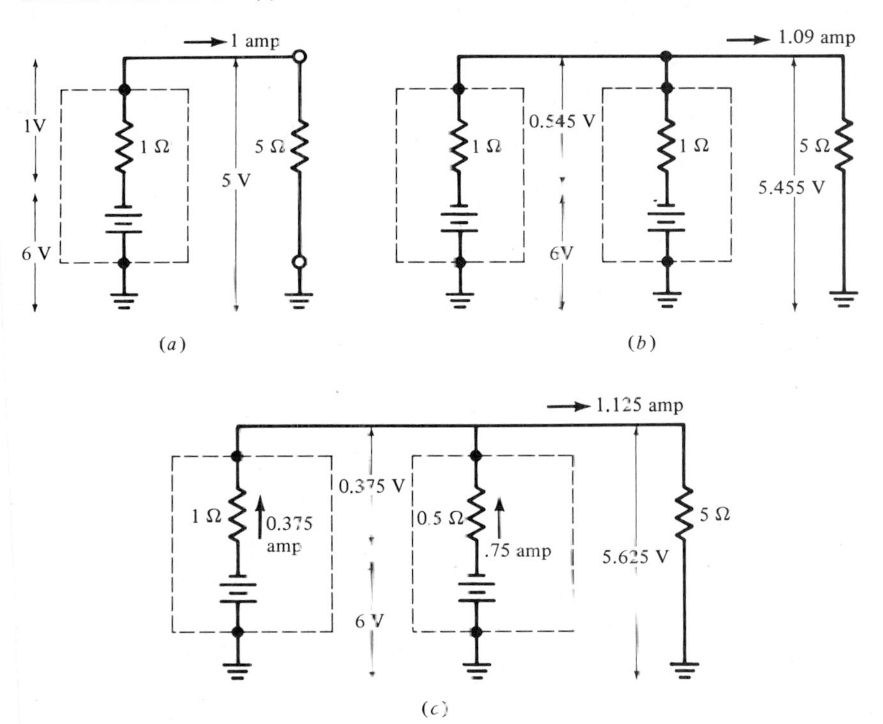

When precision devices are to be used in such a way that voltage control is important, the designer usually wishes a battery with as low an internal resistance as possible, so that load fluctuations will not produce terminal voltage fluctuations. When parallel battery systems (Fig. 9-1*b*) are used, it may be important that the batteries be "matched" as far as this is possible. That is, the internal resistance and open circuit voltage should be identical. If not, a parallel connected system, as shown in Fig. 9-1*c*, may actually discharge one cell faster than another, thereby producing a serious mismatch in the cells as they run down and possibly producing an unexpected failure. Matching cells is quite simple: the designer needs only to place a known resistance across the terminals, observe the change in potential, and compute the internal resistance. In general, it is desirable in battery-electric power systems never to connect a load that will produce a significant lowering of terminal voltage, whatever the "suggested maximum" discharge rate may be. However, in the case of many storage batteries the internal resistance will increase as discharge occurs. Thus a matched set of batteries might become mismatched as they neared the end of their useful life. Perfect matching would require that the batteries be drained by the prototype load and checked periodically for terminal open circuit voltage and internal resistance.

Although they are direct-conversion devices, batteries can be compared with the thermal and compressed-air systems already discussed, using the following nomenclature:

w_b	Dry weight of battery, lb
V_b	Battery volume
ρ_w	Density of seawater
K	Stored energy, Wh or Btu
α_w	Stored energy per pound of dry weight
α_v	Stored energy per unit volume

The apparent weight of the battery in the sea, W_a, is now:

$$W_a = W_b - \frac{V_b}{\rho_w}$$

and

$$W_b = \frac{K}{\alpha_w} \tag{9-13}$$

$$V_b = \frac{K}{\alpha_v} \tag{9-14}$$

Thus

$$W_a = K\left(\frac{1}{\alpha_w} - \frac{\rho_w}{\alpha_v}\right) \tag{9-15}$$

An optimization study for the Sealab II experiment collected basic data for the several common battery types, as shown in Table 9-1.

Comparing the figures in the last column with the computations in the previous section, it is easy to see why battery-electric energy storage is a universal favorite in small ocean systems. Our fanciful hot block storage system has a theoretical storage parameter of 91,500 Btu/ft^3, over four times that of the silver-zinc battery (right-hand column in Table 9-1) but includes nothing regarding the practical harnessing of this thermal energy for propulsion or other purposes. We would be fortunate indeed to obtain even a quarter of its energy at a propeller shaft, and this might require a sophisticated turbine power plant plus some volume of working fluid. The compressed-gas storage parameter of 38,200 Btu/ft^3, based on idealized and reversible processes, is probably two to three times what a practical system would achieve in an underwater context. The batteries, on the other hand, can provide their energy as ready-to-use work, and only a compact electric motor with an efficiency that might exceed 90 percent needs to be added to obtain shaft work.

The use of Table 9-1 and Eqs. (9-13) through (9-15) can be shown in the following double example.

We wish to select the optimum batteries for (1) a small research submarine requiring 3 hp for propulsion over a 3-h period, and (2) the diver riding in this machine, who will use an electrical heated suit (see Sec. 8-6) with a storage requirement of 1 kWh. The diver may leave the submarine and swim. Let us obtain the in-water weight and volume of the Table 9-1 batteries for these two requirements: 9 hph is equivalent to 6,750 Wh. Using Eqs. (9-14) and (9-15), we can examine the size and weight of the two batteries (Table 9-2).

Taking the case of the heated scuba diver, the crucial parameter will probably be weight, since the computed volumes are not importantly large for any of the units, excluding the nickel-cadmium. With a suit, the diver will need from 15 to 30 lb of weights to achieve neutral buoyancy; thus the choice is narrowed to the three lightest batteries, since we would not wish to have to provide the diver with reserve buoyancy. However, we might wish to give the diver the maximum potential for positive buoyancy in an emergency (by dropping his weight belt) so that the best choice (and the one made for Sealab II) is probably the lightest unit, silver-zinc.

Table 9-1 Battery types and parameters

Type of battery	$\alpha_{\hat{w}}$ (Wh/lb)	← $\alpha_{\hat{v}}$ (Wh/in^3)	(Btu/ft^3) →
Silver-zinc	55	3.5	20,400
Silver-cadmium	35	2.8	19,400
Magnesium-silver chloride	30	2.5	14,500
Lead-acid	20	2.1	12,200
Nickel-cadmium	13	0.9	5,250

Table 9-2 Sizes and weights of batteries for divers and submarines

	1,000 Wh (scuba)		*9 hph (sub)*	
Type of battery	V_b, *in*^{3}	W_a, *lb*	V_b, *ft*^{3}	W_a, *lb*
Silver-zinc	286	7.4	1.12	50
Silver-cadmium	358	15.4	1.4	104
Magnesium-silver chloride	400	18.5	1.57	125
Lead-acid	476	32.4	1.85	220
Nickel-cadmium	1,110	35.9	4.34	240

As already noted, the submarine has different problems. Here we wish to minimize volume (and thus water drag). However, the first four choices in Table 9-2 do not differ too greatly in total volume (lead-acid is about a third larger than silver-zinc). Since a small sub will certainly have need of ballast, the wet weight W_a may also be of interest. On this consideration, lead-acid may be the best choice; we pay with a little more volume, but this may be more than made up by the volume saving in ballast. The lead-acid battery gives us 170 lb more of negative buoyancy than the silver-zinc battery, which represents about ¼ ft^3 of lead ballast or 0.35 ft^3 of iron ballast. Also, the lead-acid battery will be much less expensive, easier to maintain, and somewhat tougher under emergency load conditions. Actually, the final choice will depend on many other factors in addition to those considered here.

9-4 SOLAR CELLS AND OTHER DIRECT CONVERSION DEVICES

Silicon solar cells have been used with great success in various space applications, where money is no consideration and long life is crucial. Although not practical for underwater power applications, the silicon cell should find application in drifting and remote buoy systems, provided the cell faces can be kept reasonably clean. Most such applications will probably involve batteries as well, the cells being used to charge the batteries on "good" days.

Taking, as typical, a Hoffman H5C solar module, we note that such a unit has an active area of 1.44 in^2 and a 10 percent conversion efficiency. More modern units may reach 15 percent, but will show similar characteristics. The silicon cell has a perfectly linear response of current as a function of light input on short circuit, enabling it to be used as an underwater turbidity sensor (see Sec. 11-11). For remote power applications, it will normally be connected to a load, thus behaving as a voltage source. Ideally, we should like to match the external load impedance to the cell impedance at maximum power. However, this is not possible with a variable sunlight situation. With a solar constant of 155 to 250 Btu/(h)(ft^2) (reasonably typical for summer in the temperate zones), the impedance of test cells varied from 75 to 50 Ω. The design problem is more complex if batteries are being charged, since battery impedance will drop as charging proceeds. Circuits to compensate for such variations can be designed. However, an easy approach is to estimate a mean solar constant for the location and design and cell bank to provide a substantial overvoltage. Diodes will prevent leak back from the batteries to the cells when illumination falls at night. Solar cells are costly (more than $100 per watt) but might prove useful in ocean monitoring applications.

Various other direct conversion schemes (thermoelectric systems, with conventional or nuclear heating, magnetohydrodynamics, mechanical conversion of ocean wave energy, and so on) have been proposed and sometimes used in ocean systems. None of them, so far, have been competitive with battery-electric storage methods.

9-5 ENERGY TRANSFER BY COMPRESSED GAS

The previous sections have dealt with some of the basic ways in which energy can be stored in underwater or surface systems and have shown some of the limitations governing such storage. In cases where cable or umbilical connection with surface vessels is possible, or where an underwater system is to be connected to the surface, as in the case of oil platforms and habitats connected to a surface buoy, three general methods of power transfer are commonly used: electrical energy transfer by wire, mechanical energy transfer by shaft or motion of a cable, and transfer of energy by movement of a compressed gas through a hose. We will consider this compressed-gas method here, since it is the only one involving engineering theory of a special nature and is also important in diving life support systems (Chap. 8) and in the transfer of natural gas from underwater wells.

The common Fanning equation for flow of an incompressible fluid in uniform conduits is

$$P_1 - P_2 = f_s \frac{L}{r_h} \frac{\rho V^2}{2g_c} \tag{9-16}$$

where $P_1 - P_2$ is the pressure drop in the conduit between stations 1 and 2 of length L and hydraulic radius r_h of a fluid of density ρ and velocity V. f_s is the so-called "small friction factor" which, as Fig. 9-2 shows, is dependent on the Reynolds number of the fluid and the roughness of the conduit. In some texts, L/r_h is replaced by L/D_h, where Dh is the hydraulic diameter. In such a case, we must use the "large friction factor" f_1 such that

$$f_1 = 4f_s \tag{9-17}$$

since the hydraulic radius is one-quarter of the hydraulic diameter. If g_c is in the units foot-pound force per second squared per pound mass and the other parameters in corresponding units, the pressure drop will be in pound force per square foot.

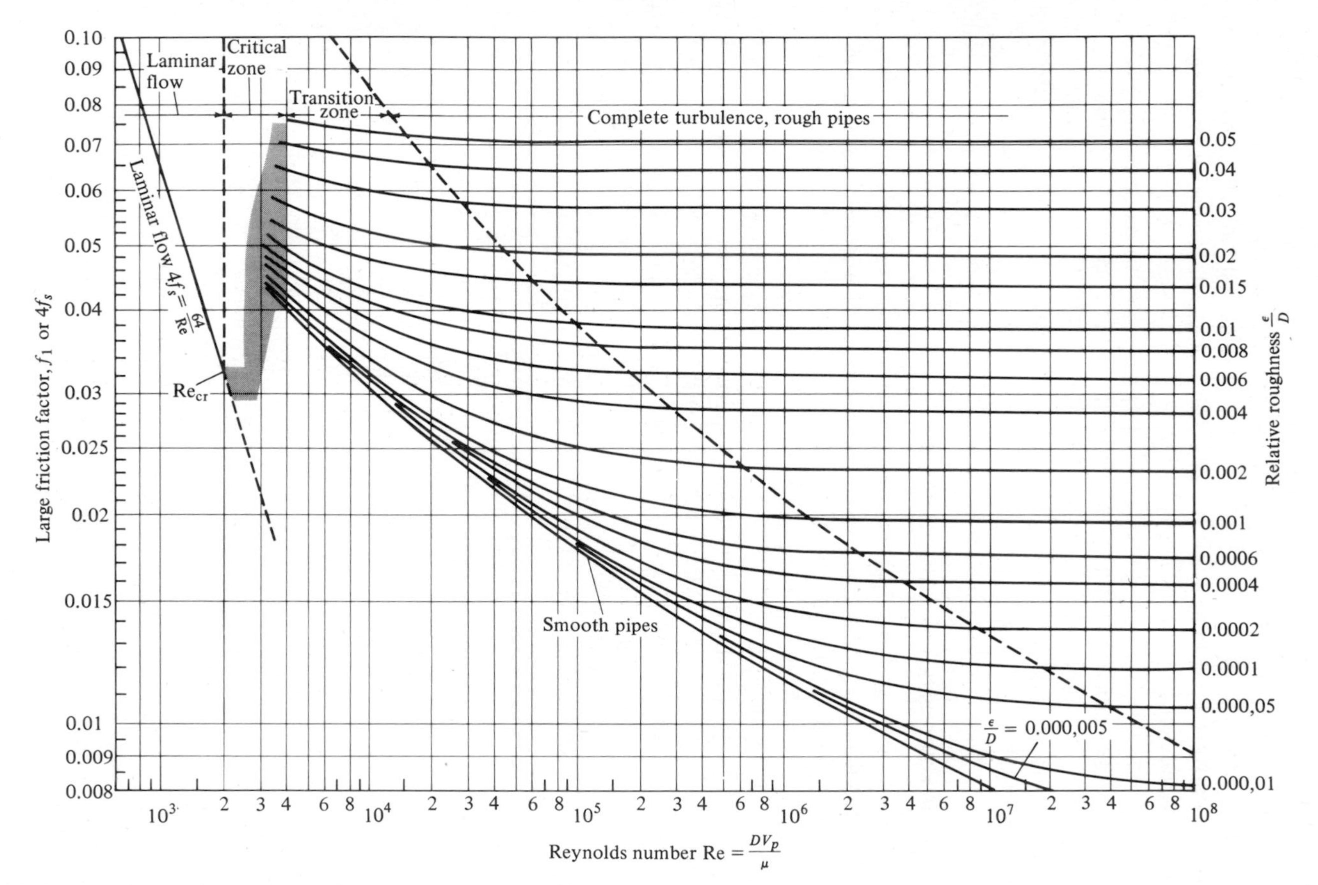

Fig. 9-2 Friction factor as a function of Reynolds number and roughness. Roughness height is ϵ, hydraulic diameter is D.

Equation (9-16) is correct only when the gas flows are incompressible. When air is pumped to a compressed-air tool held by a diver, or a deep diving suit, compressibility effects will often be present. To solve for pressure drop in such a case, we must know, first, whether the flow is isothermal, adiabatic, or subject to some intermediate condition involving both heat flow to or from the gas plus gas temperature change. In most ocean systems, where a hose of 100 ft or more immersed in water is involved, the assumption of isothermal flow is a good one. Actually, it is usually possible to predict whether the isothermal assumption is appropriate by analyzing the hose as a heat exchanger losing heat to a heat source of constant temperature. If the gas temperature approaches the water temperature within the first 10 percent of the hose, then the entire hose can be treated as isothermal with reasonable prediction accuracy. This will usually be the case since seawater is an excellent heat-transfer agent and the high gas velocity and surrounding water combine to give a high overall heat-transfer rate. When such is the case, the equation for compressible flow of a perfect gas in uniform conduits becomes (see, for example, Shapiro 1956):

$$G^2 = \frac{g_c}{RT}(P_1{}^2 - P_2{}^2)\left[f_s\frac{L}{r_h} + \log\left(\frac{P_1}{P_2}\right)^2\right]^{-1} \tag{9-18}$$

where G is the mass velocity, with typical units pounds mass per second per square foot as defined by

$$G = V\rho \tag{9-19}$$

R is the gas constant and T the constant fluid temperature. The incompressible Fanning equation (9-16) can be written

$$G^2(\text{incompressible}) = 2\rho g_c(P_1 - P_2)\frac{f_s L}{r_h} \tag{9-20}$$

for comparison.

Both these equations will give us the delivery rate per unit area of hose (G) if the pressure drop is known. This computation will, of course, be iterative, since we must assume a flow to obtain the Reynolds number to enter Fig. 9-2 to get f_s, then correct the Reynolds number and friction factor for the found G value (note that $N_{re} = GD/\rho$), and so on. G is constant in a constant area hose and the gas viscosity is a function only of temperature. N_{re} will thus be constant also.

In many practical problems, the term ln $(P_1/P_2)^2$ is small compared with $f_s L/r_h$. When this is so, we can obtain the approximate ratio of $G^2{}_{\text{compressible}}/G^2{}_{\text{incompressible}}$ by forming the ratio of Eqs. (9-18) and (9-20) and canceling like terms. The result is

$$\frac{G^2_{\text{compressible}}}{G^2_{\text{incompressible}}} = \frac{P_1 + P_2}{2P_1} \tag{9-21}$$

When this pressure ratio is significantly different from unity, we should use the compressible flow equation for accurate flow estimations.

In the case of compressed-air power transmission to the sea floor, the surface compressor must overcome three different pressure drops: the hose drop predicted by Eq. (9-18) or (9-20), the pressure excess due to the location of the tool in a region where the hydrostatic head is greater than at the compressor location, and the pressure drop requirement to operate the tool or otherwise provide work.

PART 2 STRUCTURAL FORMULAS

9-6 COMBINED STRESSES

Most practical externally pressurized systems in ocean engineering involve us in *combined stress* situations. A variety of ways of combining the three principal stresses exists (Roark 1964); we will suggest only one:

$$\bar{S} = \frac{1}{\sqrt{2}} [(S_1 - S_2)^2 + (S_2 - S_3)^2 + (S_3 - S_1)^2] \tag{9-22}$$

where $\bar{S}$ is the combined stress and S_1, S_2, and S_3 are the maximum, intermediate, and minimum principal stresses respectively.

9-7 SPHERICAL BODIES

The spherical form finds wide use in ocean engineering where large external pressures are to be resisted. For a perfectly elastic (Hooke's law) material the stress in the radial direction S_r is given by (Roark 1964):

$$S_r = -P\left[\frac{1 - (a/r)^3}{1 - (a/b)^3}\right] \tag{9-23}$$

and the tangential stress (S_t) is given by

$$S_t = -P\left[\frac{1 + \frac{1}{2}(a/r)^3}{1 - (a/b)^3}\right] \tag{9-24}$$

where a is the inner sphere radius, b the outer radius, and r some radius at or between a and b. P is the pressure on the external surface; and we assume that the inside pressure is relatively small.

Taking a small, interior section, we can see that S_1 and S_2 each equal S_t, and S_3 is S_r. Combining Eqs. (9-23) and (9-24) with Eq. (9-22) gives

$$\bar{S} = P\frac{(3/2)(a/r)^3}{1 - (a/b)^3} \tag{9-25}$$

Equation (9-25) predicts a maximum $\bar{S}$ at the inner surface ($r = a$). If we thus set $\bar{S}$ equal to S_{yp}, the tensile yield stress of the sphere material, and $r = a$, we will obtain an equation setting the yield point limit on P, and thus a criterion of maximum sphere depth.

For unmanned spheres used for buoyancy applications, this criterion may result in spheres with very thick walls. An approximate solution for the case of combined elastic and plastic deformation of hollow spheres can be obtained if we idealize the material stress-strain behavior as shown in Fig. 9-3. Let us place the radius at which the material changes from elastic to plastic behavior at $r = t$. Then for the elastic region ($t \leqslant r \leqslant b$) a new solution of the differential equation for a uniformly loaded spherical shell (Drucker 1967) provides S_r and S_t values that give a combined stress of

$$\bar{S} = \left(\frac{t}{r}\right)^3 S_{yp} \tag{9-26}$$

For the *plastic region* ($a \leqslant r \leqslant t$) the solution for the principal stresses becomes (Drucker 1967):

$$S_r = -2S_{yp} \ln \frac{r}{a} \tag{9-27}$$

$$S_t = -S_{yp}\left[2 \ln \left(\frac{r}{a}\right) + 1\right] \tag{9-28}$$

Combining these two equations with Eq. (9-22) shows for the plastic part

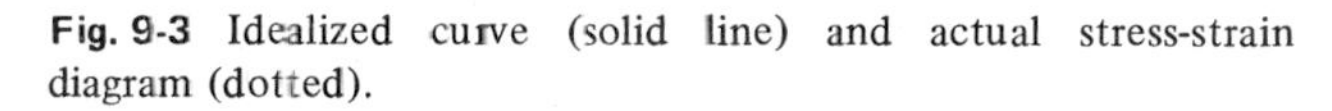

Fig. 9-3 Idealized curve (solid line) and actual stress-strain diagram (dotted).

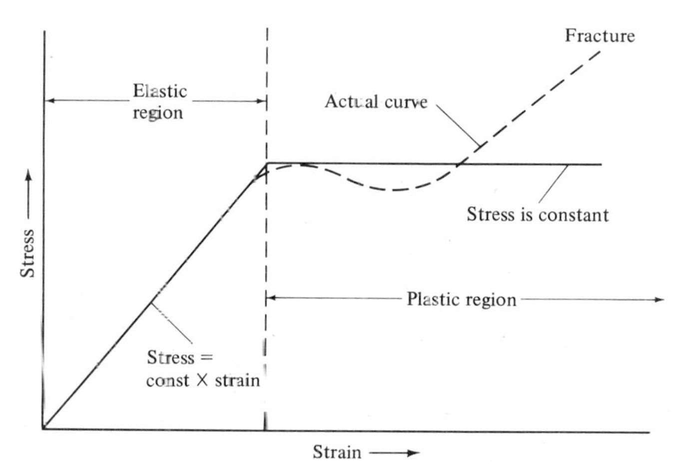

$$\overline{S} = S_{yp} \tag{9-29}$$

as we would expect from Fig. 9-3.

We can establish the location of the interface between the two zones t by equating the two equations for radial stress S_r, Eqs. (9-23) and (9-27), since there can be no discontinuity of stress in the radial direction. Setting these equations equal for the case of $r = t$ gives

$$\ln\left(\frac{t}{a}\right) + \frac{1}{3}\left[1 - \left(\frac{t}{b}\right)^3\right] = \frac{P}{2S_{yp}} \tag{9-30}$$

When yielding has progressed completely through the shell, collapse is imminent. For this case ($t = b$), Eq. (9-30) gives

Fig. 9-4 Failure criteria for a sphere based on yield and collapse equations.

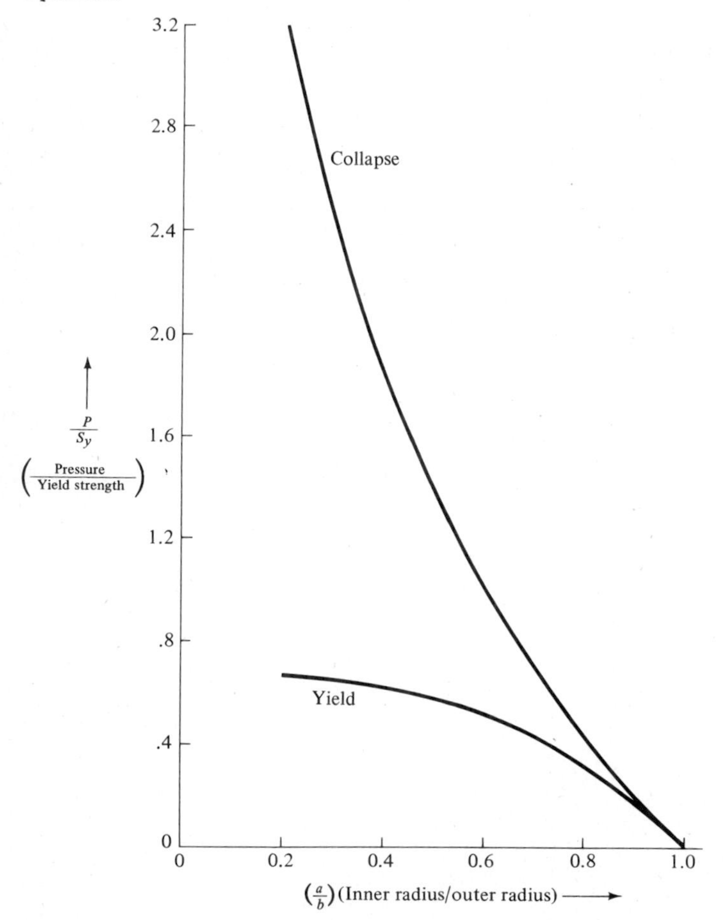

$$\frac{P}{S_{yp}} = 2 \ln \frac{b}{a} \tag{9-31}$$

Equations (9-25) (with $r = a$) and (9-31) define two limiting design conditions: the start of plastic yielding and collapse. Figure 9-4 shows how the dimensionless pressure P/S_{yp} behaves for the two criteria as a function of the a/b ratio. For very deep (thick-walled) applications, we can obtain a much lighter sphere if we are willing to allow yielding. Indeed, with some materials at great depth, it may be impossible to use a hollow sphere unless some yield occurs. Equation (9-31) will be somewhat conservative for many substances since some work hardening may occur after yielding begins. For such a material, the horizontal line on Fig. 9-3 will show an upward trend until rupture occurs.

Example 9-1 What is the greatest depth in seawater to which a neutrally buoyant, hollow sphere of ferrocement can go, based on the compressive strength?

Solution From Table 6-12 we note that ferrocement has a specific gravity of 2.49 and a compressive strength of 6.5/ksi, at which point we assume yielding begins. We first obtain the neutral buoyancy criterion. The weight of the displaced water equals $\frac{4}{3}\pi b^3 \rho_{\text{water}}$. The weight of the hollow sphere equals $\rho_{\text{concrete}}\frac{4}{3}\pi(b^3 - a^3)$. Setting these equal gives

$$\left[1 - \left(\frac{a}{b}\right)^3\right] = \frac{\rho_{\text{water}}}{\rho_{\text{concrete}}} = \frac{1}{2.49}$$

and a/b becomes 0.844, the condition for neutral buoyancy.

Putting this value in Eq. (9-25), with $r = a$, gives

$$\frac{\bar{S}}{P} = \frac{S_{yp}}{P} = \frac{3}{2}[1 - (0.844)^3]^{-1} = 3.75$$

Then P is 6.5 ksi/3.75 or 1,730 psi, a depth of about 3,840 ft.

9-8 CYLINDRICAL BODIES

A capped-end, thick-walled cylinder has the following principal stresses far from the ends (Roark 1964):

$$S_1 \text{ (longitudinal)} = -P\left(\frac{b^2}{b^2 - a^2}\right) \tag{9-32}$$

$$S_2 \text{ (circumferential)} = -P\left[\frac{b^2(a^2 + r^2)}{r^2(b^2 - a^2)}\right] \tag{9-33}$$

$$S_3 \text{ (radial)} = -P\left[\frac{b^2(r^2 - a^2)}{r^2(b^2 - a^2)}\right] \tag{9-34}$$

where a is the inner radius, b the outer radius, and P the large external pressure. Putting these equations in the combined stress formula (9-22) gives, for the case of maximum stress ($r = a$):

$$\bar{S} = P\sqrt{3}\,\frac{b^2}{b^2 - a^2} \tag{9-35}$$

and setting $\bar{S}$ equal to S_{yp} will give us a design equation based on yielding for a thick cylinder. Note, however, that when a cylinder is capped by two hemispheres, these equations will not be strictly correct at the junction. If one plots the radial stress as a function of radius for the cylinder, Eq. (9-34), and the same for a sphere, Eq. (9-23), it will be apparent that two different curves result. Thus, where these two shapes join, some adjustment in the stress field must occur and the resulting equations will be much more complicated. However, it should be noted that a spherically capped cylinder will usually fail near the center of the cylindrical part since the spherical ends impart considerable additional stiffness. Thus Eq. (9-35) might be a reasonable design equation for inexpensive missions where human safety is not involved. In the case of submarines and large tanks, a much more careful analysis is called for, since bending stresses may be set up at the sphere-cylinder junction. Plastic deformations involving cylinders are complex and will not be considered here.

9-9 FATIGUE FAILURE

As we saw in Chap. 2, ocean wave data are given as a statistical distribution of heights, lengths, and energies. These various waves, in turn, impart various forces to structures and anchored devices which may eventually fail by fatigue, even though the forces have never exceeded the yield strength of the structure at any time. The subject of fatigue is a complex one, covered in many books and papers in great detail and involving vast amounts of test information. Our approach here will be to introduce the topic with special reference to the analysis of ocean systems under wave loadings.

Virtually every fatigue problem in ocean engineering involves both static and oscillating loads. For example, a drill platform will support the large weight of the structure and machinery as a static compressive stress on the columns. Wave loading will produce an oscillating stress that may vary from compressive to tensile, or may be entirely one or the other. One would intuitively expect that a structure that is already loaded would be able to support less of an oscillating load than one that is unloaded, and such is generally the case. Various means of combining static and oscillating loads exist; we shall discuss only one, the so-called *Soderberg design criterion.*

If we designate the maximum, long-term, reversed or oscillating stress as S_r and the maximum static or average stress as S_{av}, then the Soderberg criterion is based on remaining within the envelope formed by plotting S_r in the vertical direction and S_{av} in the horizontal direction, and connecting the two points by a straight line; see Fig. 9-5. When the failure limit for pure reversed stress S_r is not known for a material, a common approximation is to take

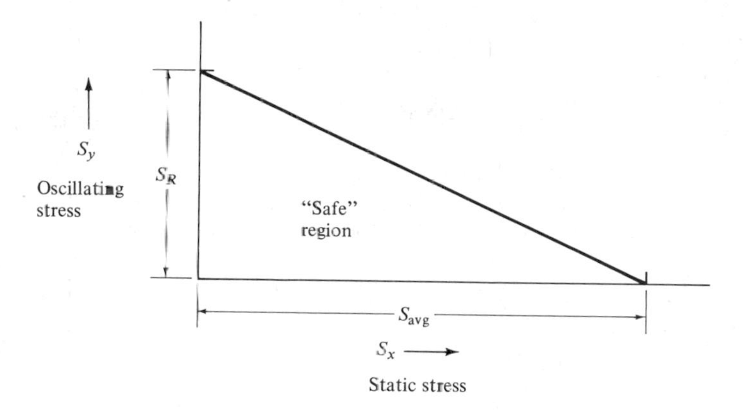

Fig. 9-5 Design criterion for combined stresses.

$$S_r = \tfrac{1}{2} S_{av} \qquad (9\text{-}36)$$

This approximation can be judged by noting Fig. 6-7, which shows typical data relating ultimate static strength to ultimate pure fatigue strength at various numbers of cycles. The Soderberg criterion can be used with either ultimate or yield point stresses, and in either case the S_r value is usually taken as half the static value. From Fig. 9-5 we can relate the general oscillating stress S_y and the static stress S_x by

$$S_y = S_r - \frac{S_r}{S_{av}} S_x \qquad (9\text{-}37)$$

where compressive stresses are taken as negative and tensile stresses as positive.

Example 9-2 A steel beam having a yield point tensile stress of 80,000 psi is subjected to reversed stress loading of 40,000 psi in tension to 20,000 psi in compression. What static tensile stress can the beam support in addition to this reversed stress?

Solution The reversed stress (S_y in Fig. 9-5) is equal to [40,000 − (−20,000)]/2 or 30,000 psi. The average stress due to reversed loading is (−20,000 + 30,000) or +10,000 psi. From Eq. (9-37) and using Eq. (9-36) to estimate S_r,

$$30{,}000 = 40{,}000 - \tfrac{1}{2} S_x$$
$$S_x = 20{,}000 \text{ psi}$$

Since the average stress due to reversed loading only is 10,000 psi, the beam can support an additional (20,000 − 10,000) or 10,000 psi in tension without exceeding the long-term yield stress in combined loading. Figure 9-6 shows the stress relations in this problem.

In "dry" engineering, reversed loading is often uniform over many cycles. Waves, however, follow some probability law over any period of time, and thus there will be a wide variation of stresses imposed by wave loading on a structure during its lifetime.

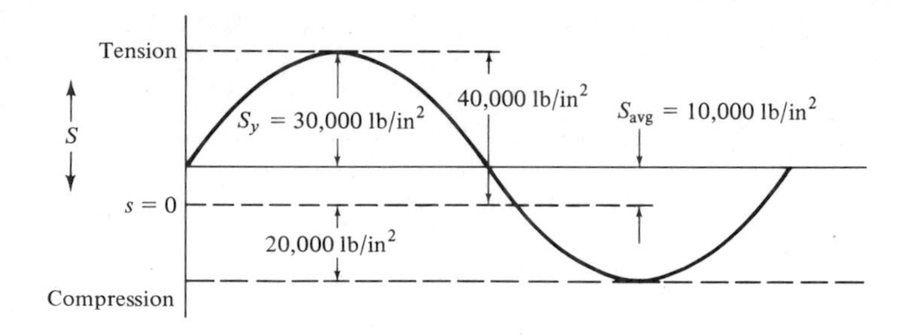

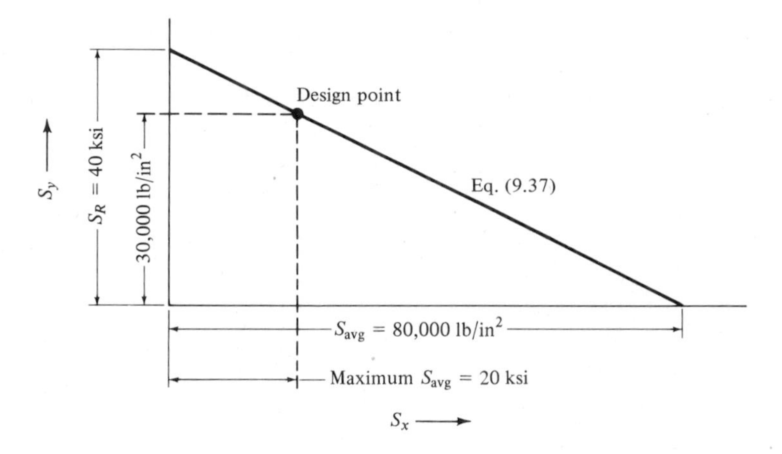

Fig. 9-6 Example 9-2 system. Since the S_{av} due to the oscillating load is 10 ksi and a static load of 20 ksi is allowed, 10 ksi added static load is permitted.

To solve this type of problem, we require data on the magnitude of alternating stress to failure as a function of the number of cycles. Figure 6-7 is an example of such data. If we have a member subjected to reversed stresses S_1, $S_2, S_3, \ldots, S_n$, for number of cycles $N_1, N_2, N_3, \ldots, N_n$, we can represent the fraction of life "lost" by each stress group by dividing the actual number of cycles N by the maximum number of cycles M which the material could support before failure if the reversed stresses were all at that level. Then the total life of the material is the sum of all these fractions or:

$$1.0 = \frac{N_1}{M_1} + \frac{N_2}{M_2} + \frac{N_3}{M_3} + \cdots + \frac{N_n}{M_n} \tag{9-38}$$

Example 9-3 A high-strength steel structure is in a seaway where waves of varying size produce reversed stresses. Computations and wave records give the following data:

Reversed stress, ksi	170	140	120	110	100
Life at this stress, cycles	10^5	10^6	10^7	10^8	Indefinite
Percent of waves inducing this stress	1.0	3.0	6.0	20.	70.

How many waves can the structure withstand without failure?

Solution Let N be the total number of waves until failure. Then Eq. (9-38) gives

$$1.0 = \frac{0.01N}{10^5} + \frac{0.03N}{10^6} + \frac{0.06N}{10^7} + \frac{0.2N}{10^8} + 0.7$$

and $N = 2.17 \times 10^5$ waves.

These few stress equations in Part 2 of this chapter only introduce the serious and complex design problems inherent in ocean structural problems. In addition to the many materials and properties discussed in Chap. 6, the ocean engineer must consider the effects of corrosion attack discussed in Chap. 5 and learn to apply safety factors to cover this form of material deterioration. Complex submarine and housing shapes are difficult to design from the simple shape formulas shown here. The designer wishes to minimize material thickness where it is not needed, and it may be actually possible to fail with an externally pressurized system by making one part of it too stiff. When human occupants are involved in deep vehicles, the cost and responsibility aspects of design are heavy and all designers depend on past experience with successful systems.

PROBLEMS

9-1. An underwater "pulse jet" for divers operates in the following manner: A volume (take 1 ft^3) is allowed to half fill with water. Air at a pressure of 100 atm rushes into the empty half, and the water is driven out as a jet under pressure. Note that the air only does work as it expands from 100 to 50 atm, at which point the water is all gone. Assume the flow is slow enough so that the process occurs at the dead state temperature of 500°R. What is the absolute maximum work obtainable from this air during such an expansion at 500°R with a dead state pressure of 1 atm? Compare with battery-stored energy.

9-2. An inventor proposes the following energy storage systems: A deep-diving midget sub will have an enclosed space (assume 1 ft^3 for simplicity) containing air at some elevated pressure P. The sub will go to its design depth of 5,000 ft in seawater and compress the air to some smaller volume, following the perfect gas law. Now when the sub returns to the surface, this compressed gas will do work at 1 atm. We will assume that both compression and work-producing expansions are so slow as to be isothermal. (Water temperature is 500°R.) With the dead state at 500°R and 1 atm:

(*a*) Decide what the optimum starting pressure of air should be to produce the maximum total change in available energy (which is the product of the mass of air times the unit change in A). Assume a closed system.

(*b*) Compare this optimum with the U/w VSP parameter for the other systems in the chapter.

(*c*) Would a gas other than air be better? Which gas, and how much better would it be?

9-3. A supercritical nuclear steam station using a flash-steam system provides water (steam) at 1,000 psia and 1100°R. Condensation following the turbine process can occur at two dead states: (*a*) shallow, estuarine water with T_0 of 75°F and (*b*) deep water below the thermoline at 42°F. The dead state pressure in each case is the appropriate vapor pressure of water at the low temperature since the condensation system is closed. How much additional available energy (per pound of steam) is possible when expanding (in a flow system) to the lower T_0?

Assuming that we might realize half this added available energy as work and that for each pound of steam 100 lb of water must be processed through the condensers, what pumping head of water might we obtain with no degradation of station performance? That is, if we use the added available energy to pump the cool water to the condensers, what head might we obtain? Comment on the numbers you get in regard to the siting of such plants so as to minimize environmental damage.

9-4. The Claude cycle proposed that a vapor power system be operated between the hot, tropic, surface waters at 80°F and the deep (2,000 ft) cool waters at 42°F. The working fluid would be a natural gas derivative similar to propane (molecular weight 44.1 lb/lb, C_p 0.39 Btu/(lb)(°F), $C_v = 0.353$). The cool water was to rise in pipes as it was displaced by the hot surface water so that the plant had a dead state pressure of 1 atm.

(*a*) What is the maximum work per pound of working fluid possible from the Claude cycle?

(*b*) What is the change in availability of the ocean water as it rises from 2,000 ft to sea level, including the compressibility effect and assuming the water at sea level reaches a temperature of 80°F, on a per pound basis?

9-5. A natural gas well on the continental shelf is bled by a 12-in-ID line 500 mi in length. We assume its inside is smooth. The surrounding water is at 60°F and the gas properties are: C_p of 0.53 Btu/(lb)(°F), $K(C_p/C_v)$ of 1.31, viscosity of 2.3×10^{-7} lb·s/ft², and molecular weight of 17.3 lb/mol.

(*a*) Compute the delivery rate (*G* times the area of the pipe) assuming a wellhead pumping station capable of 20 atm over ambient, and 1 atm at the line end.

(*b*) If we have a suction machine at the downstream end that reduces the pressure to 2 psia, what is the new delivery? Note that before accepting either answer we should check that the downstream Mach number is not choked. For isothermal flow this is given by Mach no. $= (1/K)^{1/2}$.

9-6. What is the absolute maximum work obtainable from a scuba tank having 70 ft³ of free (1 atm) air compressed to 2,200 psia at the surface of the ocean with a dead state temperature of 60°F? What is a reasonable level 2 work magnitude? If this work were spread over ½ h and used for diver propulsion, at what horsepower could the vehicle be propelled?

9-7. A 200-ft-long hose with ID of 0.75 in is to supply air from the surface to a habitat 100 ft under seawater. The surface compressor supplies air at 20 atm absolute, and we assume that the small friction factor is reasonably constant at 0.006. Assume the interior of the habitat is at bottom pressure. Find the delivery rate in cubic feet of surface air per minute assuming isothermal flow. The inlet hose condition is 20 atm and 60°F.

9-8. Design (predict size and weight of) a battery bank for a submersible requiring 1,000 hp for a 6-h period.

9-9. Design a switching circuit for a small adjustable-speed motor that can use 6, 12, 18, or 24 V to give different speeds. Four identical 6-V lead-acid batteries will be used, but we want to drain them as evenly as possible to prevent mismatch. Assume that part-load running is more or less evenly distributed over the four possible speeds.

9-10. A dangerous electric shock involving human subjects in water is a current of 70 mA for 5 s. A typical low value for body resistance from hand to hand in seawater might be 800

Ω. What is the smallest silver-zinc battery that might be capable of dangerously shocking a diver, based on these values?

9-11. Obtain a general formula for the Soderberg fatigue design line based on the assumption that pure reversed bending stress cannot exceed half the maximum pure tensile stress. Include in your equation factors of safety (FS) on both the reversed and the allowed tensile stresses that might be different in magnitude. For example, we might apply a factor of safety of 2 to the maximum allowed tensile (or compressive) stress and a factor of safety of 4 to the maximum allowed reversed bending stress.

9-12. A large buoy is supported in a seaway by a high-strength, stranded, wire cable with characteristics $S_{ult} = 90{,}000$ psi and $S_{yp} = 60{,}000$ psi. Minimum force with average waves is estimated at 20,000 lb (taut wire mooring) with a maximum force of 50,000 lb when the buoy is on the top of a wave crest. Find the wire diameter based on the yield point with a factor of safety of 1.5 for long-term use, applied to both S_{yp} and S_r.

It is found that the fittings and thimbles in the setup produce a stress concentration of 1.4 on the oscillating stress only (the factor of safety applies to both reversed and steady stresses). What is the wire area with this charge? Show Soderberg lines for these two cases.

9-13. An in-water structure made of 4340 steel is subject to wave impact forces. The following table relates the wave period to its equivalent maximum stress in the structure and its relative frequency as predicted by wave monitors.

Mean period, s	1.0	2.0	3.0	4.0	5.0
Percent of time observed	15.	40.	25.	15.	5.
Maximum reversed stress, ksi	100	125	160	180	200

Using Fig. 6-7, assume that the fatigue curve for this steel represents the 50 percent failure point and that when 50 percent of the supports have failed, the structure will fall over. Predict its life in seconds.

9-14. Show that the combined stress equation (9-22) predicts the correct result for a pure tension test on a uniform bar (that $\bar{S} = S_{yp}$) when yield is reached.

9-15. A material is subjected to an alternating stress of 50 ksi for 1×10^6 cycles. At this stress, fatigue failure would be expected at 1.5×10^6 cycles. The stress is now raised to 60 ksi, at which stress failure would be expected in 5×10^5 cycles. How many cycles at 60 ksi will occur before expected failure?

9-16. Obtain the general equation for effective stress in a capped, thick cylinder far from the ends and verify that Eq. (9-35) is a special case.

9-17. Taking the case of a cylindrical body with hemispherical end-caps of constant a and b throughout, obtain the ratio of the maximum effective stress for the spherical part to the same for the cylindrical part, as a function of a/b or whatever parameter or ratio of parameters seems most useful. Show graphically how this function behaves. Will one part always fail first?

9-18. Using material properties from Chap. 6, design the most buoyant sphere possible that will reach the deepest part of the ocean (36,000 ft) based on (*a*) thick-walled, all elastic and (*b*) thick-walled, all plastic. Can either float?

9-19. Using Table 6-2 ("Low-Cost Steels"), design the cheapest spherical floats to buoy 1 ton at 1,000-ft depth. You may use one float or several. Base your cost on material cost only.

9-20. An ocean tower is loaded as shown by wave forces. The maximum bending moment occurs at point x and is equal to $\frac{1}{3}wl^2$, where w is the force per unit length at the surface of the water. w varies sinusoidally from 100 lb/ft to 50 lb/ft and the tower column will be a steel square shape of dimensions b by b ft and solid. Based on the Soderberg design criterion with S_{yp} of 25,000 psi, what must b equal with no factor of safety?

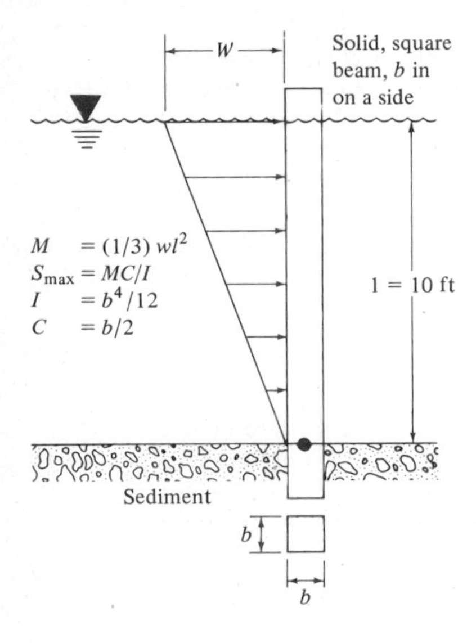

Fig. P9-20

9-21. The crossbeam shown in the figure is subjected to a varying load per foot by waves varying from zero to w pounds per foot. The beam has the dimensions shown, and the moment equation for uniform loading is given. The material can withstand a maximum yield point stress of 60,000 psi in pure tension and 30,000 psi in pure reversed bending S_r. Based on the point of maximum stress in the beam, what sustained reversed loading of 0 to w will produce yielding?

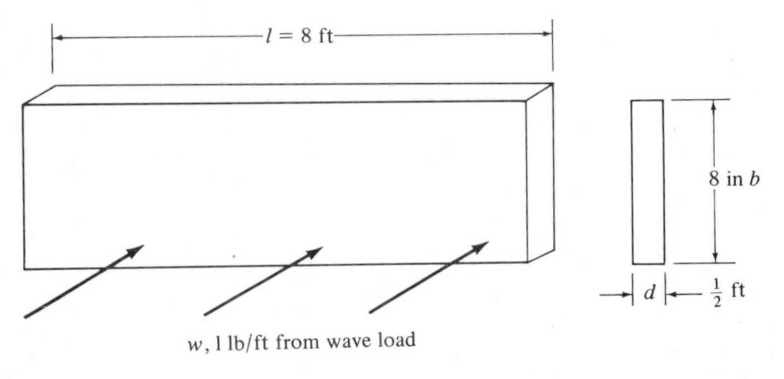

$S = MC/I \quad M = \frac{1}{2}wl[(L/2) - (l^2/4l)] \quad I = (bd^3)/12 \quad C = d/2$

Fig. P9-21

9-22. The approximate equations of principal stresses for a "thin" cylinder are: $S_1 = PR/2t$, $S_2 = PR/t$, and $S_3 = P$ or zero on the outside or inside. R is the mean radius of the cylinder and t is the thickness. What a/b ratios allow us to use the thin-walled equations? What effective stress is predicted for the thin-walled case in terms of a and b?

9-23. What is the size and thickness of a tempered glass sphere going to 18,000 ft in the ocean with an interior diameter of 5 ft based on a factor of safety of 3, and the compressive strength? What is its sinking, or buoyant, force?

The sphere is made up of two half spherical shells flanged and bolted together. What is the total force in pounds pressing the two halves together (at depth), and what is the average pressure in psi along the junction?

REFERENCES

Drucker, D. (1967): "Introduction to Mechanics of Deformable Solids," McGraw-Hill, New York.
Keenan, J. (1941): "Thermodynamics," Wiley, New York.
Kenyon, R., and H. Schenck (1964): "Fundamentals of Thermodynamics," Ronald, New York.
Obert, E. (1948): "Thermodynamics," McGraw-Hill, New York.
Roark, R. (1964): "Formulas for Stress and Strain," 4th ed., McGraw-Hill, New York.
Shapiro, A. (1956): "Compressible Flow," vol. 1, Ronald, New York.

10

Water Quality

Donald L. Sussman

10-1 INTRODUCTION

The pollution of water is indicated by changes in water quality. Levels of water quality depend on its use, such as public drinking water, bathing and recreational water, agricultural purposes, industrial and cooling water, fish and wildlife habitat, and aesthetic value.

In the past, we were concerned with such water quality indicators as bacterial count, hardness, and the appearance of waters. However, today there is an increasing awareness of the many physical, chemical, and biological substances or factors which contribute to the overall quality of water. As analytical techniques become more sophisticated and precise, there will be new factors or parameters found that will describe water quality and water pollution in more detail. This is vital to expanding industrial and domestic water use.

It has been known since the nineteenth century that water is one of the major vehicles for the transmission of disease. Such diseases as typhoid and cholera were causes of major epidemics. In all cases there was a lack of proper sanitary disposal facilities and little or no water treatment. In 1892, an explosive cholera epidemic occurred throughout Europe. It was seen that any area which treated its water, even partially, showed a significant decrease in the number of cholera cases.

A water of high quality must be free from pathogenic bacteria and viruses, waterborne disease-producing organisms, and organic and inorganic compounds. Examples of the diseases are as follows:

From pathogenic bacteria and viruses Typhoid, paratyphoid, cholera, dysentery, gastroenteritis, leptospirosis, polio, and hepatitis

From waterborne disease-producing organisms Schistosomiasis, amebic dysentery

Illnesses from chemicals Methemoglobinemia from excess nitrates, fluorosis from excess fluoride, and toxic metal ion (Cu, Pb, Hg, etc.) poisoning

Water may hold harmful substances. Generally, there are two approaches to this problem: to protect a water system from contamination and to remove or destroy any substances which may have gained access to the water system. As the domestic and industrial demands for water become greater, protection from pollution and maintenance of a high water quality are becoming more difficult unless complete examinations of waters and extensive treatment are applied.

10-2 WATER EXAMINATION

By means of quantitative water examinations a water system can be effectively monitored. The tests can determine the quality of the water and can be used as a guide for treatment. The tests carried out on water in order to determine its condition are physical, chemical, and biological. A number of different reference works present these tests in detail.

Certain physical characteristics of a water are important, as they are indications of the aesthetic qualities. Tests are based on color, odor and taste, turbidity, solids, immiscible liquids, and temperature.

10-2.1 Color

Many waters have a color due to either dissolved or suspended material. Color can be either natural or from pollution by highly colored wastewaters.

Natural pollution is caused by the decay of organic matter such as leaves, conifer needles, and vegetable extracts. The organic compounds responsible for natural color are tannates and humates. Waters that drain from swamps and deciduous forests are intensely colored owing to organic matter.

Some color may result from inorganic ions or complexes. Ferriorganic complexes and manganese compounds are examples of inorganic compounds that produce color at relatively low concentrations.

Waters may become colored by pollution with sewage or industrial wastewaters. Sewage color reflects its strength and condition; fresh sewage is gray whereas septic sewage is black.

Improperly treated industrial wastes can produce color in certain cases. Textile dye wastes, chrome-plating wastes, and paint wastes impart a color to wastewaters. Acid mine waters and drainage in mining regions often carry heavy iron discoloration, sometimes intensified by natural constituents in the water.

The fact that colored waters are conspicuous makes them objectionable. This necessitates treatment for color removal which is relatively expensive (depending on the final use of the water). Colored waters can be quite harmless, but they are aesthetically unpleasing; color may indicate the presence of more serious pollution.

Color is measured by comparing the intensity of the sample with standards prepared from potassium chloroplatinate and cobaltous chloride. The filtered sample is matched to the closest standard and reported in color units. The "International Standards for Drinking Water" states that a raw water should have a limiting color value of 300 units, on the basis that a value of less than 300 units indicates an acceptable quality for treatment. Instrumental methods have been developed for the measurement of color.

10-2.2 Odor and taste

The sources of odor and taste in water are:

1. Growth of microorganisms in water that release taste- and odor-producing substances, plus the decomposition products of these organisms when killed
2. Decomposition of leaves, grasses, and aquatic vegetation
3. Growth of slime, molds, and fungi
4. Anaerobic (without oxygen) processes in water producing odorous end products (e.g., H_2S, ammonia)
5. Industrial wastes
6. Chlorine chemicals

Odor may be characteristic of a waste, or it may develop in the receiving water by decomposition or reaction. The presence of a particular odor may indicate that a certain reaction is occurring in the water. An ammoniacal odor is indicative of the decomposition of animal life in a water. Sulfide odor characterized by the "rotten-egg smell" can be used as an indication that the water is corrosive and may chemically react with metals and concrete and discolor paints.

Taste is a sensitive but somewhat reluctant water quality parameter, as it is difficult to conceive of anyone tasting a potentially noxious or toxic waste material. Many chemicals cause objectionable tastes. Phenolic compounds which occur naturally and as industrial pollutants are the most conspicuous taste producers. The limiting concentration for phenols is 1 ppb (part per billion).

Odor is determined by a gas chromatograph which can qualitatively and quantitatively determine the odor. The most widely used method is by subjectively testing taste and odor samples by using qualified human testers.

10-2.3 Turbidity

Turbidity in water is caused by the presence of colloidal particles in water. Colloidal particles are larger than true ions and molecules dissolved in a liquid and smaller than settleable solids. A true solution can be called a homogeneous phase whereas a suspension of colloidal particles in a liquid is heterogeneous.

Colloidal suspensions have some unique properties: They scatter light, and the amount of light scattered is related to the degree of turbidity and particle surface area, among other factors. The particles have a similar surface charge which accounts for the stability of the suspension since the particles repel each other. By the addition of chemicals called destabilizing agents, the colloids can be coagulated, and they settle out rapidly.

Sources of turbidity in water are clay, silt soil, sewage, bacteria, industrial wastes, and weathered minerals. The determination of turbidity is measured by the degree of light scattered by the suspension, which increases as the concentration of colloid increases, or by the amount of light transmitted, which decreases as the concentration of colloid increases. The construction of such a turbidity meter is described in the next chapter.

10-2.4 Solids

The amount of solids in a water can be used as a guide for its use or treatment:

1. Dissolved solids: These consist mainly of inorganic salts, soluble organic matter, and dissolved gases.
2. Suspended solids: Colloidal matter (source of turbidity) as just described.
3. Settleable solids: Solids that will settle because of the influence of gravity. The rate of settling depends directly on the size of the particles and the density of the particles, and inversely on the viscosity of the liquid.
4. Volatile solids: Material that will decompose at 600°C. These include mainly organic compounds, and to a small degree some carbonate salts and ammonium salts.

Total solids are determined by evaporating and drying a measured sample in a weighed container. This sample is then heated in an oven at 600°C, cooled,

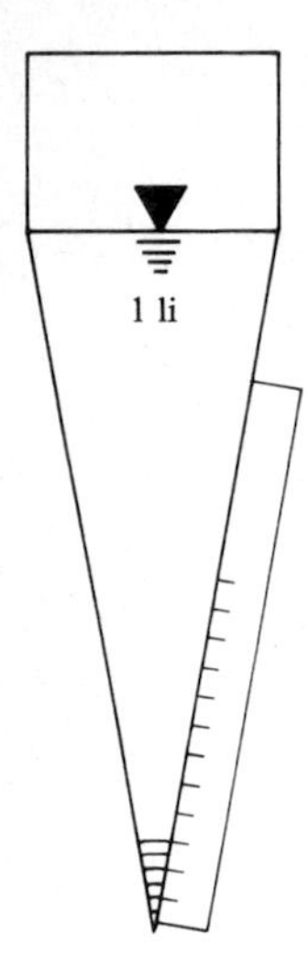

Fig. 10-1 Imhoff cone for the measurement of settleable solids.

and weighed. The loss in weight from heating is equal to the volatile solid weight.

Settleable solids are measured in an Imhoff cone (Fig. 10-1). The sample is placed in the cone and the solids are allowed to settle for 1 h. The results are measured in milliliters of settleable solids per liter. Suspended solids are measured after settleable solids by filtering a known volume of solution through a weighed microfilter (pore size of 0.45 μm). The filter is dried, then weighed.

Dissolved solid concentration can be estimated by measuring the conductance, which is a measurement of the ability of the solution to carry electric current, and it is related to the concentration and type of ionic substances in water. Another method of getting a good estimate of the dissolved solids is by subtracting the weight of the settleable and suspended solids from the weight of total solids.

10-2.5 Immiscible liquids

Immiscible liquids include oils, greases, fats, and any liquid that is insoluble or slightly soluble in water. The sources of this material are industrial wastes such as meat-packing and food-handling residues, petroleum, chemical wastes and paint wastes, and material from ships.

The major problems with these compounds are that they usually prevent the transfer of oxygen by forming a surface layer over water or by coating microorganisms and aquatic flora. The majority of the immiscible liquids are toxic to aquatic life.

10-2.6 Temperature

The temperature of surface waters of the United States varies from 0 to over 40°C (32 to over 100°F). The natural temperatures depend upon season, time of day, duration of flow, depth, latitude, altitude, and currents.

The species of aquatic life that occur naturally in each body of water are dependent upon the temperature and other variables (e.g., salinity). The interrelations of aquatic life and temperature are so intimate that even a small change in temperature may have far-reaching effects. Fish may be under stress or die when placed in a water whose temperature range is higher than where the fish are normally found. Temperature is a signal for spawning, hatching, and migration (Table 10-1), and an artificially raised temperature will have fish hatching or moving into areas with insufficient food.

Elevated temperatures affect chemical and physical processes and biological activity. The solubility of relatively insoluble compounds increases and the solubility of gases decreases as the temperature of a water increases. Physical properties such as density and viscosity are temperature-dependent.

One critical problem with elevated temperatures is the decrease in dissolved oxygen (DO) (Fig. 10-2). Many aquatic organism deaths result from

Table 10-1 Provisional maximum temperatures recommended as compatible with the well-being of various species of fish and their associated biota

93°F	Growth of catfish, gar, white and yellow bass, spotted bass, buffalo, carpsucker, threadfin shad, and gizzard shad
90°F	Growth of largemouth bass, drum, bluegill, and crappie
84°F	Growth of pike, perch, walleye, smallmouth bass, and sauger
80°F	Spawning and egg development of catfish, buffalo, threadfin shad, and gizzard shad
75°F	Spawning and egg development of largemouth bass, white and yellow bass, and spotted bass
68°F	Growth or migration routes of salmonids, and egg development of perch and smallmouth bass
55°F	Spawning and egg development of salmon and trout (other than lake trout)
48°F	Spawning and egg development of lake trout, walleye, northern pike, sauger, and Atlantic salmon

Source: "Water Quality Criteria," Federal Water Pollution Control Administration, U.S. Department of the Interior, Washington, D.C., 1968.

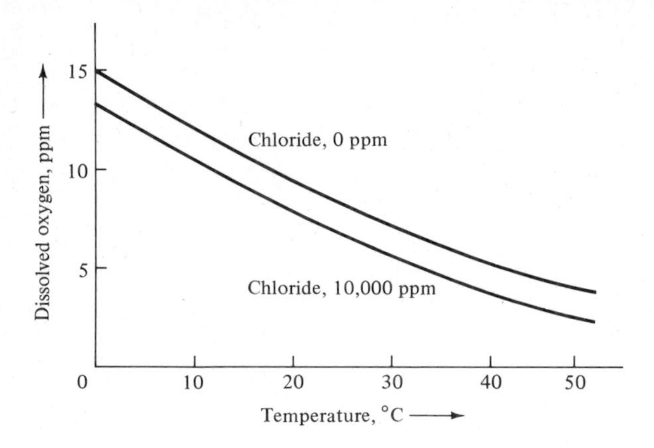

Fig. 10-2 Dissolved oxygen as a function of temperature for fresh and brackish waters. See also Eq. (11-2).

suffocation due to insufficient oxygen in the water. Natural purification processes and bacterial action are accelerated at elevated temperatures and may cause depletion of oxygen.

The sources of elevated-temperature waters include any industry or utility that uses water for cooling.

This can be overcome by the use of holding tanks, channels, or cooling towers. All these concepts involve cooling before the water is returned to a receiving water body.

10-3 CHEMICAL EXAMINATION

There are many different chemical pollutants which can affect a water quality. The following discussion will be limited to the chemical pollutants or factors that are typically found in many waters.

10-3.1 pH, acidity, and alkalinity

pH is a term used to express the acid or alkaline intensity of a solution. pH is the negative logarithm of the concentration of hydrogen ions:

$$pH = -\log H^+ = \log \frac{1}{H^+} \tag{10-1}$$

The concentration is expressed in moles per liter where a mole is the weight of the substance per molecular weight of the substance.

Example 10-1 What is the pH of a solution containing 5 mg/l (milligrams per liter) HCl (hydrochloric acid)?

Solution

HCl molecular weight = 1.0 + 35.5 = 36.5 g/mol

Then

$$5 \text{ mg/l} = 0.005 \text{ g/l}$$

and

$$\frac{0.005 \text{ g/l}}{36.5 \text{ g/mol}} = 0.000137 \text{ mol/l}$$

$$\text{pH} = -\log (1.37 \times 10^{-4})$$

$$\text{pH} = 4.00 - \log 1.37$$

$$\text{pH} = 4.00 - 0.14 = 3.86$$

The pH range in aqueous solution goes from 0 to 14 (which corresponds to an H^+ concentration from 1 to 10^{-14} mol/l). At a low pH the solution is acidic and at a high pH it is alkaline. A pH of 7 is considered neutral. The optimum pH range for aquatic life in waters is from 6.0 to 9.0, and care should be taken to make sure that no pollutants are added which would lower or raise the pH and kill aquatic life. Table 10-2 shows the pH of some substances.

pH measurement is one of the simplest to make as long as one has a properly working pH meter and electrodes. These measurements can be made in a wide variety of materials and under extreme conditions.

Acidity is the base-neutralizing capacity of a water, and it is usually a measurement of the dissolved carbon dioxide (CO_2) which is acidic, since

$$CO_2 + H_2O \rightarrow H_2CO_3 \text{ (carbonic acid)}$$

Natural acidity can be due to the presence of certain salts in solution which hydrolyze to give acids. Acids can be present as pollutants from certain industries.

Acidity is determined by titrating a sample with an alkali solution (NaOH, KOH) of known concentration and determining the end point with a pH meter or indicator.

Example 10-2 If a 200-ml sample of a water requires 10 ml of 0.01 *M* NaOH to reach the end point, what is the acidity in milligrams per liter as CO_2?

Solution

$$\text{Volume}_{\text{solution}} \times M_{\text{solution}} = \text{volume}_{\text{titrant}} \times M_{\text{titrant}}$$

$$200 \times M_s = 10 \times 0.01$$

$$M_s = \frac{10 \times 0.01}{200} = 0.0005 \text{ mol/l}$$

$$0.0005 \text{ mol/l} \times 44 \text{ g/mol } CO_2 \times 1{,}000 \text{ mg/g} = 22 \text{ mg/l}$$

Alkalinity is the acid-neutralizing capacity of a water and is due to the presence of certain anions and alkalies in water. Bicarbonates (HCO_3^-) represent the major contribution since they are formed by the reaction of CO_2 with alkaline materials. Other anions that can contribute to alkalinity include

Table 10-2 pH values in aqueous solution (25°C)

$[H^+]$	[pH]		[pOH]
1	0	1.0 *M* HCl, HNO_3	14
0.1	1	0.1 *M* HCl, HNO_3 0.1 *M* oxalic acid Acid mine wastes	13
10^{-2}	2	Limes 0.1 *M* lactic acid Swamp peats	12
10^{-3}	3	0.1 *M* acetic acid	11
10^{-4}	4	Peat water, saturated H_2CO_3	10
10^{-5}	5	0.1 *M* HCN	9
10^{-6}	6	Rainwater Lake muds; 0.02 *M* Na_2HPO_4	8
10^{-7}	7	CO_2 free distilled water	7
10^{-8}	8	0.1 *M* $NaHCO_3$ Seawater	6
10^{-9}	9	0.05 *M* $Na_2B_4O_7$ (borax) Saturated $CaCO_3$	5
10^{-10}	10	Alkali soils Saturated magnesia	4
10^{-11}	11	0.1 *M* NH_3	3
10^{-12}	12	0.1 *N* Na_3PO_4 (T.S.P.) Saturated $Ca(OH)_2$ (hydrated lime)	2
10^{-13}	13	0.1 *M* NaOH, KOH, LiOH	1
10^{-14}	14	1.0 *M* NaOH, KOH, LiOH	0

silicates, phosphates, borates, tannates, humates, carbonates, and hydroxides. The latter two anions can be found at higher pH values ($pH > 9$).

Waters of high alkalinity are found where there is a high degree of weathering.

The major constituents of alkalinity are bicarbonate (HCO_3^-), carbonate (CO^{--}), and hydroxide (OH^-). The determination is based on a titration with an acid of known concentration. The total alkalinity can be determined as well as combinations of certain anions. There are five possible combinations of the anions: (1) HCO_3^- alone, (2) CO_3^{--} alone, (3) OH^- alone, (4) $HCO_3^- + CO_3^{--}$, and (5) $CO_3^{--} + OH^-$.

The titration is based on using two indicators, phenophthalein and methyl orange or preferably methyl purple. The titration curves are shown in Fig. 10-3 and the relations in Table 10-3.

Phenolphthalein changes color at pH 8.3, and methyl orange or methyl purple changes color at pH 4.5.

Alkalinity data are used as a variable in many water purification and waste treatment processes, as in coagulation, water softening, and corrosion control. They are used to determine the capacity for boiler and heat exchanger waters to form insoluble carbonate and hydroxide scale in pipes.

Example 10-3 A solution (100 ml) requires 8.0 ml of 0.02 N H_2SO_4* to reach a phenolphthalein end point and an additional 12.5 ml to reach the methyl purple end point. What is the total alkalinity and the concentration of each component? Represent each value as "ppm as $CaCO_3$."†

Solution Total alkalinity:

$$V \times N = V \times N$$

$$(8.0 + 12.5)(0.02) = 100 \times N$$

$$N = \frac{(20.5)(0.02)}{100} = 4.1 \times 10^{-3} \text{ equiv/l}$$

$$4.1 \times 10^{-3} \text{ equiv/l} \times 50 \text{ g/equiv } CaCO_3 \times 1{,}000 \text{ mg/g}$$

$$= 205 \text{ mg/l as } CaCO_3 \text{ or } 205 \text{ ppm as } CaCO_3$$

*N represents normality, which is the number of equivalents per liter of solution. Equivalents = moles/valence.

HCl	Equivalents = moles/1
H_2SO_4	Equivalents = moles/2
H_3PO_4	Equivalents = moles/3
$CaCO_3$	Equivalents = moles/2

† A common practice is to represent concentrations at ppm as $CaCO_3$, which is really a form of equivalents. By multiplying the equivalents/liter of a substance by the equivalent weight of $CaCO_3$ in milligrams, one gets the expression equiv/l × (50,000 mg/equiv) = mg/l as $CaCO_3$.

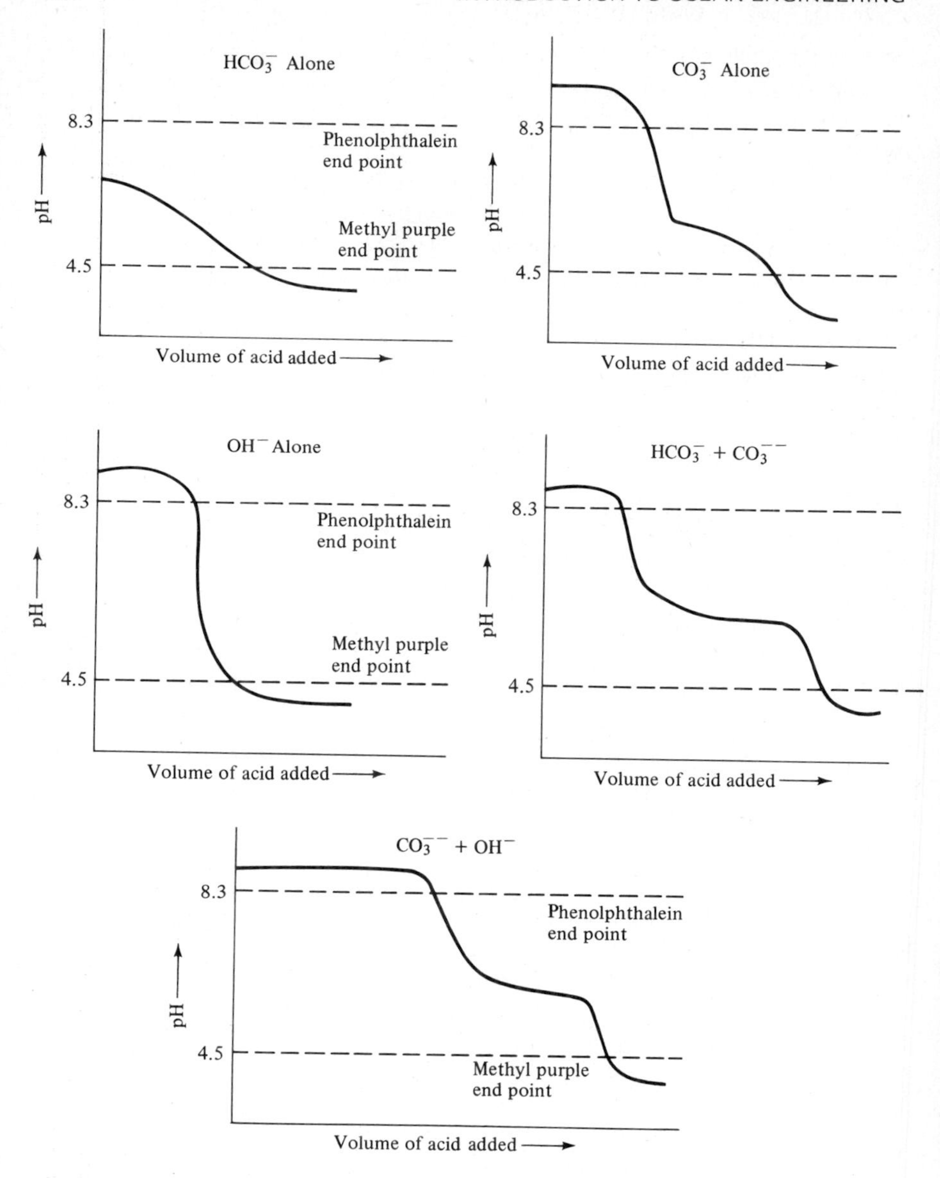

Fig. 10-3 Titration curves for phenolphthalein and methyl purple.

Components:

$P < \frac{1}{2}T$; therefore it must be CO^{--} and HCO^{--} (from Table 10-3). Then,

$CO_3^{--} = 2P$

and

$$V \times N = V \times N$$

$$8.0 \times 0.02 = 100 \times N$$

Thus

$$N = \frac{8.0 \times 0.02}{100} = 1.6 \times 10^{-3} \text{ equiv/l} = 80 \text{ ppm as } CaCO_3$$

And

$$CO_3^{--} = 2P = 160 \text{ ppm as } CaCO_3$$

so

$$HCO_3 = T - 2P = 205 - 160 = 45 \text{ ppm as } CaCO_3$$

10-3.2 Hardness

A hard water is one which will cause scale formation in hot-water pipes and will precipitate soaps. Hardness is caused by divalent (++) charged metal cations in water. These include primarily calcium (Ca^{++}), magnesium (Mg^{++}), ferrous iron (Fe^{++}), and manganese (Mn^{++}). Hardness occurs when an acid or CO_2 dissolves minerals in soil or in rock formations. CO_2 can be absorbed from the atmosphere and released by soil and subsoil bacteria. Hardness can be removed by adding lime and/or soda ash to precipitate the cations, or by the use of ion-exchange resins or zeolites.

10-3.3 Dissolved oxygen

The major parameter used in describing the quality of a water is the dissolved oxygen. All living organisms depend upon oxygen to maintain the metabolic processes that produce energy for growth and reproduction.

The concentration of dissolved oxygen (DO) in water is a measure of its sanitary quality. A polluted body of water will have a relatively low (or no) DO.

Table 10-3 Alkalinity relations*

Result	OH *alkalinity*	CO_3 *alkalinity*	HCO_3 *alkalinity*
$P = T$	T	0	0
$P = \frac{1}{2}T$	0	$2P$	0
$P > \frac{1}{2}T$	$2P - T$	$2(T - P)$	0
$P < \frac{1}{2}T$	0	$2P$	$T - 2P$
$P = 0$	0	0	T

*P = phenolphthalein alkalinity; T = total alkalinity = phenolphthalein + methyl orange alkalinity.

Table 10-4 Saturation values of dissolved oxygen in fresh and sea water exposed to an atmosphere containing 20.9 percent oxygen under a pressure of 760 mm of mercury*
Calculated by G. C. Whipple and M. C. Whipple from measurements of C. J. J. Fox.

Tempera-ture, °C	*Dissolved oxygen (mg/l) for stated concentrations of chloride, mg/l*					*Difference per 100 mg/l chloride*
	0	5,000	10,000	15,000	20,000	
0	14.62	13.79	12.97	12.14	11.32	0.0165
1	14.23	13.41	12.61	11.82	11.03	0.0160
2	13.84	13.05	12.28	11.52	10.76	0.0154
3	13.48	12.72	11.98	11.24	10.50	0.0149
4	13.13	12.41	11.69	10.97	10.25	0.0144
5	12.80	12.09	11.39	10.70	10.01	0.0140
6	12.48	11.79	11.12	10.45	9.78	0.0135
7	12.17	11.51	10.85	10.21	9.57	0.0130
8	11.87	11.24	10.61	9.98	9.36	0.0125
9	11.59	10.97	10.36	9.76	9.17	0.0121
10	11.33	10.73	10.13	9.55	8.98	0.0118
11	11.08	10.49	9.92	9.35	8.80	0.0114
12	10.83	10.28	9.72	9.17	8.62	0.0110
13	10.60	10.05	9.52	8.98	8.46	0.0107
14	10.37	9.85	9.32	8.80	8.30	0.0104
15	10.15	9.65	9.14	8.63	8.14	0.0100
16	9.95	9.46	8.96	8.47	7.99	0.0098
17	9.74	9.26	8.78	8.30	7.84	0.0095
18	9.54	9.07	8.62	8.15	7.70	0.0092
19	9.35	8.89	8.45	8.00	7.56	0.0089
20	9.17	8.73	8.30	7.86	7.42	0.0088
21	8.99	8.57	8.14	7.71	7.28	0.0086
22	8.83	8.42	7.99	7.57	7.14	0.0084
23	8.68	8.27	7.85	7.43	7.00	0.0083
24	8.53	8.12	7.71	7.30	6.87	0.0083
25	8.38	7.96	7.56	7.15	6.74	0.0082
26	8.22	7.81	7.42	7.02	6.61	0.0080
27	8.07	7.67	7.28	6.88	6.49	0.0079
28	7.92	7.53	7.14	6.75	6.37	0.0078
29	7.77	7.39	7.00	6.62	6.25	0.0076
30	7.63	7.25	6.86	6.49	6.13	0.0075

*For other barometric pressures the solubilities vary approximately in proportion to the ratios of these pressures to the standard pressures.

When DO is absent, biological changes are brought about by anaerobic organisms and noxious end products such as methane, ammonia, and hydrogen sulfide will form. With a good DO content in the water, aerobic conditions will occur and result in innocuous end products.

The concentration of DO (and most gases in water) depends on the temperature (Fig. 10-2), ionic concentration (Table 10-4), atmospheric pressure, and organic matter.

DO is determined by the Azide Modified Winkler Method or by the use of DO electrodes. In the Azide Modified Winkler Method, three solutions are required:

1. $MnSO_4$
2. KI, KOH, and NaN_3 (sodium azide)
3. Concentrate H_2SO_4

The reactions are:

$$Mn^{++} + 2OH^- + \frac{1}{2}O_2 \rightarrow \underline{MnO_2} + H_2O$$

$$MnO_2 + 2I^- + 4H^+ \rightarrow Mn^{++} + I_2 + 2H_2O$$

The elemental iodine that forms is then titrated with 0.025 N $Na_2S_2O_3$ (sodium thiosulfite), using starch as an indicator. If a 200-ml sample of the I_2 solution is used, then each milliliter of 0.025 N $Na_2S_2O_3$ is equivalent to 1 mg/l DO. The DO electrode consists of a membrane which allows DO to pass and measures the current as a function of the DO. This electrode has to be calibrated with waters which have been analyzed for DO previously by using the Winkler method.

Field samples of DO should be "fixed" as soon as possible to stop all biological reactions. Fixing involves the adding of the three solutions in the Azide Modified Winkler Method. Fixed solutions are good for 6 h.

10-3.4 Biochemical oxygen demand (BOD)

BOD is a measure of oxygen required to biodegrade organic matter in solution. As such it is used as an indicator of pollution in bodies of water.

Any oxidizable contaminants introduced into a natural waterway will be oxidized by chemical and bacterial processes, therefore consuming O_2. Dependent on the rates of oxygen depletion and replenishment, the DO will decrease to a minimum and increase to a maximum value within a certain time period (Fig. 10-4).

Oxygen depletion is caused by natural pollution from surface waters carrying salts and silt, decaying plants, human and animal excreta, chemical

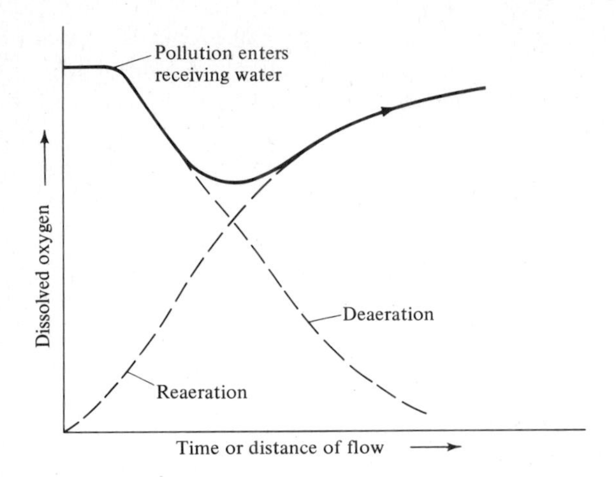

Fig. 10-4 Dissolved oxygen "sag" curve.

industrial pollution containing reducing agents such as nitrates, ferrous iron, sulfides and organics, and biochemical pollutants such as proteins, amino acids, and carbohydrates.

The bacterial reaction is generalized by

Oxidizable compounds + bacteria + oxygen + nutrients →
oxidized (stable) compounds + CO_2 + H_2O

The nutrients consist of carbon and nitrogen compounds, energy, inorganic ions, and growth factors.

The BOD reaction is primarily a first-order kinetic reaction. That is, the rate of biodegradation depends on the concentration of biodegradable material present

$$-\frac{dc}{dt} = kC \tag{10-2}$$

where k is the reaction rate constant, C is the concentration of BOD at the start of time t, and $-(dc/dt)$ is the rate of biodecomposition. C equals the initial BOD (L) less the BOD (y) at any time (t). Then:

$$\frac{d(L-y)}{dt} = -k(L-y)$$

$$\frac{d(L-y)}{(L-y)} = -k\,dt$$

Integrating,

$$\int_{y=0}^{y=y} \frac{d(L-y)}{(L-y)} = -\int_{t=0}^{t=t} k\,dt$$

$$\ln \frac{L-y}{L} = -k\,dt$$

$$\frac{L-y}{L} = e^{-kt} = 10^{-k't}$$

where

$$k' = \frac{k}{2.303}$$

and

$$y = L(1 - e^{-kt}) = L(1 - 10^{-k't}) \tag{10-3}$$

This equation is complicated by the fact that k and L are usually unknown. L and k can be found from a series of measurements (Fig. 10-5).

Theoretically infinite time is required for complete biochemical oxidation, but for all purposes it can be considered complete in about 3 weeks. This is usually too long to wait for results, and other complications can occur in this length of time. Since it has been found that a reasonably large degree of the total BOD is used up in 5 days, the test is based on a 5-day incubation period. The 5-day BOD includes from 40 to 90 percent of the total BOD. The 5-day BOD depends on the material decomposing–that is, certain materials decompose faster than others–and the nature of and viability of the bacteria which are reacting.

The BOD test is based upon the determination of DO. If the BOD is low (< 7 mg/l), fill two bottles with the water, determine the DO in one as soon as possible, and the DO in the other after incubating at 20°C for 5 days.

$$\text{BOD} = \text{DO}_0 - \text{DO}_5 \tag{10-4}$$

If the BOD is greater than 7 mg/l, the dilution method will be used. This is based on the idea that the rate of biodegradation depends on the concentration of material, and the rate at which O_2 is used is proportional to the waste in the dilution. Dilution water has to be prepared and extreme care has to

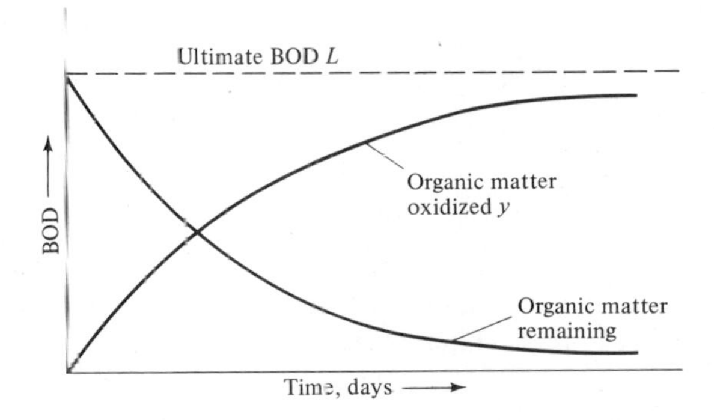

Fig. 10-5 Biological oxygen demand curve showing the nomenclature of Eq. (10-4).

be taken in preparing it. There should be no toxic materials such as heavy metal ions, chlorine and chloramines, the pH should be near neutral, good bacteria should be present, and the water should be close to saturation with dissolved oxygen. Nutrients are added. These include phosphates, calcium and magnesium salts, ferric chloride, and ammonium chloride.

BOD data can be used to show where pollution is entering bodies of water and where or when the worst conditions occur. It can be used as a method for monitoring streams, rivers, and estuaries.

10-3.5 Chemical oxygen demand (COD)

This test is used to measure the total pollution strength of a water. The COD is determined by reaction with a high chemical oxidizing agent such as dichromate ($Cr_2O_7^{--}$) or permanganate (MnO_4).

10-4 METAL IONS

There are many metal ions that can and do exist in waters. This discussion will be limited to lead (Pb), iron (Fe), manganese (Mn), copper (Cu), and chromium (Cr).

1. *Lead.* Lead occurs naturally in rocks, but is more likely to be introduced into water as an industrial or mining pollutant. Corrosion of lead pipes and joints can introduce lead into waters. Lead forms precipitates with many anions in natural water; thus it will not remain long in solution.
2. *Iron and manganese.* These elements are among the most abundant found in rocks and soils and, therefore, are found in natural waters in varying concentrations. Other sources include anaerobic waters such as organic-rich or polluted groundwaters, wells, and stratified lake bottoms. Iron can be introduced by acid-mine drainage, by iron bearing industrial wastes, and as corrosion products. Iron- and manganese-containing waters have a high degree of turbidity owing to the formation of colloidally dispersed compounds.
3. *Copper.* This is usually introduced into a water as a pollutant from industries engaged in the plating, fabricating, and finishing of copper.
4. *Chromium.* The hexavalent form of this ion is extremely toxic to aquatic life. These ions are found in waters near industries which use hexavalent chromium.

The most important mechanism of toxic action on aquatic life is thought to be the poisoning of enzyme systems. For this type of toxicity the order of the following metal ions occurs:

Mercury > copper > lead > cadmium > zinc > iron > manganese

Other modes of toxic action are

1. Antimetabolites: arsenic, boron, selenium, beryllium, tungsten
2. Substances that precipitate or tie up (chelate) essential metabolites: aluminum, beryllium, titanium, barium
3. Substances that catalyze the decomposition of essential metabolites: lanthanum and other lanthanide cations
4. Substances combining with cell membranes and affecting their function: mercury, copper, lead, cadmium, and uranium
5. Substances that replace essential elements in the cell, which then causes malfunctions

10-5 NITROGEN COMPOUNDS

These compounds are important, as all life processes depend on them. These compounds are nutrients for aquatic life, and at the same time they are waste products.

The nitrogen compounds of importance in water are

1. Organic nitrogen compounds: proteins, peptides, amino acids, amines, and amides
2. Ammonia: NH_3 (ammonia) in alkaline solutions and NH_4^+ (ammonium) in acid and neutral solutions
3. Nitrites: NO_2^-
4. Nitrates: NO_3^-

By bacterial oxidation, reduction, or decomposition, all these forms are cycled to different compounds.

Man and animals use protein as food and produce fecal and urine wastes (organic N). Then, fecal and urine wastes decompose or hydrolyze to produce ammonia N. Finally, ammonia is oxidized to nitrite and nitrate, which is used as fertilizer for plants.

Under anaerobic conditions, nitrite and nitrate can be reduced to nitrogen gas and ammonia.

Nitrogen compounds can be used to determine whether a body of water has been freshly polluted. A freshly polluted body of water will have a relatively high organic N and ammonia and no, or relatively little, nitrate or nitrite. A water that has been polluted a long time or far from the source will have a relatively high nitrate and very low organic N and ammonia (Fig. 10-6). The tests for the different forms of nitrogen can indicate that DO levels are low or have

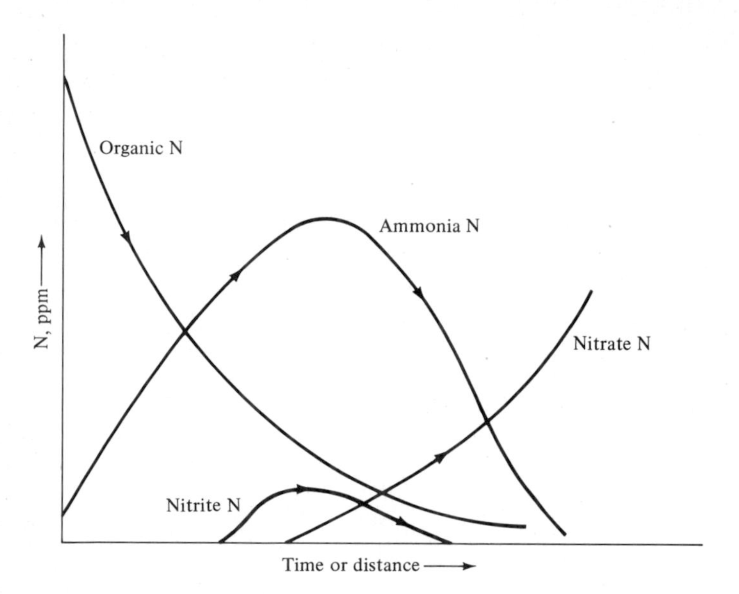

Fig. 10-6 Changes occurring in forms of nitrogen present in polluted water under aerobic conditions.

been low, and it should be noted whether the rate of reaeration is sufficient to prevent anaerobic processes from occurring.

One other source of nitrates is fertilizers.

10-6 PHOSPHORUS COMPOUNDS

These appear in waters by means of human and animal wastes, from fertilizers, and from detergent formulations. There are phosphorus compounds that are used for corrosion control, and for the control of scale in boiler waters.

Phosphorus compounds along with nitrogen compounds will increase the biological productivity of a water. Such substances are essential nutrients for the growth of phytoplankton (small aquatic plant organisms). These are predominantly algae which, if left to produce, can convert a lake into a marsh.

A body of water with very small quantities of nutrients is referred to as *oligotrophic*. As the nutrient concentration increases, it goes to the *mesotrophic* and then to the *eutrophic* state, which has a relatively high nutrient and aquatic plant life concentration.

10-7 CYANIDES

The occurrence of cyanides in water is from industrial wastes. Industries which use metal finishing and plating solutions containing cyanides will have these compounds in their effluents if they are not treated properly.

Cyanides are extremely toxic to aquatic life and aquatic standards should be based on this fact. Cyanides will oxidize slowly in natural waters to produce nontoxic materials.

10-8 PESTICIDES AND HERBICIDES

These are chemicals that are used to control, prevent, or destroy undesirable insects and plants. They can be classified into three types:

1. Chlorinated hydrocarbons: DDT, chlorodane, aldrin, dieldrin, endrin, lindane, heptachlor, etc.
2. Cholinesterase inhibitors: organic phosphorus and carbamates, parathion, malathion, naled, dicapthon, sevin, captam, IPC.
3. Herbicides: 2,4-D (2,4-dichlorophenoxyacetic acid), 2,4,5-T (trichlorophenoxyacetic acid), diquat, paraquat, and dachtal.

Pesticides may enter ground and surface waters by direct application or runoff from treated waters. Nearly all of the organophosphorous compounds and the carbamates are toxic to mammals, and some have even higher toxicity to fish. Ingestion of small quantities of these compounds over a long period of time causes damage to the central nervous system. Many of these pesticides hydrolyze rapidly in the environment to form less toxic materials.

Chlorinated hydrocarbon pesticides are extremely toxic and persist in the environment over long periods. This would indicate that they will be concentrated biologically in the food chain. These chemicals are concentrated in aquatic plants and animals to values several thousand times that occurring in the water in which they live. The herbicides 2,4-D and 2,4,5-T are chlorinated hydrocarbons.

10-9 DETERGENTS

About 85 percent of the cleaning agents sold in the United States are synthetic detergents. These detergents vary greatly in their biochemical degradation behavior. This depends on their chemical structure.

Alkyl benzene sulfonate (ABS) detergents are resistant to biological attack and have contaminated surface and ground waters. The soap and detergent industry changed over to the more readily degradable linear alkyl sulfonates (LAS).

Detergents and other surfactants (surface active agents) are readily noticed as pollutants, as they produce foaming at points of turbulence.

10-10 RADIOLOGICAL EXAMINATION

Pollution of water supplies by radioactive materials represents an increasing hazard to water quality. The effects on aquatic life can be harmful, and any radioactive material in water should be avoided. It is necessary therefore to prevent excessive levels of radiation from reaching any organism. Radioactive substances are similar to any other chemicals with the exception that they emit ionizing radiation. For example, the radioactive isotope Sr^{90} will undergo the same chemical reactions as nonradioactive Sr^{88}.

Upon introduction into an aquatic environment, radioactive wastes can remain in solution or suspension, precipitate and settle to the bottom, or be taken up by plants or animals.

When radioactive materials are added to water, they can be diluted or concentrated. The factors that dilute and disperse these materials are turbulence, transport and diffusion, and currents. Radioactive materials can be concentrated by biological uptake or by chemical or physical reactions.

Radioactive wastes can be cycled through water, sediments, and aquatic life. Sediments remove much of the radioactivity from the water. The radioactive sediments may expose benthic (bottom) species to radiation at some later time.

Although radioactive contamination has been a problem for only a short time, radioactive materials have always been part of that environment. The major sources are ores of uranium and thorium. Some of the long-lived natural radioactive isotopes are radium 226, carbon 14, cerium 142, rubidium 87, and potassium 40.

Potential sources of radioactive wastes are from chemical processing of ores of uranium, reactor fuels and blanket materials (materials placed around the reactor cores to form additional fissionable materials from neutrons), and from radioactive fallout.

10-11 MICROBIOLOGICAL EXAMINATION

The purpose of the bacteriological examination is to determine the quality of a water so that the likelihood of contamination by pathogenic bacteria can be estimated. Most waters can be polluted at some stage by human and animal excreta, and so disease may be spread. The diseases which use water as a vector are typhoid, cholera, dysentery, leptospirosis, and schistosomiasis. Time and the degree of contamination are factors in the degree of hazard.

When pathogenic microorganisms are present in a water, they are usually outnumbered by the normal excremental organisms, and these intestinal organisms are easier to detect in water. If these organisms are not found in the water, it can be inferred that pathogens are also absent. Intestinal organisms are used as indicators of fecal pollution. The class of bacteria used as indicators is called coliforms, of which *Escherichia coli* is one found in the intestines of man and animals. The bacteria *Aerobacter aerogenes* are found in vegetation and soils and to a smaller degree in fecal discharge.

The portion of the total coliforms in water that are of fecal origin may range from less than 1 percent to more than 90 percent. The presence of high fecal coliform would indicate recent pollution, and the water would constitute a threat to health.

The search for fecal streptococci, the anaerobic spore-forming *Clostridium perfringens* and *C. welchii*, and *Pseudomonas* bacteria may be of value in confirming the fecal nature of pollution. They usually occur in much smaller numbers than coliforms. *Clostridium* bacteria are capable of surviving in water and are resistant to chemical treatment and natural purification. The presence of clostridia suggests that fecal pollution has occurred, and in the absence of the coliforms, suggests that the pollution occurred a long time ago.

Over the last 20 years, there has been increasing evidence of viruses which can be found in human feces. Although the concentration of viruses is much less than bacteria, the viruses are usually resistant to most treatment processes. Virus diseases are hepatitis and polio.

Water samples for bacteriological examination should be collected in clean sterile bottles. The sample should be examined immediately or refrigerated. Tests for bacterial contamination are based on the coliforms, as these are relatively simple to detect and quantify. Tests for the coliform group may be carried out in either of three ways:

1. By inoculation, after proper dilution of the sample in a culture medium, usually gelatin or agar, and incubation for 24 or 48 h with confirming tests.
2. Ability to ferment lactose with gas formation in nutrient lactose broth.
3. Passage of a measured amount of water through a membrane filter that will retain the bacteria and permit the identification and counting of the coliforms by means of pads containing enrichment and differential mediums.

The 24-h +37°C plate count will determine the numbers of bacteria having their origin in the intestines of warm-blooded animals. The 48-h 20°C plate count includes those living within the usual range of water temperatures.

The determination of the number of microorganisms present in a sample is made by the Most Probable Number (MPN) test. The value obtained is the bacterial density which would occur most frequently. "Standard Methods for the Examination of Water and Wastewater" has the detailed procedure and the tables for determining MPN.

Other microorganisms that can exist in water are algae, fungi, protozoa, rotifers, insects, and worms.

Algae are classified by shape and color. They contain chlorophyll and thus utilize sunlight to convert carbon dioxide into organic carbon and oxygen. Algae impart an unaesthetic quality to a water, as they usually have an unpleasant odor.

The presence of algae and bacteria can serve to purify water by helping one another. Bacteria require energy, oxygen, and food, the latter two of which are produced by algae. Algae require carbon dioxide, which is a metabolic product of bacteria. The algae also remove nitrogen and phosphorus compounds from solution. If the algae are harvested frequently to prevent overgrowth, continuous purification can occur. This method of treatment is used for domestic wastewaters.

Fungi do not contain chlorophyll and obtain their food from other organisms, living or dead. Fungi can be parasites which live off other organisms or saprophytes which carry on an independent existence, finding their own food supply and adapting themselves to the conditions of their environment. Fungi can be found in polluted or nonpolluted waters, and in the latter they are often found growing in gray-colored cottony masses which attach themselves to walls and clog pipes and screens. Fungi produce odors in waters.

Protozoa are considered to be higher forms of microbiological life which are frequently found in polluted waters. They are mobile and their main source of food is bacteria. Protozoa impart a fishy or oily odor to water.

Rotifers are multicelled microorganisms with hairlike appendages that provide locomotion and create a current that draws food to the organism. They are usually found in polluted waters.

Many worms and insect larvae are found in waters. Worms are usually parasitic and can live in plant and animal tissues.

The identification of many of the larger microorganisms is made by observing water samples through a microscope. "Standard Methods" has many pictures and photographs of aquatic microorganisms.

PROBLEMS

10-1. What is the pH of a solution whose hydrogen-ion concentration is 4.6×10^{-5} *M*?

10-2. What is the hydroxyl-ion (OH^-) concentration of the solution in Prob. 10-1?

10-3. What is the pH of a solution whose hydrogen-ion concentration is 2×10^{-8} *M*?

10-4. 100,000 gal of an acid waste at pH 2.7 will be treated with a natural water source that contains 60 ppm of sodium carbonate. How much water will be required to neutralize the acid to a pH of 6.3?

10-5. What is the acidity of a solution expressed as ppm CO_2, and expressed as ppm $CaCO_3$, that requires 12.4 ml of 0.02 *N* NaOH to reach the phenolphthalein end point?

10-6. What are the molarities and normalities of the following aqueous solutions?

(*a*) 14 g HCl in 1 l
(*b*) 49 g H_2SO_4 in 1 l
(*c*) 17 g H_3AsO_4 in 400 ml
(*d*) 148 g of $Ca(OH)_2$ in 628 ml

10-7. The reaction between sulfuric acid and bicarbonate is

$$H_2SO_4 + 2HCO_3 \rightarrow 2H_2O + 2CO_2 + SO_4^{--}$$

How much water containing 50 ppm of alkalinity (expressed as ppm of $CaCO_3$) will be required to neutralize 2,000 gal of an acid-pickling waste that contains 100 ppm of H_2SO_4?

10-8. What is the total alkalinity of a solution that requires 37.5 ml of 0.02 *N* NaOH to reach the methyl orange end point if the solution volume is 50 ml?

If in the above solution it took 12.2 ml of 0.02 *N* NaOH to reach the phenolphthalein end point, what are the alkalinity components?

10-9. How would the alkalinity be affected by the addition of the following? (*a*) $Ca(OH)_2$, (*b*) acetic acid, (*c*) trisodium phosphate, (*d*) alum, (*e*) crude oil, (*f*) NaCl

10-10. In the Winkler test for dissolved oxygen, show why every milliliter of 0.025 *N* $Na_2S_2O_3$ used to titrate 200 ml of the liberated iodine is equivalent to 1 ppm of dissolved oxygen.

10-11. A sample has a 5-day BOD of 203 ppm and K' of 0.2. What is the ultimate BOD?

10-12. If a sample has a 2-day BOD of 70 ppm, a 5-day BOD of 127 ppm, and a 7-day BOD of 140 ppm, what is the ultimate BOD? What is the reaction rate constant (K')?

10-13. If an estuary has a relatively high ammonia and organic nitrogen concentration as compared to the nitrate concentration, what can be inferred about the condition of the estuary?

10-14. The major alkalinity components are OH^-, CO_3^{--}, and HCO_3^-. These can exist individually or in combinations as OH^- and CO_3^{--} and HCO_3^- and CO_3^{--}. The combination of HCO_3^- and OH^- cannot exist. Why?

10-15. What is the final pH of a solution that is made by mixing equal amounts of 0.01 *M* and 0.00001 *M* HCl?

CITED REFERENCES

American Public Health Association, American Water Works Association, Water Pollution Control Federation (1971): "Standard Methods for the Examination of Water and Wastewater," 13th ed., American Public Health Association, Inc., New York.

American Society for Testing Materials (1959): "Manual on Industrial Water and Industrial Wastewater," 2d ed., American Society for Testing Materials, Philadelphia.

Environmental Protection Agency (1971): "Methods for Chemical Analysis of Water and Wastes," Water Quality Office, Cincinnati, Ohio.

World Health Organization (1963): "International Standards for Drinking Water," 2d ed., World Health Organization, Geneva, Switzerland.

Other references

Camp, T. R. (1963): "Water and Its Impurities," Van Nostrand Reinhold, New York.

Clark, J. W., W. Viessman, Jr., and M. J. Hammer (1971): "Water Supply and Pollution Control," 2d ed., International Textbook, Scranton, Pa.

"Cleaning Our Environment–The Chemical Basis for Action" (1969): American Chemical Society, Washington.

Fair, G., J. Geyer, and D. Okun (1968): "Water and Wastewater Engineering," vol. 2, Wiley, New York.

Federal Water Pollution Control Administration, U.S. Department of the Interior (1968): "Report of the Committee on Water Quality Criteria," U.S. Government Printing Office, Washington.

Hem, J. D. (1970): "Study and Interpretation of the Chemical Characteristics of Natural Water," 2d ed., Paper 473, U.S. Geological Survey, Washington.
Sistrom, W. R. (1962): "Microbial Life," Holt, Rinehart and Winston, New York.
Stumm, W., and J. J. Morgan (1970): "Aquatic Chemistry," Wiley-Interscience, New York.
Warren, C. E. (1971): "Biology and Water Pollution Control," Saunders, Philadelphia.
White-Stevens, R. (1971): "Pesticides in the Environment," vol. 1, Marcel Dekker, Inc., New York.

11

A Laboratory in the Sea

Michael McKenna and Hilbert Schenck, Jr.

11-1 INTRODUCTION

In Chap. 10 we briefly explored the parameters involved in pollution and water quality studies in estuarine and nearshore areas. Oceanographers use these, and many other measurements in water, most of them devoted to projects and hypotheses of a scientific nature. Their purposes tend to be large in scale, involving deep regions of the ocean basins and vast distances across the seas. The oceanographer's concerns often involve the geology of underwater features, motion of biological species across the deeps, large-scale current, drift, and hydrodynamic effects, and sediment deposits related to eons gone by.

Ocean engineers, on the other hand, are more likely to be concerned with the nearshore, shallow areas where man's activities and impacts are evident. Thus, while many more measurement and data collection tasks exist than will be covered here, we have tried to organize a laboratory chapter of particular interest to engineers whose employment opportunities are most likely to involve evaluations of shallow water, man-created problems closely related to the needs (and excesses) of our modern society.

Since this is primarily a textbook and its readers will be students, most typically at the undergraduate level, we shall provide some suggestions and plans for self-built instruments that will give basic data of the more essential estuarine parameters related to water motion, sediment character, and dissolved materials. Commercial instruments are readily available to do these jobs, but the instructional benefits of building and using one's own measuring tools are too great to be ignored. Also, we hope these last two chapters may find some use in traditional engineering laboratory work, where limited funds have always governed the content of course materials. Inland schools with nearby lakes, rivers, or quarries can utilize most, if not all, of this material.

11-2 GEOGRAPHIC POSITION

Regardless of the simplicity of an ocean measuring program, it is evident that instrument position will be a basic piece of the data in almost every case. The primary question to be decided during ocean measurements is the degree of error to be accepted in position fixing, a matter of judgment and experience. In the case of current profiles across a tidal estuary or river, for example, fairly crude fixes may be acceptable, if the purpose of the measurement is to obtain net flow of water. Since relatively few flow measurements are usually possible, and each will be applied to a relatively large extent of moving water, the exact location of the meter in its zone of moving water is usually not critical, providing extreme velocity gradients do not exist. On the other hand, if one is taking a core sample of the bottom and wishes to reproduce the data at some later time, position accuracies within a few feet may be necessary.

For the beginning student working in nearshore waters, the following positioning methods will be considered, all of them assuming that a nautical chart or topographic map of the region is available.

1. Visual fix by compass bearings to two or more fixed objects
2. Use of range lines
3. Estimation of distances to fixed objects
4. Depth measurement and comparison with chart depths

A. Visual fix The most flexible technique, and the one which is the most popular, is the use of compass bearings. Unfortunately, the magnetic compass used with this technique is subject to deviation errors which result from the proximity of ferrous materials or magnetic fields associated with unshielded electrical cables. Because the ferrous materials on a boat are usually distributed along its fore-and-aft axis, deviation is generally greatest when the boat is on a course perpendicular to magnetic north and decreases to minimal values on heading of magnetic north or south. Most boats which have regularly installed marine compasses have provision for correction of deviation, but residual errors of from 5 to 10° are not uncommon, particularly if there has been any change in the distribution of ferrous materials on board or alterations to electrical systems after compensation has been performed.

The position error that results from compass deviation can be essentially eliminated if compass bearings are taken to objects which are separated by roughly 120°. In Fig. 11-1, the intersection of three such bearings, each of which includes a deviation error θ_D of −10°, forms a triangle with sides about 1 mi in length when the distance to those landmarks is 5,500 yd. It will be noted, however, that the bisectors of the triangle's three angles will intersect in a point which will coincide with actual ship's position. Unfortunately, it is not always possible to have landmarks which are placed so fortuitously with respect to the boat.

The more commonly encountered situation exists when the maximum bearing difference between landmarks is less than 180°, as when operating off a coastline with no landmarks to seaward, or when operating in an estuary under conditions of reduced visibility. In this situation, significant position errors can result. The mathematical calculation of the magnitude of this error is complex and beyond the scope of this chapter. It is intuitively obvious, however, that position error will increase with an increase in distance from the landmarks used, and with an increase in the bearing difference between bearing lines. These relations are illustrated in Fig. 11-2, which depicts a situation in which position is to be established using compass bearings to objects *A, B,* and *C* at distances of 5,000 yd from the boat. A compass deviation error of −5° is assumed.

Fig. 11-1 Visual fixed position with objects at 120° from each other, with constant compass error.

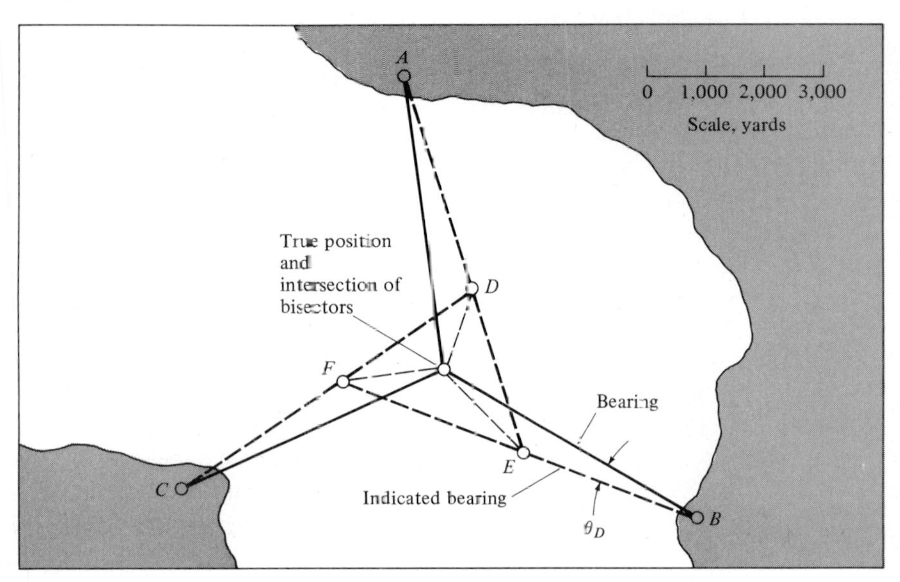

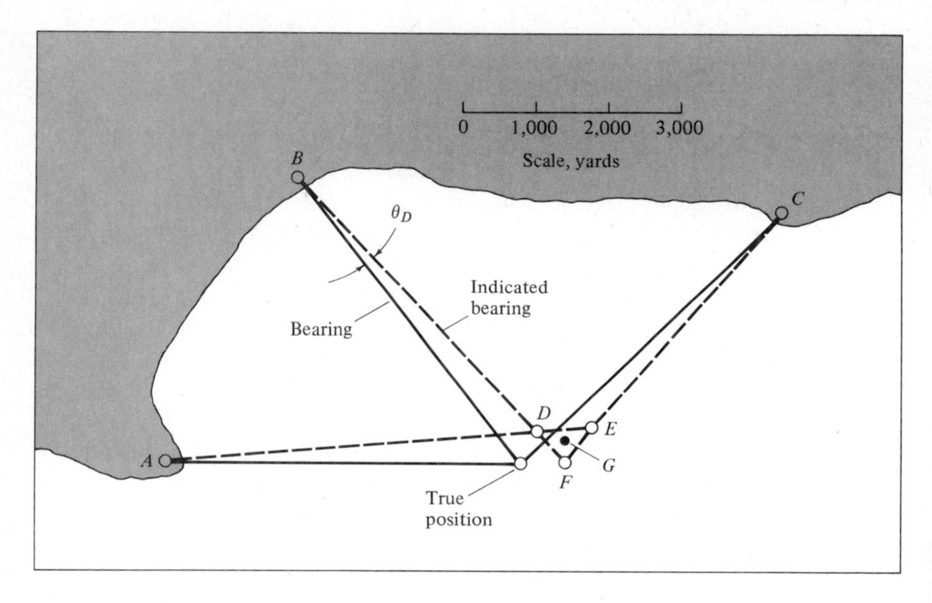

Fig. 11-2 Visual fix when operating with no marks to seaward showing apparent position (*G*) compared with true position.

If bearings to *A* and *B* are used, indicated position will be at point *D*, about 500 yd from true position. If the bearing difference is increased, as when *A* and *C* are used, the error increases to about 1,000 yd (point *E*). If all three landmarks are used, triangle *DEF* will result, the center of which (*G*) is about 700 yd distant from the actual position.

B. Ranges If one is to eliminate the errors inherent in compass deviation, it is necessary to eliminate the use of the compass in determining position. This can be accomplished by maneuvering the boat so as to bring two landmarks into "range," i.e., so that an imaginary straight line from the boat will pass through the two objects. This immediately establishes the boat's position along a line drawn on the chart through those two objects. The accuracy is dependent only upon the observer's skill in determining when the two objects are in fact aligned. The actual position of the boat along that line can be established definitively if there is another range available which intersects the first. But, because of the limited number of conventional landmarks (such as lighthouses, church spires, radio towers, etc.) which are depicted on the charts, it is seldom possible to obtain such a crossing of ranges in a specific area of interest. This difficulty can be overcome by using combinations of other identifiable objects, such as tangents to islands or headlands, prominent geographic features or structures, or other objects whose position is indicated on the chart being used. It is even possible, but more difficult, to employ objects whose position is not marked on the chart, but which can be established with reference to others which are established.

A range line can be particularly useful during a towed instrument run, such as may be possible with temperature, turbidity, or bottom profile measurements. If one can start and stop the run at known points and hold constant vessel speed over the range, then the distance along the range can be determined by the amount of time that has elapsed since the start of the run. Alternatively, compass bearings can be taken periodically on objects while the range is run.

The use of navigational buoys as objects in a range may cause problems because of a positional uncertainty of about ±20 yd, which results from the fact that the buoys will "ride" to their anchors in response to currents and wind.

C. Distance estimation Estimations of distances to objects are usually made inshore using optical range finders of the split-image type. The range of usefulness of such instruments depends on the distance between the two objective lenses. A hand-held unit with a separation of 8 to 12 in is not likely to be trustworthy for objects more than 200 ft away, and wider units may become awkward in use. When boat-shore cooperation is possible, surveyor's transits on shore can have baselines of hundreds of yards and will give very accurate position lines, provided the baseline length is exactly known. Each transit simply measures the angle between the other transit and the boat or drifting object in the water, and the two position lines plus the baseline can be plotted on a chart. However, this method will usually require three walkie-talkie units, several people, and extensive data reduction.

A single transit has been used on a high bridge by University of Rhode Island students (Teeson et al. 1970) to give both an azimuth and bearing angle on drift sheets used in a study of surface drift. Since the exact height and location of the transit were known, these data were sufficient to solve the trigonometric problem and locate a sighted object within a reasonable range of the bridge. With a transit height of 150 ft, an object within a radius of several hundred yards could be accurately tracked.

A special but important problem occurs when currents are measured using drifting poles or drogues, since the beginning and end locations of a drift must be known with high accuracy. With a typical tidal velocity of 0.5 ft/s, a 5-min drift produces a track only 150 ft long. To attempt to pinpoint the starting and ending locations by compass bearings will not be practical. Rhode Island students have solved the problem by anchoring the boat on a taut, two-point moor and then allowing the drifter to carry out a fine, floating line of known length. Current direction is easily found by a compass bearing along the drift track. Various photographic tracking methods will occur to the reader, but they may involve formidable data reduction problems.

D. Depth measurement Where charts of depth are available, fathometer data can often be used to help fix a measurement location. This is particularly true when one wishes to return to a sampled location and when depth changes are frequent, giving the bottom a distinctive topography.

It is only workmanlike to present location data in reports and papers with estimates of position error. The easiest approach is to utilize a graphic analysis, as in Figs. 11-1 and 11-2. First, determine or estimate the appropriate instrument error bands, then plot the triangles or areas of confusion directly on the working chart from these error data, and finally scale off the resulting position errors.

11-3 WATER DEPTH AND SAMPLING DEPTH

Overall water depth in most small research programs will be obtained by either an acoustic fathometer or a lead line. Most small fathometers, if they are calibrated and mounted properly, will give depths to within 2 ft or better. The lead line is simply a marked line with a 4- or 5-lb lead weight on the end that is allowed to touch bottom. Various line-marking schemes have been used for this, and we have found a color code to be practical. A lead or lowering line is marked in 5-ft intervals with white, red, yellow, green, and blue tape, which then repeats for the second and third 25-ft sections. Interpolation within each 5-ft interval is easily done.

In the early days of oceanography, the determination of sampling depth on a multisample single-line cast in the deep ocean was accomplished by recording the length of suspension line paid out, the distance between sampling devices, and the slope of the suspension line at the water's surface. Solution of the basic trigonometric relation yielded estimated depth for the various samplers. This method was inexact because of the unpredictability of the slope of the suspension line below the surface and variations in suspension line elasticity. The later development of the reversing protected and unprotected thermometer permitted accurate measurement, at each sampler, of the hydrostatic pressure, which could be converted to sampling depth. Interpretation of the readings of the thermometers, however, is laborious and time-consuming. More recently, increasing use has been made of electromechanical pressure transducers at each sampler. This permits the telemetering of electrical signals which are related to hydrostatic pressure and can be converted into *in situ*, real-time values of sampling depth.

In estuarine waters, and specifically in waters of about 100-ft depth or less, it is practicable to employ the older, length-of-suspension-line technique. The assumption is made that the slope of the suspension line remains constant along its entire length. This permits an estimate of sampling depth through conversion of slant distance, or length of suspension line L, to vertical depth D from

$$D = L \cos \theta_i \qquad (11\text{-}1)$$

where θ_i is the inclination of the suspension line from the vertical.

Another approximate method is to lower an instrument slowly until the bottom (of known depth) is just reached. Then all line lengths can be corrected by multiplying by the ratio of depth to line length at bottom. For example, if a salinometer in a current reaches a known bottom at 50 ft with 60 ft of line, then with 30 ft of line it would be assumed to be at 30 × (50/60) or 25 ft.

11-4 WATER SAMPLE COLLECTION

Biological and chemical parameters of natural waters are often measured on a water sample brought aboard the boat or back to the lab. The actual parameters, such as dissolved oxygen, biological oxygen demand (BOD), or various concentrations, may require rather time-consuming and specialized tests, some of which have been mentioned in the previous chapter. Various inexpensive kits, such as those sold by the Hach Chemical Co. and others, are readily available for a variety of such tests, and the specialized techniques are completely explained in the accompanying literature.

Various ways of obtaining an uncontaminated water sample from a known depth have been devised: continuous pumping systems, Nansen and other reversing bottles, and a variety of closure systems that can be activated by pressure, sonic means, or line messengers. For shallow-water use involving moderate sampling rates, the Van Dorn type of sampling bottle is lowered in the water to the desired depth and then has both ends closed by caps connected by a piece of stretched rubber through the body of the sampler. Closure is effected by a weighted messenger that is dropped down the line.

Several Van Dorn samplers are available through oceanographic supply firms. The one described here can be constructed in a few hours with a minimum of tools, works dependably, and allows the user to obtain water temperature at the sampling depth along with the sample itself.

Figure 11-3 is a diagram of the assembled water sampler, the detailed figures of which are described below.

A. Barrel The water sampler barrel is made of 2½-in-OD acrylic plastic tubing, 16 in in length. Wall thickness of the tubing is 1/8 in. The ends of the barrel are milled perpendicular to the longitudinal axis of the tubing and then internally cambered 45° to facilitate seating of the closure balls described below. Near the ends of the bottom side of the barrel, two 3/8-in holes are drilled and tapped to receive small, threaded copper tubes (standard plumbing fittings) which, after the attachment of flexible rubber tubing, serve as vent and drain fixtures for the sampler. Watertight sealing of these fixtures is achieved through the application of vacuum wax to the threaded ends of the copper tubes.

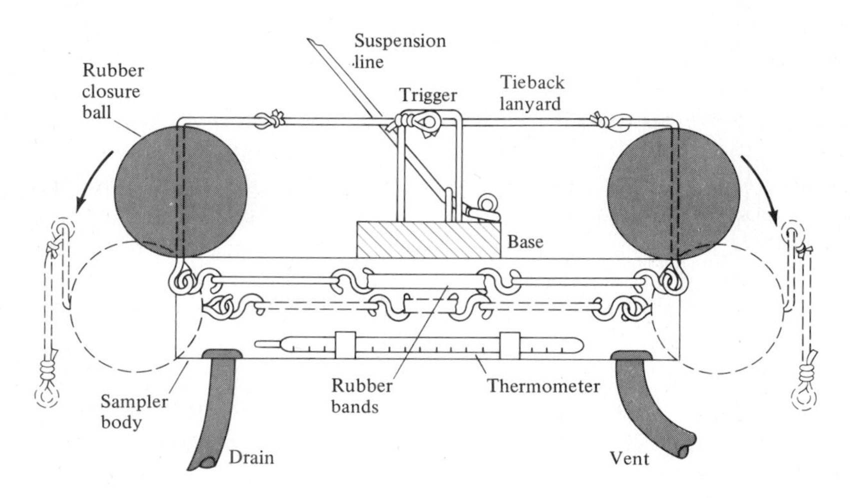

Fig. 11-3 General view of home-built Van Dorn water sampler.

The mounts for the thermometer are cut and shaped from a cork bottle stopper, and drilled out to receive the thermometer stem. The mountings are glued to the inside wall of the bottom of the barrel. This method of mounting permits easy insertion and removal of the thermometer, while still holding it firmly in place during operation of the sampler.

B. Closure system The closure devices are two 3-in-diameter sponge rubber balls which were obtained from a toy store. Each ball is pierced by a 5½-in copper stem (designed for use in a toilet tank), which is then bent 90° where it protrudes from the ball. The two balls are held snugly in the closed position against the ends of the barrel by an assemblage of four brass S hooks, two braided nylon lanyards, and five ordinary rubber bands that pass within the barrel and connect the two balls. Tied to the other end of each stem is a short lanyard of braided nylon cord which is whipped to form an eye. These lanyards serve as tiebacks for cocking the closure system.

C. Trigger mechanism (see Fig. 11-4) The trigger mechanism is mounted on a wooden base which is held against the barrel with two hose clamps. Pieces cut from an inner tube serve as cushioning devices between the wooden base, hose clamps, and the sampler barrel. Two hooks screwed into the base act as restraints for a trip pin which is used to hold the ends of the two tieback lanyards when the sampler is cocked. The suspension line for the sampler is attached to the wooden base and passes within about 3/16 in of the trip pin when the sampler is cocked. A messenger, made up of an ordinary thread spool and wrapped with lead flashing to provide negative buoyancy, rides down the suspension line to knock the trip pin free, thereby releasing the tieback lanyards. The tension supplied by the rubber bands then causes the balls to close the sampler barrel. The trip pin is tied with a length of sail twine to the base piece to prevent its loss after actuation of the trigger.

D. Sampler ballast weight The assembled water sampler, because of the buoyancy of the rubber balls and the wooden base piece, will float. To make the sampler negatively buoyant, a 4½-lb weight is attached by means of a lanyard and snap hook to the loop of sash cord which is affixed to the sides of the base piece. The length from the center of the sampler to the ballast weight is adjusted to 3 ft to allow the collection of samples at 3 ft above the bottom.

E. Suspension line A 100-ft length of ordinary 3/16-in-diameter braided sash cord is employed as a suspension line for the water sampler. The suspension cord is wrapped with colored plastic tape at 5-ft intervals to permit estimation of the depth of submergence of the water sampler.

Fig. 11-4 Water sampler trigger mechanism.

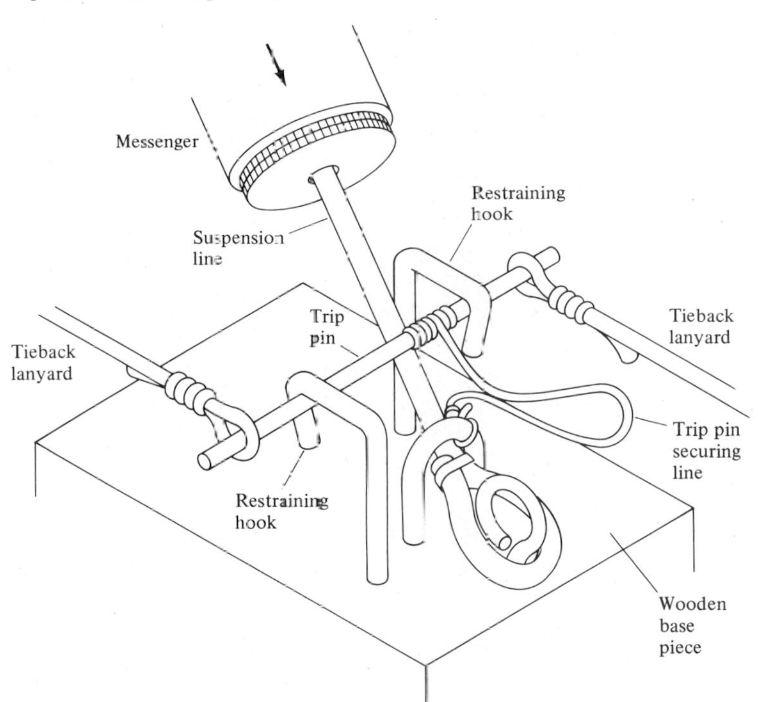

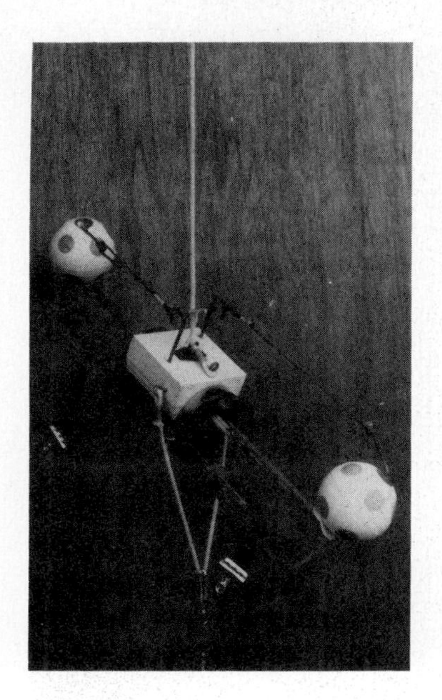

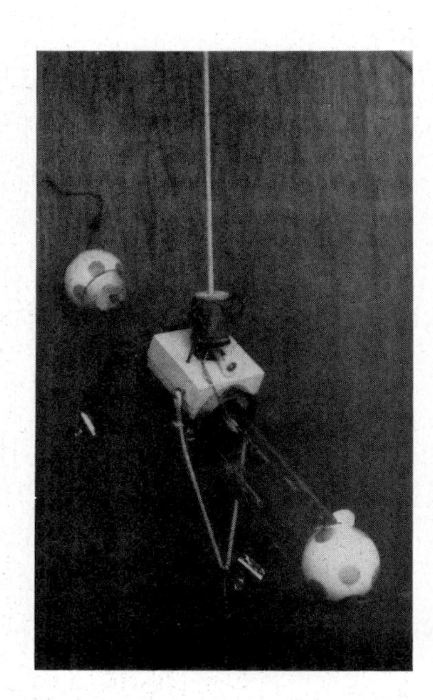

Fig. 11-5 (*a*) Sampler cocked, (*b*) messenger approaching, (*c*) closure complete.

F. Thermometer The thermometer used with the sampler is a full-immersion, −15°C to +60°C, mercury-in-glass laboratory thermometer. The thermometer is graduated in 1° increments and readability precision is about ±0.2°C.

A water sample is obtained as follows: The vent and drain fittings are closed by applying paper clamps to the ends of the rubber tubing. The sampler is cocked by pulling one of the closure balls to the open position and holding it there, placing the trip pin between the two restraining hooks, and looping the eye of the tieback lanyard over the end of the trip pin. This procedure is repeated for the other closure ball.

With the sampler cocked, it is lowered over the side to the desired depth. After a 20-s pause at depth to permit the thermometer to equilibrate with ambient temperature, the messenger is released. The impact of the messenger at the sampler can be felt in the suspension line if the depth is 85 ft or less. For greater depths, a delay of 15 s will ensure that the messenger has arrived. Figure 11-5 shows this sequence. With the sampler in the closed position, the sampler is pulled back on board, where the temperature is read and recorded immediately.

The sample can be drained into 1-qt plastic bottles and into 50-ml reagent bottles which are used in subsequent analysis.

When temperature stratification exists in the water column, the sample may suffer some temperature change before it reaches the surface. Tests showed that if this difference is 10°C or less, the error in taking a sample from 100 ft will not exceed 0.017°C per degree of water column differential. In other words, a cold bottom sample at 10°C pulled up through a 20°C water column will read no more than 10.17°C at the surface.

11-5 TEMPERATURE AND SALINITY

In addition to the temperature measuring system just described, various other water temperature measuring devices are available. The mechanical bathythermograph utilizes a bellows and bimetallic element to trace a temperature-vs.-depth curve on a smoked plate. Various thermistor units are available for sending temperature data up a cable. The design and construction of such an instrument is an excellent small project for an undergraduate electrical engineer.

Salinity is an important parameter in nearshore areas where freshwater flows are present. It can be determined in a number of ways, most of which require that temperature be measured at the same time. Electrical methods involve inductance or resistance bridge circuits that basically measure conductivity of the water sample. Optical methods involve refraction measurements that can be related to salinity. Chemical methods usually involve a titration and color change. Of these three types, the chemical test is inexpensive and quite easily done but has limited accuracy; the other methods involve instruments that cost many hundreds of dollars.

An alternative method is proposed here for those with limited resources. A sensitive hydrometer with a range of 1.000 to 1.0500 specific gravity units and Hydrographic Office publication 615, "Tables for Sea Water Density" (HO 615), are needed, along with some sort of transparent container to hold the water sample. Sample temperature must be recorded as the specific gravity is determined.

If possible, a hydrometer referenced to a reading of 1.0000 at 3.97°C (the reference point in HO 615) should be obtained. When other referenced hydrometers are used, they must be calibrated as follows (the calibration described here refers specifically to a Nurnberg brand hydrometer referenced to a reading of 1.0000 specific gravity units at 15.55°C or 60°F).

A sample of seawater of known salinity in the middle of the anticipated range of salinities is obtained, either from a scientific supply house or by test against a known standard method. The sample is cooled to just over 0°C and allowed to warm slowly while hydrometer and temperature readings are made up to, say, 18°C. Figure 11-6 shows the resulting temperature-vs.-salinity curve for a sample of known salinity of 29.2 ppt (parts per thousand). When compared with

Fig. 11-6 Calibration of a hydrometer using a known seawater sample of salinity 29.2 ppt.

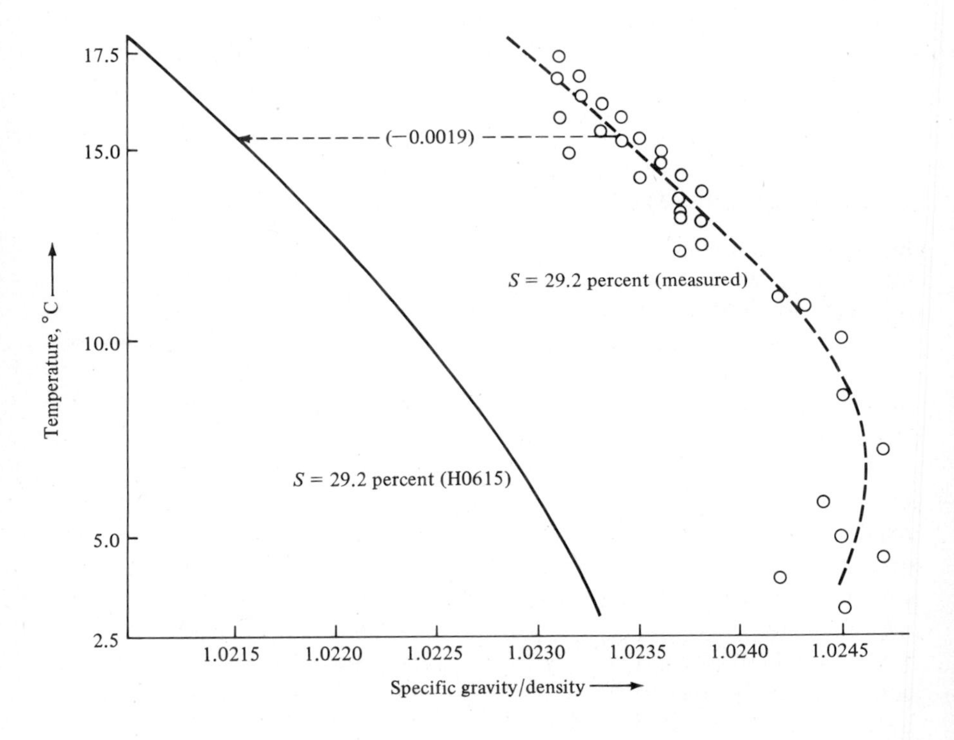

the curve predicted by HO 615, two facts are evident: (1) less scatter occurs above 10°C, so that determinations should be made on samples in the 10 to 18°C range and (2) the hydrometer reads an average of 0.0019 specific gravity units too high over this temperature range on this sample. This offset could be checked at other neighboring salinities if samples were available. We should thus subtract 0.0019 specific gravity units from this hydrometer reading before entering HO 615 to obtain salinity.

Because of the extreme sensitivity of the hydrometer, it is mandatory that a standard measurement procedure be used. It was found, for example, that the hydrometer would float about 0.0005 specific gravity units deeper if the above-water portion of its stem were wetted, and that the adherence of air bubbles to its submerged body could cause it to float about 0.0002 units shallower. Prior to each reading, therefore, air bubbles were spun off by twirling the hydrometer stem, the stem was wiped dry, and the hydrometer was caused to bob up and down about 0.0010 units before its equilibrium position was read. All readings should be made from the bottom of the meniscus.

In taking measurements of samples of unknown salinity, data should be collected, by using the procedure described above, until two readings agree within 0.0002 units. Those values, and their corresponding temperatures, are then averaged, corrected to density values, and used in entering HO 615.

The precision of this hydrometer, as indicated by the data scatter in Fig. 11-6, is approximately 0.0002 specific gravity (and density) units. In the estuarine salinity range from 25 to 32 ppt, and at a temperature of about 15°C, this is roughly equivalent to a precision of ±0.3 ppt salinity.

11-6 DISSOLVED OXYGEN AND OTHER POLLUTION PARAMETERS

Dissolved oxygen (DO) is an essential measurement in polluted areas, as noted in the previous chapter. Direct-reading DO meters are available, but the least expensive, although slower, method of obtaining DO is with a chemical test. Oxygen fixing reagents (manganous sulfate and alkaline iodide-azide) are emptied into the sample bottle (60 ml in the Hach Chemical kit for this test). The bottle is then stoppered, ensuring that no air bubbles are entrapped, and vigorously shaken. A precipitate will form which combines with the dissolved oxygen present in the sample. After this precipitate has settled, dry acid is added and the bottle restoppered as before. This will dissolve the precipitate and cause the formation of iodine in proportion to the amount of oxygen which was present in the precipitate. A measuring tube is then filled from the sample bottle and poured into the mixing bottle. This yellowish sample is then titrated with an indicator solution (PAO), using a dropper. The number of drops required to change the color of the sample from yellow to colorless will be equal to the milligrams per liter (parts per million) of dissolved oxygen in the water sample.

The precision of the analysis can be increased by using larger samples in the titration process. For example, by using five measuring tubes of sample and titrating as before, each drop will be equivalent to 0.2 ppm of dissolved oxygen.

Pollution parameters such as BOD (biological oxygen demand) and coliform count are rather more difficult to make and may require immediate sample refrigeration, plus considerable treatment and waiting time.

11-7 USE OF WATER SAMPLING SYSTEMS IN NARRAGANSETT BAY

The systems described in the previous sections (modified Van Dorn water sampler with thermometer, hydrometer, and Hach Chemical DO kit) were applied to a "broad brush" survey of Narragansett Bay, Rhode Island, during the months of November and December of 1971. Twenty-four water samples were taken over a 15-mi range from the Providence River south, with some locations involving sampling at the surface and near the bottom. Figure 11-7 shows a sketch map of the area with the stations indicated; most work was done in the west passage of the bay. No attempt was made to control the sample time with respect to a particular point in the tidal cycle.

A. Temperature The temperature data showed a slight positive gradient with depth at the majority of stations. This gradient was fairly strongly established in the area of the west passage near the Jamestown bridge. At other stations, the profiles were isothermal within measurement system precision. The general trend toward isothermal or a lightly positive gradient is in agreement with the findings of Hicks (1959), who reported that the bay was "approximately isothermal from the surface to the bottom throughout most of the year" and that "slight increases in temperature with depth were found in November and January." Hicks also found that bottom temperatures along a line from the Providence River to the area in the vicinity of Dutch Island were, at corresponding times in the tidal cycle, "essentially constant" at about 3.5°C during February. (He furnishes no data on bottom temperature variation for November or December.)

B. Salinity A marked decrease in both surface and bottom water salinity was noted between mid-bay and the lower reaches of the Providence River. With one exception (where there was a salinity inversion), bottom waters were about 0.5 ppt more saline than surface waters were. Hicks's data for February (the closest month for which he presents salinity results) is in general agreement; it is presented together with experimental results in Fig. 11-8. The wide divergence between Hicks's values for bottom salinities and those obtained in this study for the area near Conimicut Point (latitude 41°45′) is probably attributable to the fact that Hicks's readings were obtained at time of high water, while those reported here were obtained at time of maximum ebb, resulting in a sharp drop in bottom salinity. The other variations in salinities shown in Fig. 11-8 also may

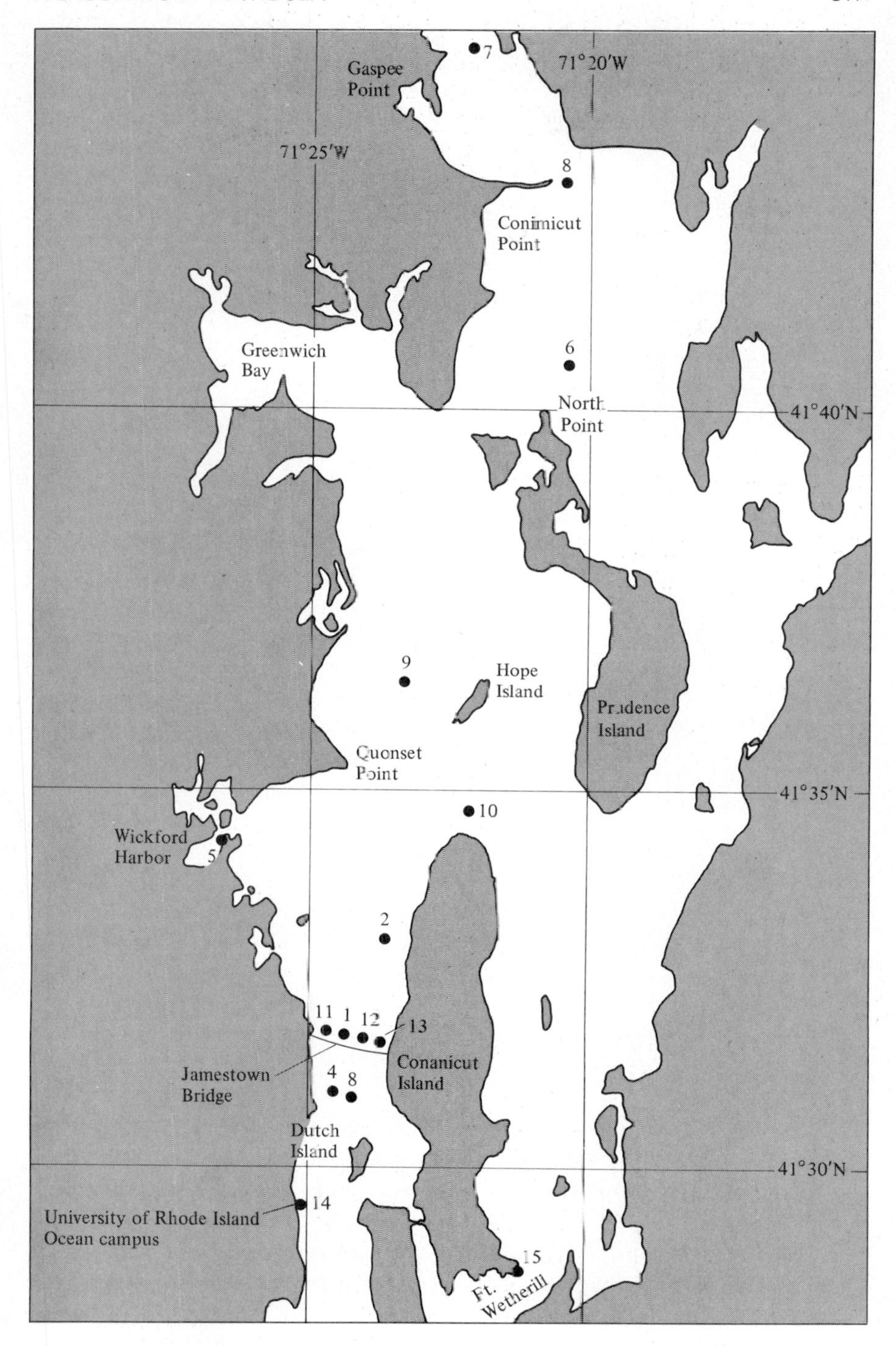

Fig. 11-7 Narragansett Bay, Rhode Island, showing sampling stations.

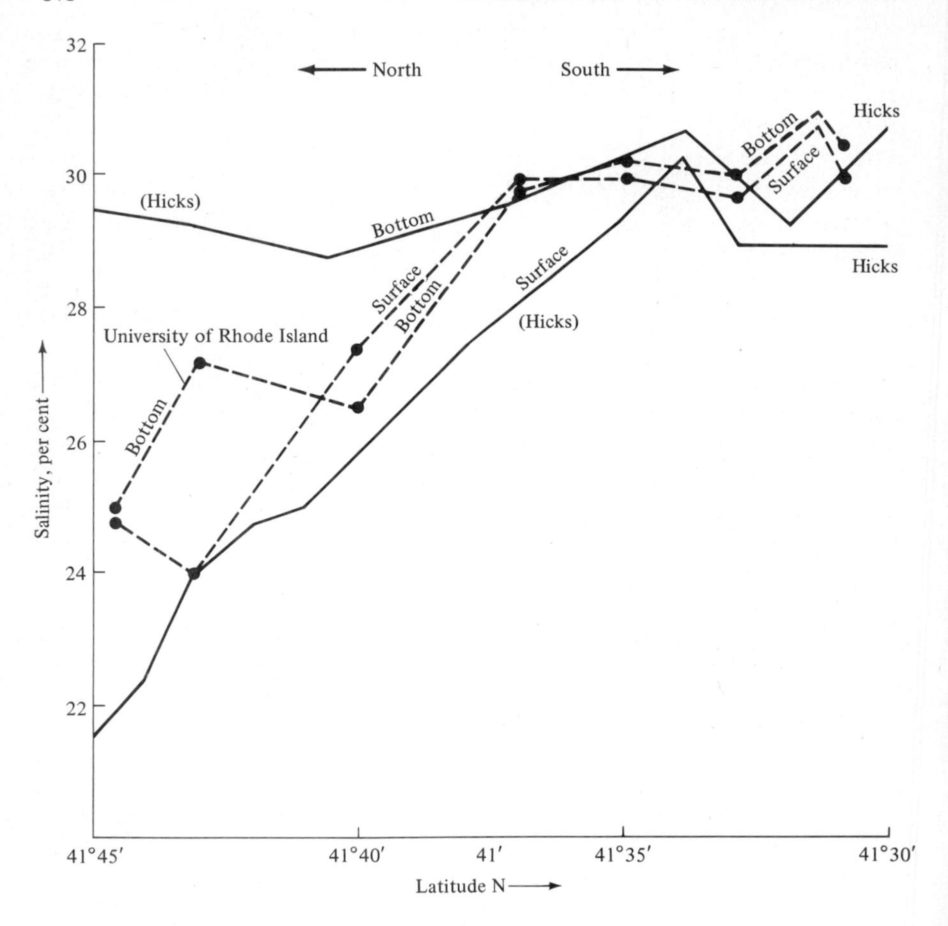

Fig. 11-8 Comparison of the north-south salinity profiles for Narragansett Bay as found by Hicks (1959) (solid lines) and University of Rhode Island (dotted lines) using apparatus described in this chapter.

be caused by differences in the state of the tide at which the readings were obtained.

C. Dissolved oxygen Most values of dissolved oxygen content of samples collected were obtained using the one-drop-per-ppm technique with the Hach kit, and no examination of dissolved oxygen variability about the 10.0 ppm values was attempted.

The most reliable information on dissolved oxygen content of Narragansett Bay waters is that obtained in the summers of 1970 and 1971 by investigators from the University of Rhode Island (unpublished) using direct-reading instruments. Those studies are of limited usefulness for comparison, however, since they were not conducted during the winter months and were concentrated mostly in the northern reaches of the bay. The results of those measurement programs and of this study are plotted vs. latitude in Fig. 11-9. The markedly higher levels obtained during this study would appear to be caused by the high solubility of oxygen in the colder water found during winter months. Oxygen solubility vs. temperature for water at a given salinity can be found from the relation

$$C_S = \frac{475 - 2.65S}{33.5 + T} \tag{11-2}$$

where C_S is the solubility (in parts per million) of oxygen in water of salinity S (in ppt) at temperature T (in °C) (U.S. Department of Health, Education, and Welfare 1958).

The point of these graphs and comparisons is to indicate that professional and significant environmental studies can be made with these inexpensive systems.

Fig. 11-9 North-south dissolved oxygen profiles for Narragansett Bay for summer as established by a Beckman electronic meter and early winter using the devices described in this chapter.

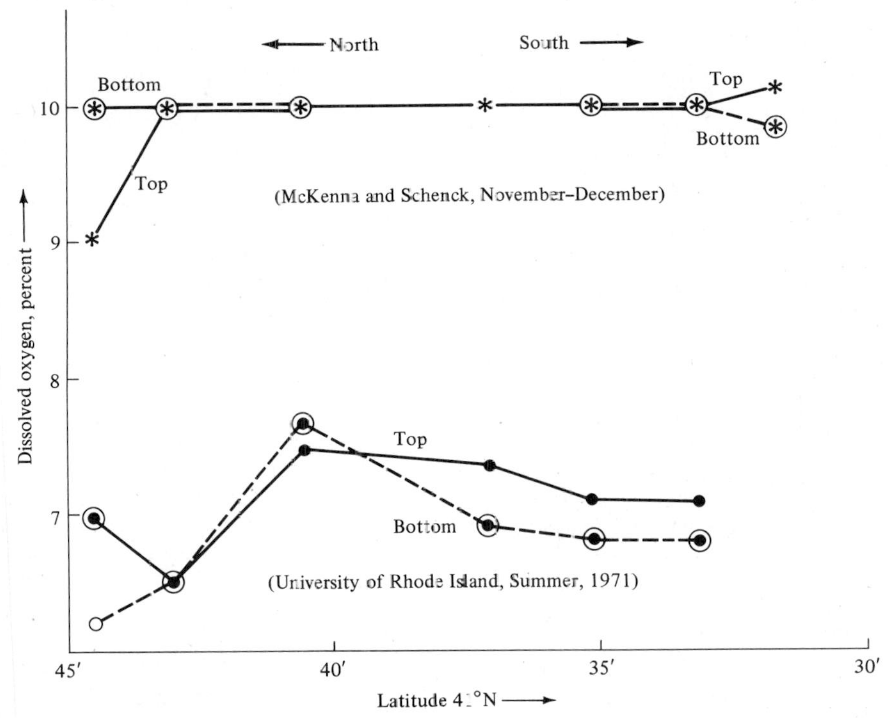

11-8 CURRENT AND FLOW MEASUREMENTS

Flow measuring systems for ocean and estuarine work based on a bewildering array of physical principles are continually being invented and marketed, proving the durable character of this problem. Systems based on sound speed and doppler effects, motion of seawater in magnetic and electrical fields, constant-depth floats with sonic telemetry, rotors, propellers, click generators, and many other devices are now available.

We will consider only two systems, those which have proved most practical in our instructional laboratories for a modest, low-funded program of current measurement: pole drifters and a novel variation of the Carruthers leaning bottle meter.

Pole drifters are probably the easiest to make. They can be constructed of hollow aluminum poles with lead shot in the bottom to adjust flotation depth or from wooden rods with lead flashing around their bottom ends. Poles have the disadvantage that they do not give local current velocity but an average based on the velocity-squared summed over the depth of submergence. Each pole increment responds to a drag force given by

$$dF = \frac{C_d dA \rho (V - V_P)^2}{2g_c} \tag{11-3}$$

where C_d is the (relatively constant) drag coefficient for the section of pole cross section of incremental area dA, V is the local water velocity (density ρ) at the section, V_p is the pole velocity, and g_c in the English unit system is 32.2 $\text{ft}\cdot\text{lb}_f/(\text{s}^2)(\text{lb}_m)$. The pole must assume a mean velocity V_p such that the algebraic sum of all the forces is zero.

Example 11-1 A 20-ft pole is drifting in water that has a linear velocity increase, from surface to 20 ft, of zero to 1 ft/s. What will be the pole velocity V_p?

Solution We can write the function of V with depth as $V = KD$, where K is equal to 0.05 ft/(s)(ft). Then Eq. (11-3) can be written

$$\frac{C_d d\rho}{2g_c}\int_0^x (KD - V_p)^2\, dL = \frac{C_d d\rho}{2g_c}\int_x^{D_{\max}} (KD - V_p)^2\, dL$$

where d is the pole diameter, D_{max} is the total depth of the pole, and x is the unknown distance down the pole at which $V = V_p$. If this equation is integrated and solved for V_p (noting that $x = V_p/K$), we find that the only real root gives $V_p = V_{max}/2$. That is, the pole will drift at the arithmetic mean velocity of the current. Such a result can be inferred by inspection of the system. Since the incremental force is a function of the local velocity difference between pole and water, equal summed differences above and below result only when x is in the center of the underwater part of the pole. Thus a pole drifter in a linearly varying velocity field will give the true average of the current over the pole range.

On Oct. 29, 1971, at station 12 (Fig. 11-7) a nonlinear velocity profile with value 2 ft/min at the surface and increasing to 46 ft/min at 20 ft was observed using subsurface drogues (point measurements). A 10-ft pole moved at 14 ft/min whereas the "true" velocity at 5 ft was 12.3 ft/min, an error of 13.8 percent. A 30-ft pole moved at 35 ft/min whereas the actual 15-ft value was 32.5 ft/min, an error of 7.7 percent with both pole values being too high.

Such a result always occurs with a velocity profile that increases from surface to depth faster than at a linear rate.

Drogues are simply cloth cones with a circular stiffener, several pounds of weight, and a line to an observable surface float. Diameters of 3 ft have proved handy and a trip line to collapse the cone speeds retrieval. It might be assumed that, since they do not have the averaging problems of poles, drogues would be preferred. Unfortunately, the lines and float slow the drogue motion and must be corrected for. By using a fine line and a very small float, this error can be minimized, but handling problems will increase. We suggest poles for modest programs, especially if some idea of the current profile is available to allow the correction noted above.

Any drifting current indicator must have its track and direction known. Many methods have been tried at Rhode Island. At present we suggest that the boat be taut-moored fore and aft to eliminate swing, that the drifter be tethered by the lightest floating line available, that clocking not start until the drifter has gone 10 to 20 ft from the boat, and that drifts of at least 100 ft be used. As the line comes up taut on, say, the 120-ft mark, there will be some doubt as to when the exact distance is reached, involving an uncertainty of perhaps 2 or 3 ft. Thus an inherent error of perhaps 3 percent is difficult to escape, although this is modest compared to most other tracking and sensing methods that do not involve high-precision electronics. The drift direction can be obtained with a compass sight along the drift track as noted in Sec. 11-2.

The leaning bottle current meter was devised by Dr. James N. Carruthers (1958) of the National Institute of Oceanography in England. Carruthers's meter was designed to measure near-bottom currents in either oceanic or coastal waters. Our shallow-water model is made from a Pyrex glass baby bottle half-filled with a gelatin solution, in which a buoyant, aircraft-type compass floats. From the bottle cap, a tether line is attached to an anchor line. After the bottle is heated to melt the gelatin solution, the device is launched and allowed to settle to the bottom. The buoyant bottle inclines under the drag force of local currents and the compass will align itself to magnetic north. As the gelatin is cooled by the surrounding water, it solidifies, freezing the compass in position. On recovery, the slope of the gelatin-air interface is measured and the magnetic direction of tilt is noted from the compass.

The buoying technique used with this system, a submerged, taut-wire array, was selected because it showed the greatest potential for reducing errors in current measurement.

With either an array which is freely suspended from a surface vessel, or an array suspended between a surface vessel or buoy and an anchor, any motion of the surface suspension point will be reflected in motion of the current-sensing device attached to the suspension. Motion of the current meter will result in an error in indicated current velocity. Paquette (1963), in a discussion of sources of error in measurement of oceanic currents, lists a variety of platform motions which induce errors, including yaw and swing of the surface suspension point about its mooring, vertical motion caused by wave action, and horizontal and vertical motion resulting from elastic-cord oscillation or slack in the suspension system. These motions result from the use of a suspension point on a boat or buoy which is coupled to the surface and which transmits surface motion along

Fig. 11-10 General layout of leaning bottle current meters.

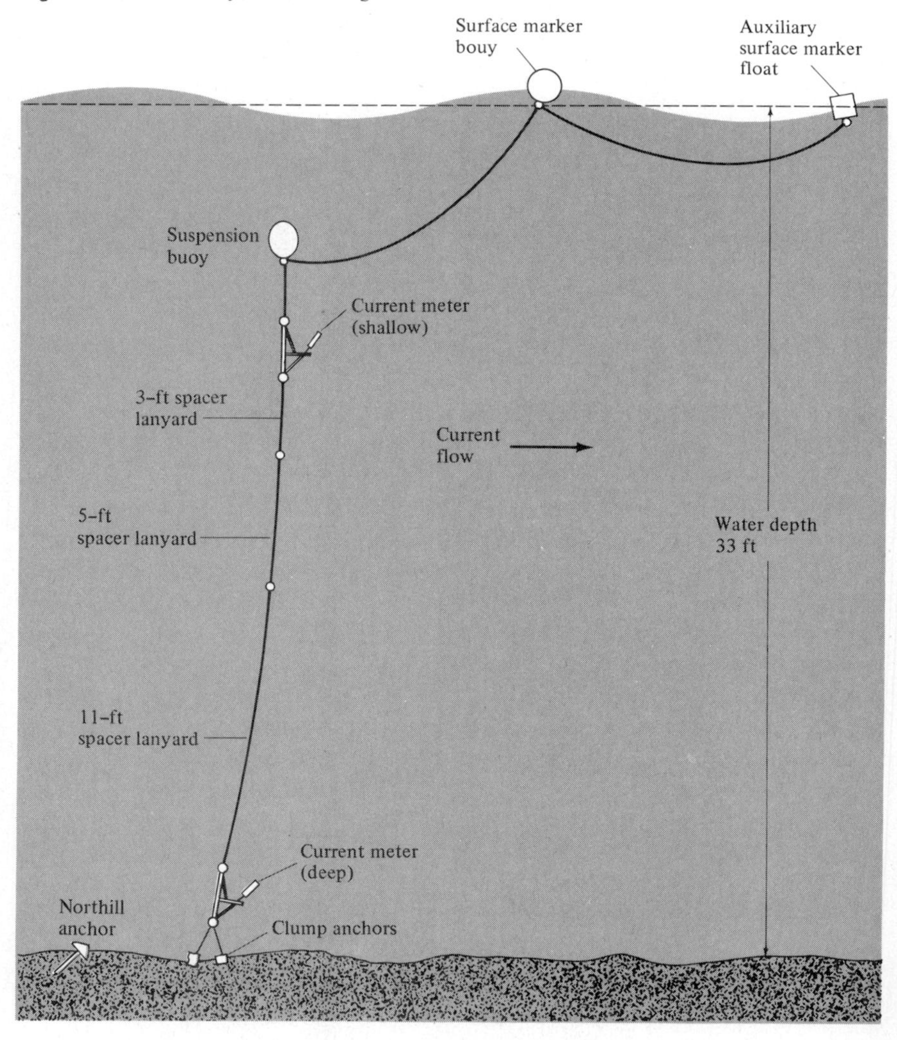

the suspension line to the measuring device. By utilizing a suspension buoy which is moored some distance beneath the surface, where wind effect is eliminated and wave action drastically reduced, the suspension point is effectively decoupled from surface motion and induced errors resulting from such motion are virtually eliminated.

Figure 11-10 depicts a typical vertical array which consists of a submerged suspension buoy with attached surface markers, a suspension line made up of interchangeable spacer lanyards which are hooked together end to end, two current meter brackets with current meters attached between spacer lanyards, and an anchor assembly made up of a clump and a Northill-type anchor.

A. Suspension buoy and surface markers An empty 1-qt plastic detergent bottle serves as a suspension buoy for each array. When fully submerged, it provides a positive buoyancy of about 2 lb. The bottle is made watertight by the application of vacuum wax to the threads of the cap. A short length of 3/16-in braided sash cord, which terminates in a plastic ring, is tied to the handle of the bottle and provides a means of attaching the buoy to the suspension line. Also made fast to the handle is a 20-ft length of 1/16-in braided nylon cord which connects the suspension buoy to the surface marker buoy.

The surface marker buoy, an empty 1-gal plastic bottle, is painted international orange for high visibility.

B. Suspension line spacer lanyards The length of the vertical array is determined by the number and length of spacer lanyards and the number of current meter brackets employed. Spacer lanyards are made up in lengths of 3, 5, 11, 22, and 44 ft. One end of each lanyard is made up to a 1-in-diameter plastic ring, and the other end to a brass snap hook, permitting the joining of any combination of lanyards and current meter brackets between the suspension buoy and the anchor assembly to achieve array lengths that vary from 6 to 105 ft.

C. Current meter brackets (see Fig. 11-11) Bracket assemblies are composed of a vertical, 2-ft piece and a horizontal 1-ft piece of 7/8-in-diameter wooden broom handle. Into the upper and lower ends of the vertical member are screwed brass eyes, through which passes a 3-ft length of 3/16-in sash cord, made up in much the same manner as a spacer lanyard, to permit the attachment of the bracket assembly between lanyards of the suspension line. The bracket is free to rotate in a horizontal plane about its lanyard and will orient itself with the prevailing current flow. In the upper end of the horizontal arm is screwed a brass eye, to which is attached a brass S hook which attaches to the cap of the current meter. The entire bracket assembly, including the current meter bottle, is ballasted to neutral buoyancy and trim by wrapping lead flashing weighing about 0.15 lb around both the vertical and horizontal arms of the bracket.

Fig. 11-11 Current meter bracket with leaning bottle in deployed position.

D. Current meter The bottle used is a standard Even-Flo brand Pyrex glass baby feeding bottle. The plastic two-piece cap assembly is used in lieu of the specially made brass cap used with the Carruthers device.

The current meter compass is a handmade device consisting of a balsa wood float-and-compass-rose portion, to which is glued an ordinary Alnico permanent magnet, available from most toy or hardware stores. The "skirt," or compass-rose section, of the float is marked in 15° increments with the four quadrants color-coded for ease of reading. The entire assemblage is given five coats of hot-fuel-proof dope to seal the wood. The compass is designed to float with its uppermost surface barely awash and with its skirt totally submerged in order to eliminate the possibility of surface tension that might cause the compass to adhere to the sides of the bottle and result in erroneous direction indication.

To prevent leakage of seawater into the bottles, rubber gaskets, cut from an inner tube, are used in the cap assemblies, and vacuum wax or silicon lubricant is applied to both faces of the gasket.

The gelatin charge used in each bottle is prepared by adding 3/4 oz of ordinary unflavored cooking gelatin powder to 24 fl oz of hot tap water. Four fluid ounces of this solution constitute a standard filling for each bottle.

E. Thawing and ready-service containers At a temperature of approximately 22°C, the standard gelatin solution used with the current meter bottles begins to solidify. In order to keep the gelatin liquefied and the bottles ready for use with current meter arrays, the capped bottles can be stored in an insulated Styrofoam cooler which is kept about half-filled with water at a temperature of at least 30°C. This temperature is maintained by periodically adding small amounts of hot water.

A separate insulated container, about half-filled with hot water (at least 40°C) can serve as a "thawing" container. As soon as arrays are recovered and current readings are recorded, the bottles can be unscrewed from the cap assemblies attached to the current meter brackets. The bottles are then recapped and placed in the thawing container. After the gelatin has reliquefied, the bottles are transferred to the ready-service container to await reuse later.

Prior to arrival of the boat at the position at which the first array is to be deployed, current meter brackets and spacer lanyards can be hooked together between the anchor assembly and suspension buoy in the appropriate configuration for the predicted water depth at that station. For example, if water depth were 33 ft, the following configuration might be used: anchor assembly, current meter bracket, 11-, 5-, and 3-ft lanyards, current meter bracket, and suspension buoy-marker system. This would provide measurements at 3 ft and 25 ft above the bottom, viz., at water depths of 8 ft and 30 ft. (See Fig. 11-10.)

About 30 min after deployment the boat can commence maneuvers for retrieval of the first array. The approach is made from the direction toward which the surface markers are tending in order to ensure that the boat does not overrun the array or the line connecting the surface markers and the suspension buoy. The auxiliary float, or its nylon cord, is snagged with a boathook and taken aboard and the array is pulled up by hand.

After the entire array is aboard, the current meter bottles are successively placed on an inclination readout device (see Fig. 11-12), and the angle of inclination is determined by aligning the gelatin-air interface plane with the lines on the readout device.

Current direction is determined by estimating the magnetic direction in which the slanted gelatin-air interface was inclined (see Fig. 11-13). To this

Fig. 11-12 Solidified gelatin-air interface placed on angle-measuring device.

reading is applied a correction factor to compensate for any magnetic variation, which in the Narragansett Bay area is about $-15°$.

11-9 CALIBRATION AND USE OF THE LEANING BOTTLE METER

Calibration and test of the Carruthers system involves three aspects: direction, velocity, and solidification time. Since many readers will not have facilities to perform all these calibration jobs themselves, we present sufficient data to permit using the bottles with little or any calibration effort.

A. Current direction The magnets used in making the compasses for the current meters were designed for adhering to ferrous materials, not for incorporation into current meters. It was therefore desirable to determine whether the magnetic axes of these bar magnets coincided with their physical longitudinal axes. This was accomplished with a simple testing arrangement. A balsa wood raft, floating in a water-filled aluminum pan, served as a buoy for the magnet, which was positioned with its longitudinal axis aligned with two sighting pins on the buoy. After the raft-and-magnet had oriented itself to magnetic north, the point was marked where an imaginary straight line through the sighting pins intersected a sighting board set up 6 ft away. This procedure was repeated for six additional magnets. A portable compass was then used to establish the actual direction of magnetic north. Knowing the adjacent and opposite legs of the resulting right triangles permitted solution of the angular errors between the longitudinal and magnetic axes of the magnets. The average accuracy was about ±4°, with the maximum discrepancy of axes being 5°.

Magnetic deviation, or the difference between compass-indicated magnetic azimuth and actual magnetic azimuth, is caused by the presence of ferrous materials or electromagnetic fields in the vicinity of the compass which cause distortion of the earth's magnetic field. Deviation errors were eliminated through the exclusive use of nonmagnetic materials in the current meter arrays. Other deviation errors can be introduced by deploying the arrays in the immediate vicinity of fixed or semifixed objects which can be expected to distort the

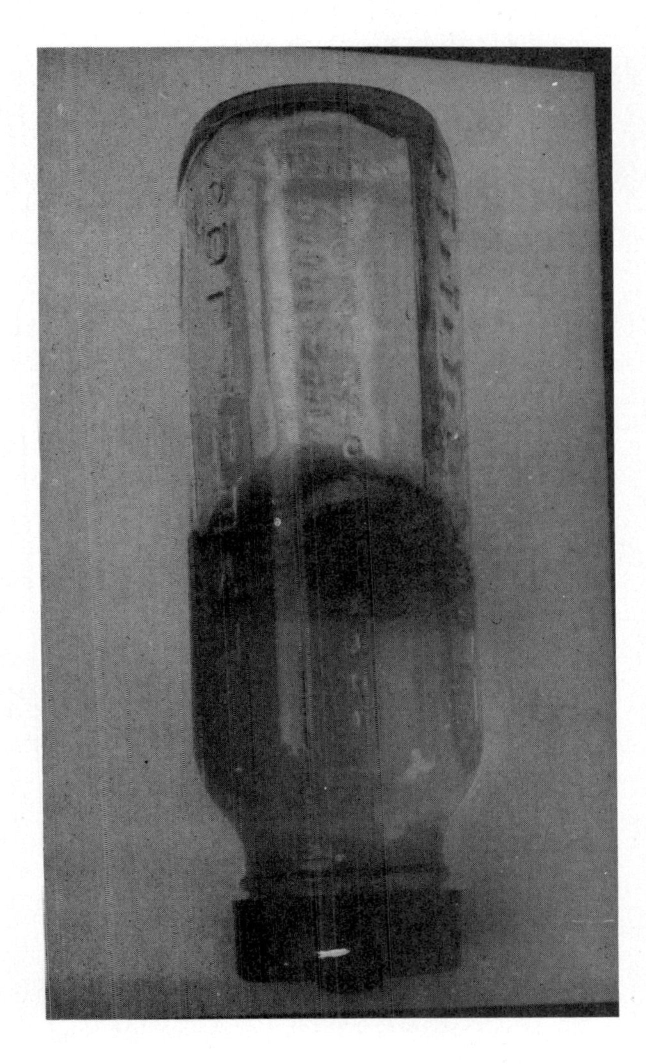

Fig. 11-13 Compass fixed in solidified gelatin.

earth's magnetic field, such as bridges, piers, or buoys. Even the presence of the research boat itself during the time that the gelatin solution is solidifying might cause an erroneous indication of direction.

The degree of precision with which the compass can be read is limited by the size of the "skirt" on which are scribed the directional markings; in this case it appears to be about ±5°. The precision is further reduced, however, by the difficulty of assessing qualitatively the exact direction in which the gelatin-air interface plane slopes (particularly when the slope is of small magnitude), and the inherent parallax errors introduced by the requirement to view the compass through glass. We estimate that the maximum precision with which direction of inclination can be read is ±10°, even with the most scrupulous attention to detail. The accuracy of the indicated direction obtained with the compass is thus considered to have an error of about ±15°.

B. Current velocity In order to establish a response curve for bottle inclination versus water velocity, a calibration experiment was carried out in a testing flume in which velocity could be varied from about 0.7 ft/s to 2.75 ft/s. A graph of bottle inclination versus water velocity is presented in Fig. 11-14. The 4-oz gelatin charge was used in all underway operational tests of the current meter because of the nearly linear response characteristics.

C. Gelatin solidification It is desirable that the gelatin solution in the bottles should solidify in at most 30 min after the time of submergence in waters of any temperature within the normal range of estuarine temperatures.

Fig. 11-14 Flow-vs.-inclination calibration curve for an 8-oz Even-Flo baby bottle with 4-oz of gelatin water charge.

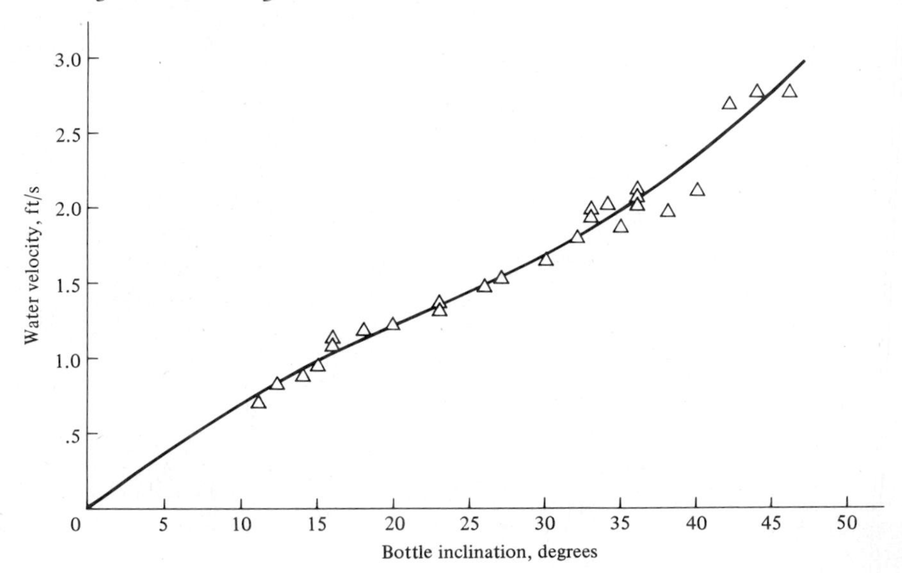

A series of tests was carried out to determine the time required from submergence to solidification, in which the bottles were heated and then immersed in a water bath. A log of gelatin temperature and water bath temperature versus time was maintained. Both the standard gelatin solution (1/8 oz of gelatin powder per 4 fl oz of water) and a concentrated solution, twice as strong, were tested over a water bath temperature range from near freezing to about 24°C.

Although the standard solution will solidify eventually at an immersion temperature of about 22°C (71.6°F), the time required is excessive. To ensure solidification within 30 min, ambient water temperature should be no greater than 17.5°C (63.5°F).

Tests with the concentrated solution yielded solidification times of under 30 min at water temperatures of 24.0°C (76.2°F) and lower.

In extremely cold waters (about 5°C, or 41°F), solidification time appeared to be virtually independent of solution concentration and initial temperature of the solution at time of immersion. Solidification took place with both solutions, and at varying initial temperatures, within 12 min.

When a bottle array, such as that shown in Fig. 11-10, is placed in a current, it will lean over and cause depth errors. That is, the bottles will be at greater depths than a taut, erect line would be. With an array length of 48 ft and a 1.2-kn current, the inclination was computed to be 17° from the vertical with a depth error of about 4 percent. For a 96-ft array in the same current the inclination becomes 30° and the depth error is now 13 percent.

In December of 1971, three pairs of Carruthers bottles were deployed across the west passage of Narragansett Bay (stations 11, 12, and 13, Fig. 11-7). In addition, water sampling and salinity determinations were made at bottom and near-surface locations. These readings were carried out by one of the authors (McKenna), two high-school students, and the Rhode Island Ocean Engineering Department boat captain. Figure 11-15 shows the current and salinity pattern about ½ h before low slack. Substantial surface water is moving down the bay, but the deeper motion is almost stopped. Surface water is saltier in the deep channel and less saline in the shallows. Figure 11-16 shows conditions about 2 h past low slack. All motion is now up the bay, and all surface waters are less saline than the deeper flows. Flow from both shallow flats is angling toward the deep channel, a phenomenon verified at this location by extensive drift experiments with poles and drogues. Four data sets of this sort were obtained in a 6-h on-station period, plus data on temperature and dissolved oxygen, all with instrumentation costing less than $100.

11-10 TIDE AND WAVE MEASUREMENTS

Although government publications give tide times and heights at many locations along the coastlines of the world, the actual measurement of tidal changes may be an important need when ponds, marshes, and rivers are involved. In the case

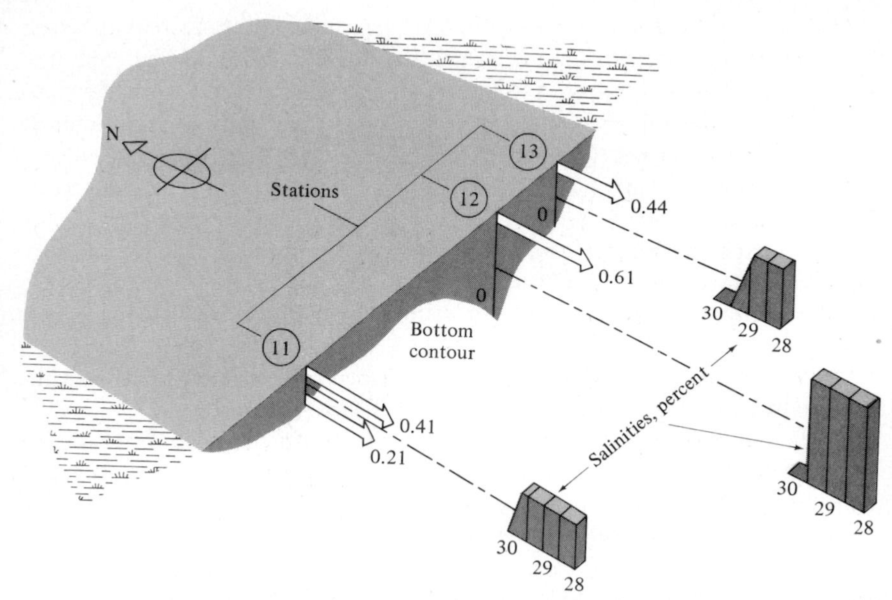

Fig. 11-15 Current (in feet per second) across the west passage of Narragansett Bay ½ h before low slack, and salinities.

Fig. 11-16 Current (in feet per second) and salinities across the west passage of Narragansett Bay 2 h after low slack.

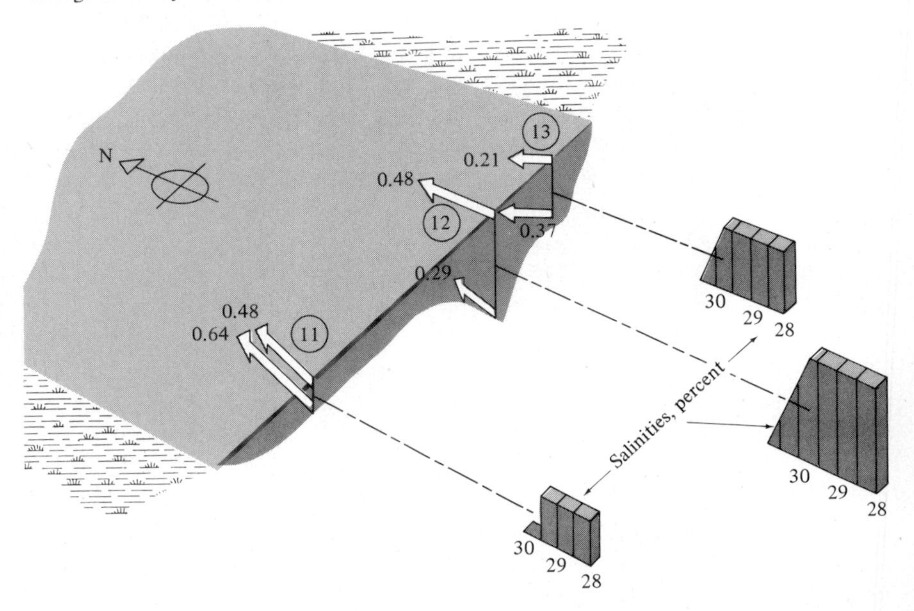

of a pond or harbor connected to the sea by a channel, the volume of flushing water over each tidal cycle can be computed if the water-height change is multiplied by the surface area of the pond. Where inlets may go dry at low water, this flushing volume may be heavily affected by the sun and moon positions. Indeed, waste disposal might be suspended during low ranges of tides. Similarly, the distance a tidal bore will reach up a river can best be determined by tide gauging along the river.

A very successful tide gauge, designed by Rhode Island students, is shown in Fig. 11-17. A float counterbalanced by a weight moves a bicycle wheel, which is directly coupled to a potentiometer which forms part of a voltage divider circuit. Unfortunately, it is necessary to use a recorder of some sort, unless the location is such that it can be visited every half hour and to take readings. We have found that a 5,000-Ω impedance Rustrak unit, with a chart speed of 1 or 2 in/h and battery-operated, will serve nicely and is in the $100 range. If the potentiometer is 100 Ω, the recorder will not load the circuit appreciably and the output will be almost linear. Because vandalism is a big problem in estuarine monitoring, the system must be carefully located and protected.

Figure 11-18 shows data taken with this unit on the Pettaquamscutt River in southern Rhode Island. Such data detect the extent of tidal action at various

Fig. 11-17 Tide gauge for sheltered locations. Since friction can produce jerky records, make the float with an area of 1 ft^2 or more to ensure smooth motions.

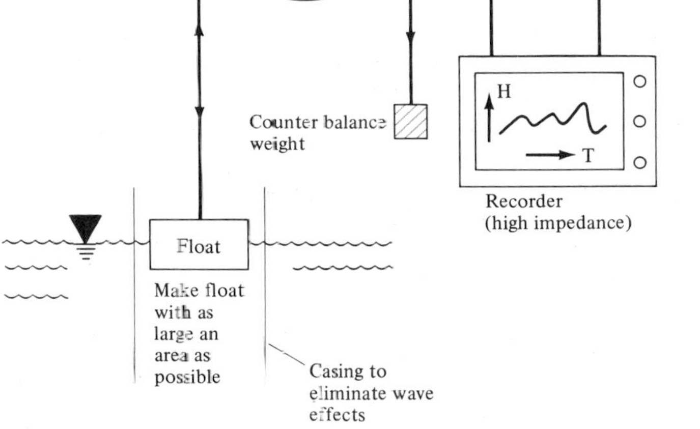

times of the year; several such stations will allow estimates of riverbed friction and other specialized parameters.

Wave data are among the most difficult of ocean parameters to obtain, and many types of instruments exist to attempt the job. In the management of estuarine waters, wave data may be useful for a number of studies. Design of instruments, buoys, and other devices may be based on some estimate of the primary wave periods and heights. In shallow water, the conditions that produce bottom scour by wave surge and the percentage of time this will occur may be desired. With the material from Chap. 2 it is possible to compute lengths between crests if wave period and water depth are known (and the assumptions noted there can be accepted).

We offer a single, crude wave measuring system developed under a grant from the National Science Foundation and shown schematically in Fig. 11-19. Again, this linear potentiometer can be connected to a Rustrak recorder, but now the recorder must operate at its maximum rate, which is a little better than three recorded points per second. The observer must be at an immovable location: the shore, a tower, or a bridge. The target pole can be fixed to the bottom or can be a spar buoy, anchored and reaching to within a few feet of the ocean bottom with a damping disk on its bottom. Clearly, the pole motion must be at a minimum with the waves rising and falling along its length. The observer

Fig. 11-18 Maximum tidal excursions at an upstream location on the Pettaquamscutt River compared with excursions at the river's mouth. Data made with a bicycle wheel tide gauge as shown in Fig. 11-17. This sort of measurement allows standard tide charts to be corrected for special locations.

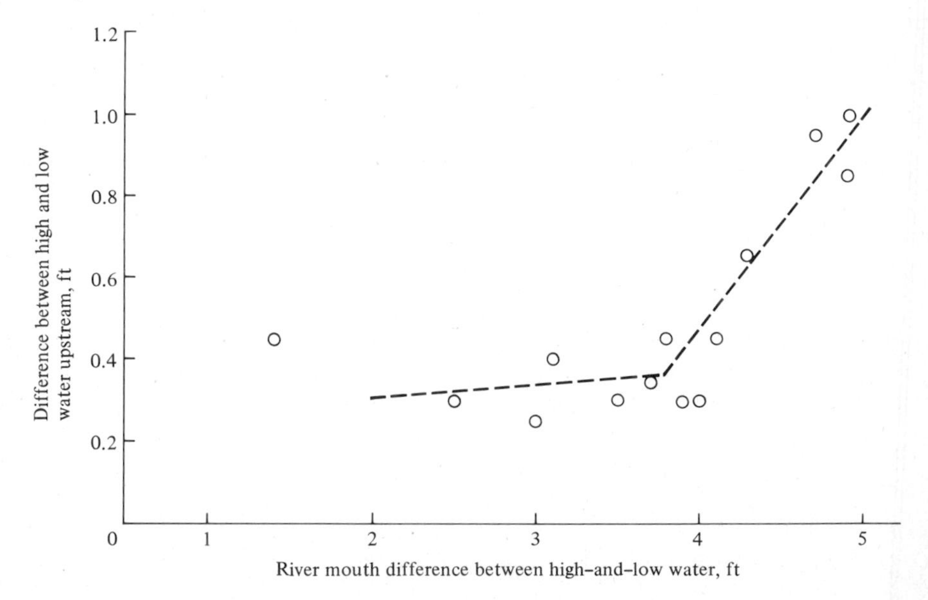

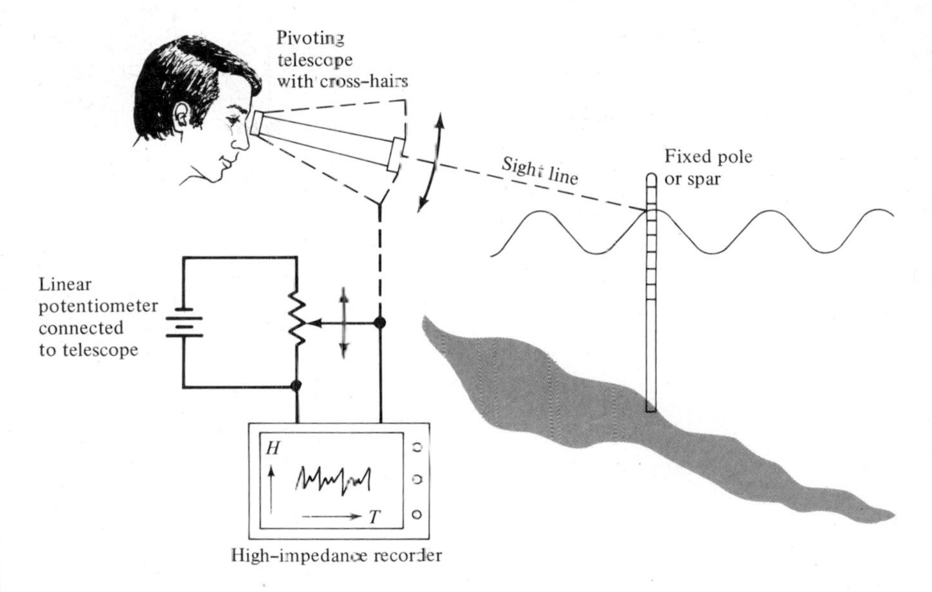

Fig. 11-19 Wave measuring system using a telescope decoupled from the waves which tracks the air-sea interface rising and falling along a pole fixed to the bottom or a spar that does not resonate to wave motion. (See Chap. 4 for design equations.)

sights the pole-water interface and moves the sighting telescope (an eight-power surplus elbow telescope in our unit) which in turn varies the linear potentiometer. It is better to use a lever and linear pot to magnify the motion than to connect the telescope to a circular pot, since the angular motion is likely to be only a fraction of a degree. Calibration is easily done if the target pole is color-coded for each 6-in segment. Tracking takes patience, but waves of ½ s or longer period can be recorded. A few minutes of operation give a great deal of information, but data reduction takes a long time.

Using this device off Point Judith, the southern tip of Rhode Island, we were able to determine that 1.4 to 1.6 s was the range of the principal period of waves coming into the surf zone under average wind conditions, a value verified for the lower bay as a whole with an elaborate 60-ft spar and capacitance wave-staff system developed by a doctoral student.

We make no apologies for this rudimentary instrument with all its obvious weaknesses and restrictions. For serious wave work we suggest consultation with specialists and a study of available systems from commercial houses, many of them running to tens of thousands of dollars.

11-11 EXTINCTION COEFFICIENTS

The most important optical parameter in natural waters is the *scalar irradiation coefficient* K. This measurement is made by lowering a linear-response photocell to various known depths, plotting the best straight line through the points on a log I_d/I_0-vs.-D plot, and solving for K in the equation

$$\frac{I_d}{I_0} = e^{-KD} \tag{11-4}$$

where I_0 is the intensity of the surface light and I_d is the intensity at depth D.

Besides predicting visible and photographic distances, K may be of considerable interest to marine biologists who will need to know the level of solar energy at particular depths and under various conditions of tide, weather, and season.

The instrument described here is entirely satisfactory and can be built for less than $20.

A. Solar cells Two Hoffman brand model 52CL silicon solar cells were used as light sensors. Cell dimensions were 0.2 by 0.8 in with an active area of 0.12 in. Spectral response of the cells is from 0.4 to 1.15 μm, with peak response at 0.85 μm in the near-infrared region of the electromagnetic spectrum.

To waterproof the submerged cell, it was encapsulated between a window of rectangular plexiglass (dimensions: $1\frac{7}{8} \times 1\frac{1}{4} \times \frac{1}{4}$ in) and a backing of clear casting resin of the same dimensions. Encapsulation was accomplished by laying the solar cell (active side downwards) on the plexiglass window and taping temporary plastic cofferdams around the edges of the plexiglass. Into the space formed by the cofferdams, the casting resin was poured and allowed to set for 36 h. The cofferdams were then stripped off and the sides and bottom of the encapsulated unit were painted with two coats of oil-base black paint to exclude all light except that passing through the window.

A 60-ft length and a 6-ft length of ordinary, rubber-coated household two-conductor wire were used to connect the submerged cell and the reference cell, respectively, to the readout unit. Junctions between the cell output wires and the two-conductor cabling were coated with silicon rubber to make them watertight.

B. Submerged cell mounting A three-armed frame made of wood and ballasted with 2.5 lb of lead was used as a mounting for the submerged cell. The cell was glued to a wooden backing piece which was screwed onto the frame. Three 3-ft lanyards of 3/16-in sash cord connected the outer ends of the three arms to a 50-ft sash cord lowering line. The lowering line was marked with colored plastic tape at 1-ft intervals to permit determination of submergence depth of the cell. The cabling from the cell to the readout unit was taped to the lowering line every 5 ft. Figure 11-20 depicts the general arrangement, without the ballast weight.

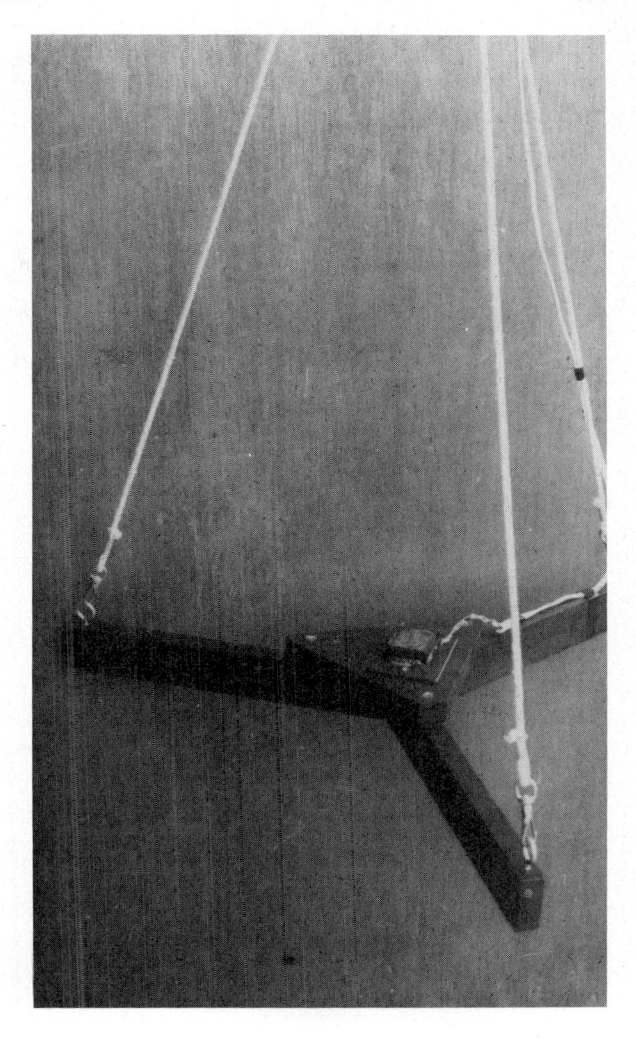

Fig. 11-20 Submerged irradiance meter.

C. Cell readout unit The cables from the two cells terminated in jack plugs which were inserted into receptacles at the readout unit. Switches permitted sending the output signals to selected milliammeters. The output of the subsurface cell could be monitored on meters with ranges of 0 to 50 mA, 0 to 10 mA, and 0 to 3 mA. The three meters were all U.S. government surplus items.

Light measurements were made by first suspending the submerged cell just above the surface of the water and placing the reference cell on a horizontal surface and recording the subsurface cell output B and the reference cell output A. The subsurface unit was then lowered into the water in 1-ft increments beneath the surface. The subsurface cell readings B_D and the reference cell readings A_D were recorded at each depth. The subsurface unit was monitored on

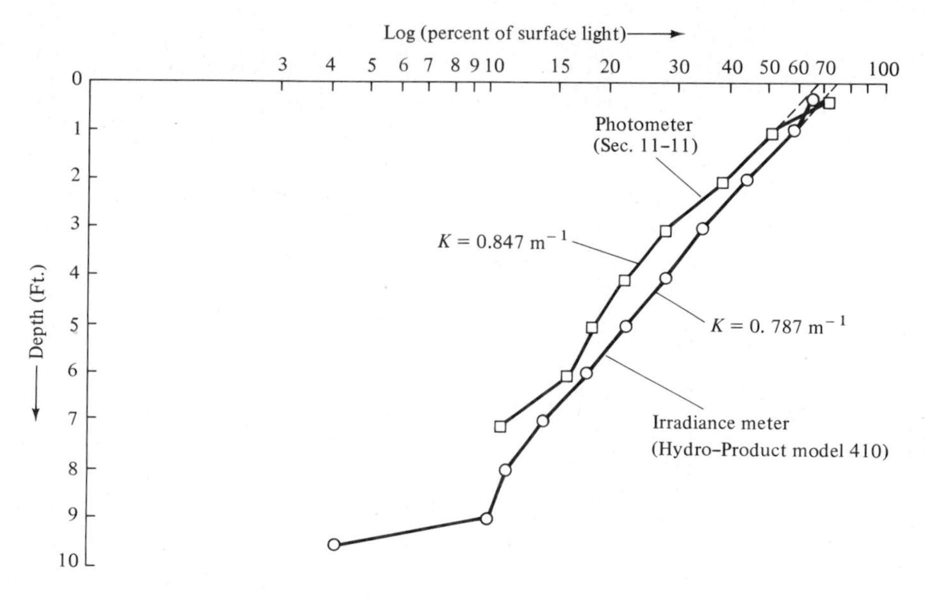

Fig. 11-21 Values of K at station 15 as found by both home-built and commercial meters. Note that only the line *slope* is of interest.

the 0 to 10 mA meter until readings had decreased to less than 3 mA, at which point subsequent readings were taken with the 0 to 3 mA meter.

The percentage of surface light that reached submergence depth D was determined by multiplying the ratio of the subsurface cell output B_D to the reference cell output A_D obtained at each depth by the initial ratio of subsurface-to-reference cell output B/A which was obtained with the subsurface cell out of the water:

$$\text{Percent of surface light at depth } D = \frac{B}{A} \times \frac{B_D}{A_D} \tag{11-5}$$

By plotting the log of the percent of surface light values thus obtained versus depth, the value of the extinction coefficient of light can be determined using Eq. (11-4).

After encapsulation and painting of the cells, they were calibrated to determine their response characteristics. The calibrations were conducted in sunlight, using perforated neutral filter plates, with one cell being used to monitor the intensity of incident light. Readings were taken only when incident light remained constant within 0.1 mA as shown by the monitor cell. The response curves were linear above about 0.05 mA, but below that output level the curves began a gradual upturn to intersect the zero output–zero percent light point.

It is therefore evident that the response of the solar cells when encapsulated and painted is usable from output levels from 0.05 mA to at least 10.0 mA and is probably linear up to the rated cell output of 20 mA. We would expect that any silicon cell with this general output level would perform in a similar manner, but each pair should be calibrated.

The home-built solar cell meter was compared with a commercial irradiance meter (Hydro-Product Model 410). A typical comparison, shown in Fig. 11-21, was made in March at station 15 (Fig. 11-7). It appeared that the two instruments gave essentially the same line slopes (and thus K values) within the inherent scatter usually found in this type of determination.

11-12 SEDIMENT TESTS AND CORING

As Chap. 3 showed in some detail, the prediction of sinkage, bearing load, and pull-out forces in marine sediments requires some knowledge of the character of the sediment. This can be studied by a diver manipulating the vane shear apparatus described in Sec. 3-6, or by obtaining a sample and returning to a soils laboratory to perform tests there. Clearly any removal of soil from the bottom will more or less change its character.

A standard sample-gathering technique involves gravity coring, for which devices are readily built and used from boats having some sort of hoist, davit, or A-frame. Figure 11-22 shows a small gravity corer. The thin-walled inner plastic tube should fit snugly in the steel-pipe barrel. The core catcher can be purchased or built of thin spring brass. Its fingers fold against the plastic tube wall as the sediment is driven into the tube, then fold out as the corer is pulled out of the sediment to hold the sample in the tube. For a nominal 2-in barrel, 100 lb of adjustable weights is reasonable.

In operation, the corer assembly is lowered to within a few feet of the bottom, then allowed to drop freely into the muck. With skill and/or luck, the unit will strike in an upright position and catch a core from a few inches to 2 or 3 ft long. A winch of some sort will help in pulling the core out of a sticky bottom.

Two samples were obtained by Rhode Island students north of station 2 with a 100-lb corer. When taken to a well-equipped soils laboratory, these cores yielded the following data:

Velocity of sound in the sediment	1,610 m/s
Sediment specific gravity	132.2 lb/ft^3
Water content of sample	55.9%
Acoustic reflectivity for sediment-water impedances	0.172
Shear strength (cone penetrometer)	132.3 lb/ft^2
Shear strength (Farnell vane shear)	102.5 lb/ft^2

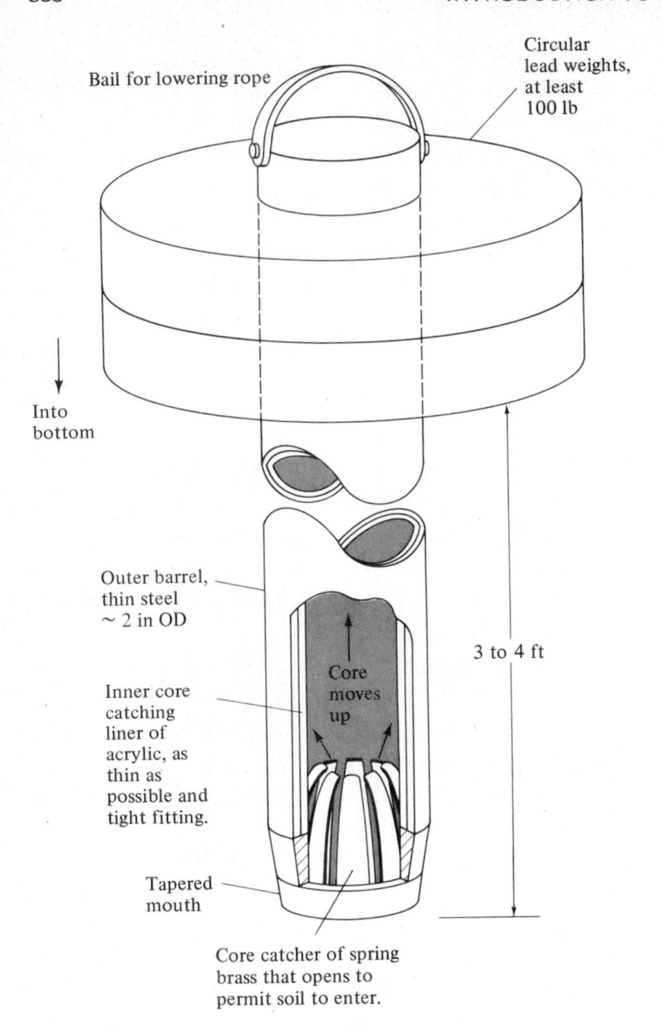

Fig. 11-22 Small gravity corer. The addition of sheet metal fins at the weighted end may aid the corer in entering the bottom in a vertical orientation.

Such data are of interest in pipe and cable sinkage and in anchoring and foundation design, and also important to acoustic signaling, bottom profiling, and underwater communications.

PROBLEMS

11-1. A magnetic compass has a 5° error in a westerly direction (when pointing north it reads 355° magnetic). If two landmarks each 1,000 yd away are to be used to establish position, make a plot of position error magnitude versus angle between the two marks in the range 20 to 160°. Use the graphical method noted in the text for several intermediate angles and scale the errors from the graph.

11-2. A range consists of a spire and a channel-marking buoy 1,500 yd apart. The buoy has a 10-yd position uncertainty, and we are 1,000 yd along the range (2,500 yd from the spire). What will be the worst position error if the second position line consists of a bridge pier 500 yd away and at right angles to the range and the compass has an uncertainty of 8°? With a ±10-yd uncertainty in the buoy and a ±8° uncertainty in the compass, what is the shape of the "area of confusion" and how large is it?

11-3. A transit on a 150-ft-high bridge makes an angle of —20° with the horizontal. How far is the sighted object from the transit? If the transit has a 0.5° error, what is the error in the range estimation?

11-4. Make a sketch design to modify the water sampler in Figs. 11-3 and 11-4 to allow a second weight to be released to trip a second unit suspended at a greater depth underneath the first. Assume the second weight is carried down with the upper sampler and is released when the upper sampler closes.

11-5. If we are measuring salinity in a range where 0.0002 specific gravity units equals 0.3 ppt salinity, what salinity error will a 2°C error in temperature of the sample produce? Assume the sample is around 29.2 ppt.

11-6. If a water sample is at 4°C and 30 ppt salinity, what is the maximum dissolved oxygen level possible if the sample is not supersaturated?

11-7. Assuming the bay was more or less isothermal when the data shown on Figs. 11-8 and 11-9 were obtained, what can we say about the *relative* amount of oxygen saturation as we move from the Providence River south?

11-8. Obtain the general equation for the average velocity of a pole drifter in a linearly increasing velocity field starting at V at the surface (finite, nonzero velocity) and increasing to V_d at the bottom of the pole. Use the derived equation to obtain the pole drift velocity when V surface is 0.5 ft/s and V_d is 1.0 ft/s.

11-9. A pole drifter 20 ft long is in a velocity field with 0 to 10 ft at 0.5 ft/s and 10 to 20 ft at 0.8 ft/s. What will be the pole velocity?

11-10. With 3-ft bottle sections and spacer lanyards of 3, 5, 11, 22, and 44 ft, design an array that puts one bottle 3 ft above the bottom and a second bottle 7 ft from the surface, with water depth of 54 ft.

11-11. Design with dimensions and weights a tide gauge with a maximum tidal excursion of 2.5 ft, using a potentiometer with a 240° total turn (0 to 100 Ω, linear wound). The potentiometer needs 4 in-oz of torque to turn it reliably.

11-12. Referring to the Fig. 11-19 wave tracking device, how long a lever arm will we need for a linear pot of 1-in total motion to go from zero to maximum ohms while following a 3-ft-high wave if the wave is 100 yd from the tracking station?

11-13. Using Fig. 11-7 for dimensions, estimate the net flow of water in the west passage of Narragansett Bay for the conditions shown in Fig. 11-16, in cubic feet per second. Ignore direction shifts in the shallow areas. The depth at station 11 is 18 ft, at station 12, 59 ft, and at station 13, 44 ft.

11-14. Prove algebraically that an irradiance K meter will give the correct line slope on semilog paper (and thus the correct K) if the two cells are linear in output with light intensity and, further, that cell mismatch (different currents at a given illumination level) will not affect K. What other sources of error might exist in a K measurement?

11-15. A K-measuring experiment gives the following ratios of I_d/I_0 versus depth.

I_d/I_0	*Depth in feet*
0.489	5
0.362	10
0.272	15
0.208	20
0.18	25
0.148	30
0.126	35
0.11	40
0.108	45

What is K for this water?

REFERENCES

Barnes, H. (1959): "Apparatus and Methods of Oceanography," vol. 1, Interscience, New York.

Carruthers, J. N. (1958): Seine Net Fishermen Can Easily Observe Bottom Currents for Themselves, *Fishing News*, vol. 2362, pp. 6–7.

Cronin, Eugene L. (1967): The Role of Man in Estuarine Processes, in G. H. Lauff (ed.), *Estuaries*, AAAS Pub. no. 83, American Association for the Advancement of Science, Washington, pp. 667–689.

Duxbury, Alyn C. (1971): "The Earth and Its Oceans," Addison-Wesley, Reading, Mass.

Emery, K. O. (1969): "A Coastal Pond Studied by Oceanographic Methods," American Elsevier, New York.

Fanning, Odom (1969): "Opportunities in Oceanography," Universal Publishing & Distributing Corporation, New York.

Gross, M. Grant (1971): "Oceanography," 2d ed., Merrill, Columbus, Ohio.

Haight, F. J. (1938): "Currents in Narragansett Bay, Buzzards Bay, and Nantucket and Vineyard Sounds," U.S. Department of Commerce, Coast and Geodetic Survey Spec. Pub. no. 208, U.S. Government Printing Office, Washington.

Hicks, Steacy D. (1959): The Physical Oceanography of Narragansett Bay, *Limnol. Oceanog.*, vol. 4(3), pp. 316–327.

Hoerner, S. F. (1965): "Fluid-Dynamic Drag," published by the author, Midland Park, N.J.

Horne, R. A. (1969): "Marine Chemistry," Wiley Interscience, New York.

Horrer, Paul L. (1967): Methods and Devices for Measuring Currents, in G. H. Lauff (ed.), *Estuaries*, AAAS Pub. no. 83, American Association for the Advancement of Science, Washington, pp. 80–89.

Ippen, A. T. (ed.) (1966): "Estuary and Coastline Hydrodynamics," McGraw-Hill, New York.

Isaacs, John D. (1952): Considerations of Oceanographic Instrumentation, in John D. Isaacs and C. O'D. Iselin (eds.), *Oceanographic Instrumentation*, NAS-NRC Pub. 309, National Academy of Sciences, Washington, pp. 1–10.

Jerlov, N. G. (1968): "Optical Oceanography," Elsevier, New York.

Levy, E. M. (1970): The Significance of the Third Figure Following the Decimal Point in Salinity Data, *Information on Techniques and Methods for Sea Water Analysis*, Conseil International pour l'Exploration de la Mer, Service Hydrographique, Interlaboratory Report N 3, Charlottenlund Slot, Denmark, pp. 37–49.

Mangelsdorf, Paul C., Jr. (1967): Salinity Measurements in Estuaries, in G. H. Lauff (ed.), *Estuaries*, AAAS Pub. no. 83, American Association for the Advancement of Science, Washington, pp. 71–79.

McMillan, D. H. (1966): "Tides," American Elsevier, New York.

Myers, John J. (ed.) (1969): "Handbook of Ocean and Underwater Engineering," McGraw-Hill, New York.

Neumann, G., and W. J. Pierson, Jr. (1966): "Principles of Physical Oceanography," Prentice-Hall, Englewood Cliffs, N.J.

Niskin, Shale J. (1963): A Low Coast Bottom Current Velocity and Direction Recorder, *ISA Marine Sciences Instrumentation*, 3, Instrument Society of America, Pittsburgh, pp. 123–132.

O'Brien, M. P., and G. H. Hickox (1937): "Applied Fluid Dynamics," McGraw-Hill, New York.

Paquette, R. G. (1963): "Practical Problems in the Direct Measurement of Ocean Currents," *ISA Marine Sciences Instrumentation*, 2, Instrument Society of America, Pittsburgh, pp. 135–146.

Ryther, J. H., and C. S. Yentsch (1957): The Estimation of Phytoplankton Production in the Ocean from Chlorophyll and Light Data, *Limnol. Oceanog.*, vol. 2, pp. 281–286.

Saucier, John W. (1968): How to Operate "Package" Wastewater Treatment Plants, *Am. City*, vol. 83 (9), pp. 92–97.

Smayda, J. J. (1957): Phytoplankton Studies in Lower Narragansett Bay, *Limnol. Oceanog.*, vol. 2, pp. 342–359.

Sverdrup, H. U., M. W. Johnson, and R. H. Fleming (1942): "The Oceans: Their Physics, Chemistry, and General Biology," Prentice-Hall, Englewood Cliffs, N.J.

Teeson, D., F. White, and H. Schenck (1970): Studies of the Simulation of Drifting Oil by Polyethylene Sheets, *Ocean Eng. V*: 1–11.

U.S. Department of Commerce (1958): "Density of Sea Water at Tide Stations: Pacific Coast, North and South America and the Pacific Ocean Islands," U.S. Coast and Geodetic Survey Pub. 31-4 (5th ed.), Washington.

—— (1963): "Tidal Current Charts for Narragansett Bay," U.S. Coast and Geodetic Survey, Washington.

—— (1970): "Tide Tables: High and Low Water Predictions 1971, East Coast of North and South America Including Greenland," U.S. Coast and Geodetic Survey, Washington.

U.S. Department of Health, Education, and Welfare (1958): Oxygen Relationships in Streams, *Public Health Service Tech. Rept.* W-53-2, Cincinnati.

U.S. Hydrographic Office (1965): "Tables for Sea Water Density," H.O. 615, Washington.

Vine, A. C. (1952): Comment and discussion on contributions in *Oceanographic Instrumentation*, NAS-NRC Pub. 309, National Academy of Sciences, Washington, p. 12.

Weyl, Peter K. (1970): "Oceanography: An Introduction to the Marine Environment," Wiley, New York.

Williams, Jerome (1970): "Optical Properties of the Sea," U.S. Naval Institute, Annapolis, Md.

Index